Processor Architecture

Springer
Berlin
Heidelberg
New York
Barcelona
Hong Kong
London
Milan
Paris
Singapore
Tokyo

Jurij Šilc • Borut Robič • Theo Ungerer

Processor Architecture

From Dataflow to Superscalar and Beyond

With 132 Figures and 34 Tables

Springer

Dr. Jurij Šilc
Computer Systems Department
Jožef Stefan Institute
Jamova 39
SI-1001 Ljubljana, Slovenia

Assistant Professor
Dr. Borut Robič
Faculty of Computer and Information Science
University of Ljubljana
Tržaška cesta 25
SI-1001 Ljubljana, Slovenia

Professor Dr. Theo Ungerer
Department of Computer Design and Fault Tolerance
University of Karlsruhe
P.O. Box 6980
D-76128 Karlsruhe, Germany

Library of Congress Cataloging-in-Publication Data applied for

Die Deutsche Bibliothek – CIP-Einheitsaufnahme

Šilc, Jurij:
Processor architecture: from dataflow to superscalar and beyond/
Jurij Šilc; Borut Robič; Theo Ungerer. - Berlin; Heidelberg; New
York; Barcelona; Hong Kong; London; Milan; Paris; Singapore;
Tokyo: Springer, 1999
ISBN 3-540-64798-8

ISBN 3-540-64798-8 Springer-Verlag Berlin Heidelberg New York

Typesetting: Camera ready pages by the authors
Cover Design: Künkel + Lopka, Werbeagentur, Heidelberg
Printed on acid-free paper SPIN 10665103 33/3142 – 5 4 3 2 1 0

To my father, Alfonz,
my wife, Marjetka, and my sons, Tomaž & Jaka.

Jurij

To Marinka, Gregor, and Rebeka
for their patience, support, and love

Borut

To my wife Gertraud

Theo

Preface

Today's microprocessors are the powerful descendants of the von Neumann computer dating back to a memo[1] of Burks, Goldstine, and von Neumann of 1946. The so-called von Neumann architecture is characterized by a sequential control flow resulting in a sequential instruction stream. A program counter addresses the next instruction if the preceding instruction is not a control instruction such as, e.g., jump, branch, subprogram call or return. An instruction is coded in an instruction format of fixed or variable length, where the opcode is followed by one or more operands that can be data, addresses of data, or the address of an instruction in the case of a control instruction. The opcode defines the types of operands. Code and data are stored in a common storage that is linear, addressed in units of memory words (bytes, words, etc.).

The overwhelming design criterion of the von Neumann computer was the minimization of hardware and especially of storage. The most simple implementation of a von Neumann computer is characterized by a microarchitecture that defines a closely coupled control and arithmetic logic unit (ALU), a storage unit, and an I/O unit, all connected by a single connection unit. The instruction fetch by the control unit alternates with operand fetches and result stores for the ALU. Both fetches access the same storage and are performed over the same connection unit – this turned out to be a bottleneck, sometimes coined by latter authors as the von Neumann bottleneck.

The sequential operating principle of the von Neumann architecture is still the basis for today's most widely used high-level programming languages, and even more astounding, of the instruction sets of all modern microprocessors. While the characteristics of the von Neumann architecture still determine those of a contemporary microprocessor, its internal structure has considerably changed. The main goal of the von Neumann design – minimal hardware structure – is today far outweighed by the goal of maximum performance. However, the architectural characteristics of the von Neumann design are still valid due to the sequential high-level programming languages that are used today and that originate in the von Neumann architecture paradigm.

[1] A.P. Burks, H.H. Goldstine, J. von Neumann, Preliminary Discussion of the Logical Design of an Electronic Computing Instrument. Report to the U.S. Army Ordnance Department, 1946. Reprint in: W. Aspray, A.P. Burks (eds.) *Papers of John von Neumann.* MIT Press, Cambridge, MA, 1987, pages 97–146.

A slightly more efficient implementation than the alternating of instruction fetch and operand fetch is the overlapping of the following two steps: next PC computation and instruction fetch and decode, with operand fetch, instruction execution, and result storage. This overlapping already defines two-stage instruction pipelining.

A more consistent use of overlapping results in an instruction pipeline with the following basic steps that are characteristic of so-called reduced instruction set computer (RISC) processors: instruction fetch, instruction decode and operand fetch, instruction execution, memory access in the case of a load/store instruction, and result write-back. Ideally each step takes about the same amount of time.

However, a storage access today needs much more time than a single pipeline step. The introduction of registers on the processor chip and restricting the operands of ALU instructions to register accesses allows the pipeline to be balanced again. However, the problem of the memory accesses – the von Neumann bottleneck – is still one of the main hindrances to high performance even today. A whole memory hierarchy of cache storages now exists to widen that bottleneck.

Current superscalar microprocessors are a long way from the original von Neumann computer. However, despite the inherent use of out-of-order parallelism within superscalar microprocessors today, the order of the instruction flow as seen from outside by the compiler or assembly language programmer still retains the sequential program order as defined by the von Neumann architecture.

Radically different operating principles, such as the dataflow principle and the reduction machine principle, were surveyed very early on. The dataflow principle states that an instruction can be executed when all operands are available (data-driven) while the reduction principle triggers instruction execution when the result is needed (demand-driven). We find a modified variant of the dataflow principle, called local dataflow, in today's superscalar microprocessor cores to decide when instructions are issued to the functional units.

Since present-day microprocessors are still an evolutionary progress from the von Neumann computer, at least four classes of future possible developments can be distinguished:

- Microarchitectures that retain the von Neumann architecture principle (the result sequentiality), although instruction execution is internally performed in a highly parallel fashion. However, only instruction-level parallelism can be exploited by contemporary microprocessors. Because instruction-level parallelism is limited for sequential threads, the exploited parallelism is enhanced by speculative parallelism. Besides the superscalar principle applied in commodity microprocessors, the superspeculative, multiscalar, and trace processor principles are hot research topics. All these approaches belong to the same class of implementation techniques because result sequentiality

must be preserved. A reordering of results is performed in a retirement phase in order to conform this requirement.
- Processors that modestly deviate from the von Neumann architecture but allow the use of the sequential von Neumann languages. Programs are compiled to the new instruction set principles. Such architectural deviations include very long instruction word (VLIW), SIMD in the case of multimedia instructions, and vector operations.
- Processors that optimize the throughput of a multiprogramming workload by executing multiple threads of control simultaneously. Each thread of control is a sequential thread executable on a von Neumann computer. The new processor principles are the single-chip multiprocessor and the simultaneous multithreaded processor.
- Architectures that break totally with the von Neumann principle and that need to use new languages, such as, e.g., dataflow with dataflow single-assignment languages, or hardware-software codesign with hardware description languages. The processor-in-memory, reconfigurable computing, and the asynchronous processor approaches also point in that direction.

In particular processor architecture covers the following two aspects of computer design:
- the instruction set architecture which defines the boundary between hardware and software (often also referred to as the "architecture" of a processor), and
- the "microarchitecture", i.e., the internal organization of the processor concerning features like pipelining, superscalar techniques, primary cache organization, etc.

Moreover, processor architecture must take into account the technological aspects of the hardware, such as logic design and packaging technology.

Intended Audience

The primary intended audience of this book are computer and/or electrical engineers and researchers in the fields of computer science. It can also be used as a textbook for processor architecture or advanced microprocessor courses at the graduate student level of computer science or electrical engineering. As such it is not intended for beginners.

The book surveys architectural mechanisms and implementation techniques for exploiting fine-grained and coarse-grained parallelism within microprocessors. It starts with a review of the basic instruction set architecture and pipelining techniques, continues with a comprehensive account of state-of-the-art superscalar and VLIW techniques used in microprocessors. It covers both the concepts involved and implementations in modern microprocessors. The book ends with a thorough review of the research techniques that will lead to future microprocessors.

Using the book

Each book chapter comprises a tutorial on the specific techniques and comprehensive sections on sample processors. The reader may quickly browse the sample processor sections, if interested mainly in learning the techniques. The conversant reader may even start with Chap. 4 – the main chapter of the book, while the student is advised to read at least Chaps. 1 and 3.

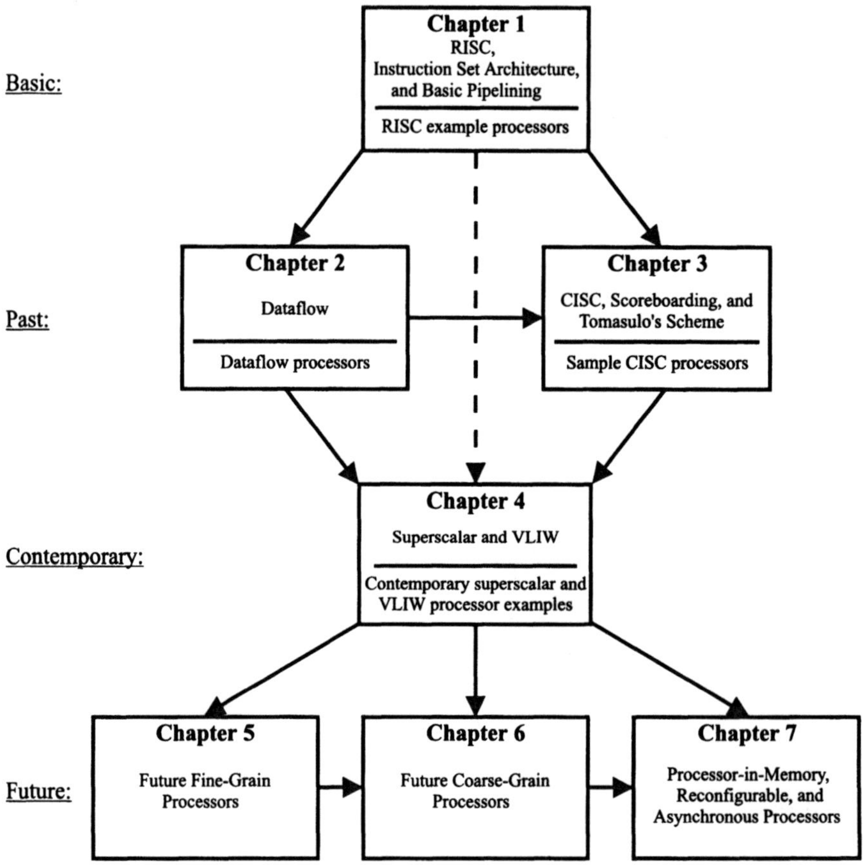

Overview of the book

Chapter 1. Basic Pipelining and Simple RISC Processors. After a period of programming in assembly language, the permanent desire for reduced software costs ultimately resulted in the appearance of high-level programming languages. However, at that time – about two decades after the von Neumann architecture had been proposed – processor design did not provide stream is a branch instruction which potentially interrupts the instruction flow through the pipeline.

VLIW processors use a long instruction word that contains a (normally) fixed number of operations that are fetched, decoded, issued, and executed synchronously. VLIW relies on a sequential stream of long instruction words, i.e., instruction tuples, in contrast to superscalar processors, that issue from a sequential stream of "normal" instructions. The instructions are scheduled statically by the compiler, in contrast to superscalar processors which rely on dynamic scheduling by the hardware. VLIW is not as flexible as superscalar and therefore has been confined to signal processors during the last decade. Recently the VLIW technique has come into focus again in the explicitly parallel instruction computing (EPIC) design style proposed by Intel for its IA-64 processor Merced.

The chapter presents all components of superscalar and VLIW-based multiple-issue processors in detail and provides descriptions of nearly all major superscalar microprocessors.

Chapter 5. Future Processors to Use Fine-Grain Parallelism. Current microprocessors utilize instruction-level parallelism by a deep processor pipeline and by the superscalar instruction issue technique. VLSI technology will allow future generations of microprocessors to exploit aggressively instruction-level parallelism up to 16 or even 32 instructions per cycle. Technological advances will replace the gate-delay by an on-chip wire-delay as the main obstacle to increase chip complexity and cycle rate. The implication for the microarchitecture is a functionally partitioned design with strict nearest neighbor connections.

One proposed solution is a uniprocessor chip featuring a very aggressive superscalar design combined with a trace cache and superspeculative techniques. Superspeculative techniques exceed the classical dataflow limit which says: Even with unlimited machine resources a program cannot execute any faster than the execution of the longest dependence chain introduced by the program's data dependences. Superspeculative processors also speculate about data dependences.

The trace cache stores dynamic instruction traces contiguously and fetches instructions from the trace cache rather than from the instruction cache. Since a dynamic trace of instructions may contain multiple taken branches, there is no need to fetch from multiple targets, as would be necessary when predicting multiple branches and fetching 16 or 32 instructions from the instruction cache.

Multiscalar and trace processors define several processing cores that speculatively execute different parts of a sequential program in parallel. Multiscalar uses a compiler to partition the program segments, whereas a trace processor uses a trace cache to generate dynamically trace segments for the processing cores.

A DataScalar processor runs the same sequential program redundantly on several processing elements with different data sets.

Chapter 6. Future Processors to Use Coarse-Grain Parallelism. The instruction-level parallelism found in a conventional instruction stream is limited. Recent studies have shown the limits of processor utilization even of today's superscalar microprocessors. The solution is the additional utilization of more coarse-grained parallelism. The main approaches are the multiprocessor chip and the multithreaded processor which optimize the throughput of multiprogramming workloads rather than single-thread performance. The multiprocessor chip integrates two or more complete processors on a single chip. Every unit of a processor is duplicated and used independently of its copies on the chip.

In contrast, the multithreaded processor stores multiple contexts in different register sets on the chip. The functional units are multiplexed between the threads in the register sets. Because of the multiple register sets, context switching is very fast. The multiprocessor chip is easier to implement, but does not have the capability of multithreaded processors to tolerate memory latencies, by overlapping the long-latency operations of one thread with the execution of other threads.

The performance of a superscalar processor suffers when instruction-level parallelism is low. The underutilization due to missing instruction-level parallelism can be overcome by simultaneous multithreading, where a processor can issue multiple instructions from multiple threads each cycle. Simultaneous multithreaded processors combine the multithreading technique with a wide-issue superscalar processor such that the full issue bandwidth is utilized by potentially issuing instructions from different threads simultaneously. Depending on the specific simultaneous multithreaded processor design, only a single instruction pipeline is used, or a single issue unit issues instructions from different instruction buffers simultaneously.

Chapter 7. Processor-in-Memory, Reconfigurable, and Asynchronous Processors. Architectural techniques that partly give up the result serialization that is characteristic of von Neumann architectures arise from an on-chip processor–memory integration and from reconfigurable architectures. Such innovations have the potential to define highly parallel chip architectures.

The processor-in-memory or intelligent RAM approach integrates processor and memory on the same chip to increase memory bandwidth. The starting points for processor and memory integration can be either a scalar or superscalar microprocessor chip that is enhanced by RAM memory rather than cache memory, or it can be a RAM chip combined with some computing capacity. Researchers at Sun Microsystems propose a processor-in-memory design that couples a RISC processor with multi-banked DRAM memory.

The Mitsubishi M32R/D is a similar processor-in-memory approach designed for embedded systems applications. Vector intelligent RAM processors couple vector processor execution with large, high-bandwidth, on-chip DRAM banks. The Active Page approach is at the other end of the design spectrum which may be characterized as smart memory approaches. Active pages provide data access and manipulation functions for data arrays integrated in a RAM chip, the processor staying off-chip.

Reconfigurable computing devices replace fixed hardware structures with reconfigurable structures, in order to allow the hardware to adapt to the needs of the application dynamically at run-time. The MorphoSys reconfigurable architecture combines a reconfigurable array of processing elements with a RISC processor core. The Raw architecture approach is a set of replicated tiles, wherein each tile contains a simple RISC-like processor, a small amount of bit-level reconfigurable logic and some memory for instructions and data. Each Raw tile has an associated programmable switch which connects the tiles in a wide-channel point-to-point interconnect. The Xputer defines a non-von-Neumann paradigm implemented on a reconfigurable Datapath Architecture.

Conventional synchronous processors are based on global clocking whereby global synchronization signals control the rate at which different elements operate. For example, all functional units operate in lockstep under the control of a central clock. As the clocks get faster, the chips get bigger and the wires get finer. As a result, it becomes increasingly difficult to ensure that all parts of the processor are ticking along in step with each other.

The asynchronous processors attack clock-related timing problems by asynchronous (or self-timed) design techniques. Asynchronous processors remove the internal clock. Instead of a single central clock that keeps the chip's functional units in step, all parts of an asynchronous processor (e.g., the arithmetic units, the branch units, etc.) work at their own pace, negotiating with each other whenever data needs to be passed between them. Several projects are presented, one of these – the Superscalar Asynchronous Low-Power Processor (SCALP) – is presented in more detail.

Additional Information

The book's home page provides various supplementary information on the book, its topics, and the processor architecture field in general. Lecture slides are available in PowerPoint, PDF, and Postscript, covering the whole book. Over time, enhancements, links to processor architecture-related web sites, corrigenda, and reader's comments will be provided. The book's home page is located at `goethe.ira.uka.de/~ungerer/proc-arch/` and can also be accessed via the Springer-Verlag home page `www.springer.de/cgi-bin/search_book.pl?isbn=3-540-64798-8`.

The home pages of the authors are **www-csd.ijs.si/silc** for Jurij Šilc, **www-csd.ijs.si/robic/robic.html** for Borut Robič, and **goethe.ira.uka.de/people/ungerer** for Theo Ungerer.

Additional information can be drawn from the "WWW Computer Architecture Home Page" of the University of Wisconsin at **www.cs.wisc.edu/~arch/www/** which provides comprehensive information on computer architecture research. Links are provided to architecture research projects, the home pages and email addresses of people in computer architecture, calls for papers, calls for conference participation, technical organizations, etc.

The National Technology Roadmap for Semiconductors **www.sematech.org/public/home.htm** is a description of the semiconductor technology requirements for ensuring advancements in the performance of integrated circuits. Sponsored by the Semiconductor Industry Association (SIA) and published by SEMATECH, this report is the result of a collaborative effort between industry manufacturers and suppliers, government organizations, consortia, and universities.

The CPU Info Center of the University of California, Berkeley, at **infopad.eecs.berkeley.edu/CIC/** collects information on commercial microprocessors such as, e.g., CPU announcements, on-line technical documentations, a die photo gallery, and much more.

An excellent guide to resources on high-performance microprocessors is the "VLSI microprocessor" home page **www.microprocessor.sscc.ru** at the Supercomputer Software Department RAS.

The newest commercial microprocessors are presented at the Microprocessor Forum, the Hot Chips Conference, and the International Solid State Circuit Conference (ISSCC). The most important conferences on research in processor architecture are the Annual International Symposium on Computer Architecture (ISCA), the biennial International Conference on Architectural Support for Programming Languages and Operating Systems (ASPLOS), the International Symposium on High-Performance Computer Architecture (HPCA), the Annual International Symposium on Microarchitecture (MICRO), and the Parallel Architectures and Compilation Techniques (PACT) conferences.

Acknowledgments

Several people in academia have provided invaluable assistance in reviewing drafts of the book. We would especially like to thank *Bruce Shriver* who greatly influenced the structure and terminology of this book by sharing his experiences from his own recent book "The Anatomy of a High-Performance Microprocessor".

Jochen Kreuzinger of the University of Karlsruhe and *Andreas Unger* of the University of Jena provided thorough and detailed critics of several chapters of the book leading to considerable improvements in the text. *Silvia Müller* of the University of Saarbrücken, *Ulrich Sigmund* of Viona GmbH, Karlsruhe, and *Ulrich Nageldinger* of the University of Kaiserslautern provided valuable improvements to the sections on scoreboarding and Tomasulo's scheme, superscalar processors, and reconfigurable computing. *Yuetsu Kodama* of the Electrotechnical Laboratory at Tsukuba provided invaluable information on threaded dataflow and, especially, on the EM-X and RWC-1 parallel computers. Several contributions to the chapter on RISC processors are due to *Veljko M. Milutinović* of the University of Belgrade. Numerous suggestions and improvements in the chapter on future processors using fine-grain parallelism have been made by *Lalit Mohar Patnaik* of the Indian Institute of Science at Bangalore. We have also received significant help from *Nader Bagherzadeh* of the University of California at Irvine, *Reiner W. Hartenstein* of the University of Kaiserslautern, and *Krishna M. Kavi* of the University of Alabama at Huntsville. Many thanks to all of them.

We also thank *Daniela Tautz* of the University of Karlsruhe who has drawn several figures, and *Gregor Papa* of the Jožef Stefan Institute, Ljubljana, who has drawn the tables.

Numerous students of the University of Karlsruhe have improved this book by their questions and ideas. In particular we have to thank *Alexander Schulz*, *Jan Albiez*, and *Patrick Ohly*.

Finally we would like to express our thanks to our editors, *Hermann Engesser* and *Brygida Georgiadis*, and the anonymous copy-editor of Springer-Verlag.

Ljubljana and Karlsruhe, March 1999

Jurij Šilc
Borut Robič
Theo Ungerer

Contents

1. Basic Pipelining and Simple RISC Processors

What is "reduced" in a RISC? Practically everything: the number of instructions, addressing modes, and formats...
... The application area of RISCs is expected to widen in the future.

Daniel Tabak
Advanced Microprocessors
(McGraw-Hill, 1995)

1.1 The RISC Movement in Processor Architecture

CISC. Conventional state-of-the-art computers in the 1960s and 1970s, exemplified by the IBM System/370, and the DEC PDP-11 minicomputer series and VAX-11/780 super minicomputer, were rack-based machines implemented with discrete logic and only rarely with microchips. A processor of such a machine used a complex instruction set, consisting of as many as 304 instructions, 16 addressing modes, and more than 10 different instruction lengths in the case of VAX.

What were the reasons for such a large number of instructions? In fact, computers before the 1960s were limited in their instruction sets. That was due to the hardware technology of that time. Computer architecture of the 1960s and early 1970s was dominated by high hardware cost, in particular by the high cost of memory. Technology at that time only allowed a small main memory and slow memory access compared to more recent technology. High-speed local memory was not yet available except for a few general-purpose registers. Instruction fetch was done from main memory and could be overlapped with decode and execution of previous instructions. As a result, CISC processors were based on the observation that the number of cycles per instruction was determined by the number of cycles taken to fetch the instruction. It was acceptable to increase the average number of cycles taken to decode and execute an instruction. To improve performance, the two principal goals of CISC were to reduce the number of instructions and to encode these instructions densely. With these assumptions, CISC processors evolved densely encoded instructions at the expense of decode and execution time inside the processor. Multiple-cycle instructions reduce the overall number of instructions, and thus reduce the overall execution time because they reduce the instruction-fetch time (*Johnson* [150]).

In addition, as long as they were programmed in assembler, the programmer's view of the machine was close to the computer architect's view. However, the permanent desire for reducing software cost by simplifying the task of software design ultimately resulted in appearance of *high-level languages* (HLL). Because the computer design did not provide any hardware-based support for HLL features (such as array management, handling of procedure parameter passing, process and memory management, etc.), this introduced a wide *semantic gap* between HLL and the machine design. Thus the programmer's view of the machine deviated from the architect's view. In the 1970s, several breakthroughs in microelectronic technology made it possible to replace software with hardware. This was done in HLL computer architectures, which attempted to incorporate HLL features in their instruction sets. A HLL instruction set provided powerful instructions with a wide range of flexibility. This resulted in *complex instruction set computers* (CISC) characterized by a complex *instruction set architecture* (ISA), i.e., a large number of instructions, addressing modes, and instruction formats. Unfortunately, these instructions often did more work than was required in the frequent case, or they did not even exactly match the requirements of the language.

RISC. In mid-1970s, researcher noticed that, instead of providing the ISA with a large number of instructions, a promising approach would be to support only the most frequently used instructions (see Table 1.1) while leaving less frequent operations to be implemented as instruction sequences. Systems having such a reduced ISA were called *reduced instruction set computers* (RISC).

Table 1.1. The ten most frequently used instructions in the SPECint92 benchmark suite for the CISC Intel x86 microprocessor

Instruction	Average (% total executions)
load	22
conditional branch	20
compare	16
store	12
add	8
and	6
sub	5
move register-register	4
call	1
return	1
Total	95

Historically, ideas of having a small number of instructions can be traced back to 1964, when the Control Data Corporation CDC 6600 supercomputer (see *Thornton* [295, 296]) used a small (64 opcodes) load/store and register-register instruction set. In the mid-1970s, when researchers at IBM developed the 801 architecture (see *Radin* [237]), they found that about 80 % of the computations of a typical program required only about 20 % of the instructions in a processor's instruction set. The most frequently used instructions were simple instructions such as `load`, `store`, and `add`. IBM 801 emphasized the importance of the cooperation between a well-chosen set of simple instructions implemented directly in hardware and an optimizing compiler. Because this approach abandoned microcode implementations of complex instructions in favor of a few simple instructions implemented with hardwired control, it was called *RISC architecture*. Some time after the IBM 801, around 1980, researchers at the University of California at Berkeley (*Patterson and Ditzel* [231]) and at Stanford University (*Hennessy et al.* [133]) began parallel efforts in RISC technology. The Berkeley team concentrated on understanding the principles for achieving the most effective use of the area on a VLSI chip for building computers. Since the limited number of instructions required a relatively small amount of on-chip control logic, more chip space could be used for other functions, thus enhancing the performance and versatility of the processor. For example, efficient procedure parameter passing was enabled by having a large CPU *register file*, whose global registers were accessible to all procedures while the so-called window registers acted both as input registers for one procedure and output registers for another. This *register window* approach is nowadays used in the Scalable Processor ARChitecture (SPARC) family of microprocessors (see Sect. 1.7.2). The Stanford team concentrated in its project, called *Microprocessor without Interlocking Pipeline Stages* (MIPS), on combining an optimizing compiler with the design of a VLSI RISC processor. MIPS used the pipeline technique to enable a number of instructions to be active at once. Since the pipeline hardware was not able to recognize so-called pipeline hazards, the compiler had to guarantee correct instruction execution in the pipeline and improve its efficiency. MIPS has been also called the *single register set* approach because it uses a small register file with no register windows. The developments of Berkeley's RISC and Stanford's MIPS initiated a number of projects that resulted in modern scalar and superscalar RISC microprocessors.

The fundamental theme of RISC architectural design is that of maximizing the effective speed of a design by performing most functions in software, except those whose inclusion in hardware yields a net performance gain. The basic design principles are:

- *Simple instructions and few addressing modes*: CISC architecture includes an extensive set of ways for addressing system memory, many of which require the processor to use several different parameters to calculate an effective address during program execution. Complex instructions and addressing modes warrant microcode or multi cycle execution and complicate the processor and compiler design. In contrast, RISC designs are able to eliminate microcode because they have a small and simple set of instructions and addressing modes. The ISA is designed so that most instructions remain only a single cycle in each pipeline stage.

- *Register-register (or load/store) design*: References to data in system memory are limited to load/store instructions. All other instructions operate on data in registers. A load instruction moves data from memory to registers, where the data can be rapidly processed and temporarily held for further access. When appropriate, a store instruction returns the data to its place in memory. In contrast, CISC architectures support arithmetic-logic instructions denoting memory operands.
- *Pipelining*: Modern processor design generally includes a multistage pipeline to increase the rate of instruction execution. Because each pipeline stage is responsible for an individual execution phase, such as instruction decoding or operand fetching, a pipelined processor is actually working on several instructions simultaneously. CISC processor pipelines are subject to various kinds of inefficiencies, such as the use of complex addressing modes, that slow down instruction execution in the pipeline. In a RISC design, however, the predictability of the time required to perform instructions allows pipelines to operate with high efficiency.
- *Hardwired control, with little or no microcode*: RISC designs eliminate microcode ROM and implement instructions directly in hardware. This means that there is no translation from a machine instruction to primitive microcoded operations, which would increase the number of cycles required to execute the instruction. In the case of a RISC implemented as a microprocessor, this also frees up chip space and gives the opportunity to use it for performance-enhancing functions.
- *Reliance on optimizing compilers*: Optimizing compilers recognize when the contents of a register can be re-used in subsequent instructions without reloading the data from system memory. When a memory reference cannot be avoided, the compiler rearranges the instructions so that useful work that is not dependent on the referenced data can be performed while the processor is waiting for the data to be loaded into a register. The large number of registers, the small, simple instruction set, and the limited number of addressing modes in RISC designs make it easier for an optimizing compiler to limit references to memory, recognize computations that can be streamlined, and reorganize instructions to ensure maximum pipeline efficiency. An optimizing compiler for RISC machines can calculate the savings of an optimization more simply because, ideally, all instructions require equal execution time.
- *High-performance memory hierarchy*: As RISC design increases the performance of the CPU, it is important to provide a fast, efficient memory hierarchy to keep pace with the processor. In a RISC system, the memory

hierarchy typically consists of a large register file (i.e., a set of on-chip registers), fast static RAMs for split *data cache* (D-cache) and *instruction cache* (I-cache), and write buffers. Usually, there is also an on-chip memory management unit. Recent advances in the cost, density, and speed of semiconductor memory have contributed to the ability to design high-performance memory hierarchies which support the speed of a RISC processor.

We will investigate ISA features and basic pipelining in the rest of this chapter in the context of RISC processors, because the relative simplicity of RISC facilitates understanding. However, most of the techniques apply equally well to either RISC or CISC processors.

1.2 Instruction Set Architecture

The instruction set architecture refers to the programmer-visible instruction set. It defines the boundary between hardware and software. Often it is identified with the *processor architecture*. The *processor microarchitecture* refers to the internal organization of the processor. The microarchitecture comprises implementation techniques like the number and type of pipeline stages, issue bandwidth, number of FUs, size and organization of on-chip cache memories, etc. All these features cannot be seen in the ISA. Yet, an optimizing compiler may also use the knowledge of microarchitecture features. Several specific processors with differing microarchitectures may share the same architecture.

The programmer's view of the machine depends on the answers to the following five questions:

- How is data represented?
- Where can data be stored?
- How can data be accessed?
- What operations can be done on data?
- How are instructions encoded?

The answers to these questions define the *instruction set architecture* (ISA) of the machine. In what follows, we describe some common features of RISC ISA. We will partly follow the presentation given by *Tabak* [286].

How is data represented? The programmer can usually declare data of different *data formats*. One of the key ISA issues is to support several data formats by providing representations for characters, integers, floating-point numbers, etc. For instance, in DEC Alpha there is byte, 16-bit word, 32-bit longword, and 64-bit quadword. Integer data formats can be signed or unsigned. There are two ways of ordering byte addresses within a word, *big-endian* (most significant byte first) and *little-endian* (least significant byte first). There are also packed and unpacked BCD numbers, and ASCII characters. Floating-point data formats can be, according to the ANSI/IEEE 754-1985 standard, basic or extended, each having two widths, single or double. Multimedia data formats are 32-bit or 64-bit words (sometimes also 128-bit) including several 8-bit or 16-bit pixel representations.

Where can data be stored? For storing data, several *address spaces* are often distinguished by the (assembly language) programmer, such as register space, stack space, heap space, text space, I/O space, and control space. Except for the registers, all other address spaces are mapped onto a single contiguous memory address space, which is accessed by the RISC processor. A RISC ISA additionally contains a *register file*, which consists of a relatively large number of general-purpose CPU registers. Early RISC processors contained thirty-two 32-bit general purpose registers, or register windowing (RISC I and SPARC processors). Contemporary RISC processors provide an additional register set with thirty-two 64-bit floating-point registers.

How can data be accessed? The way in which data can be accessed is defined by the *addressing mode*. In modern processors several of the following addressing modes can be found (Table 1.2):

- *Register* mode, which is used when the operand is stored in one of the registers.
- *Immediate* (or *literal*) mode, if the operand is a part of the instruction.
- *Direct* (or *absolute*) mode, where the address of the operand in memory is stored in the instruction.
- *Register indirect* (or *register deferred*) mode, where the address of the operand in memory is stored in one of the registers.
- *Autoincrement* (or *register indirect with postincrement*) mode, which is like register indirect, except that the content of the register is incremented after the use of the address. This mode offers an automatic address increment useful in loops and in accessing byte, halfword, or word arrays of operands.
- *Autodecrement* (or *register indirect with predecrement*) mode, where the content of the register is decremented and is then used as a register indirect address. This mode can be used to scan an array in the direction of decreasing indices.
- *Displacement* (also *register indirect with displacement* or *based*) mode, where the effective address of the operand is the sum of the contents of a register and a value, called displacement, specified in the instruction.
- *Indexed and scaled indexed* mode that works essentially as the register indirect. The register containing the address is called the index register. The main difference between the register indirect and the indexed modes is that the contents of the index register can be scaled by a scale factor (e.g., 1, 2, 4, 8 or 16). The availability of the scale factor, along with the index

Table 1.2. Addressing modes

Addressing mode	Example instruction / Meaning
Register	`load Reg1,Reg2` `Reg1 ← (Reg2)`
Immediate	`load Reg1,#const` `Reg1 ← const`
Direct	`load Reg1,(const)` `Reg1 ← Mem[const]`
Register indirect	`load Reg1,(Reg2)` `Reg1 ← Mem[(Reg2)]`
Autoincrement	`load Reg1,(Reg2)+` `Reg1 ← Mem[(Reg2)], Reg2 ← (Reg2) + step`
Autodecrement	`load Reg1,-(Reg2)` `Reg2 ← (Reg2) - step, Reg1 ← Mem[(Reg2)]`
Displacement	`load Reg1,displ(Reg2)` `Reg1 ← Mem[displ + (Reg2)]`
Indexed and scaled indexed	`load Reg1,(Reg2*scale)` `Reg1 ← Mem[(Reg2)*scale]`
Indirect scaled indexed	`load Reg1,(Reg2,Reg3*scale)` `Reg1 ← Mem[(Reg2) + (Reg3)*scale]`
Indirect scaled indexed with displacement	`load Reg1,displ(Reg2,Reg3*scale)` `Reg1 ← Mem[displ + (Reg2) + (Reg3)*scale]`
PC-relative	`branch displ` `PC ← PC + step + displ` (if branch taken)

`const, displ` decimal, hexadecimal, octal or binary numbers
`step ...` e.g., 4 in systems with 4-byte uniform instruction size
`scale ...` scaling factor, e.g., 1, 2, 4, 8, 16

register, permits scanning of data structures of any size, at any desired step.

- *Indirect scaled indexed* mode, where the effective address is the sum of the contents of the register and the scaled contents of the index register.
- *Indirect scaled indexed with displacement* mode, which is essentially as the indirect scaled indexed, except that a displacement is added to form the effective address.
- *PC-relative* mode, where a displacement is added to the program counter (PC). The PC-relative mode is used automatically with program control

instructions in many systems. Branches and jumps that use PC-relative addressing mode are advantageous, because the targets are often near the current branch/jump instruction. Therefore specifying the displacement requires fewer bits.

RISC ISAs have a small number of addressing modes, not usually exceeding four. Note that the displacement mode already includes the direct mode (by setting the register content to zero) and the register indirect mode (by setting the displacement to zero).

What operations can be performed on data? One of the key features of a computer is its *instruction set*, i.e., a specific set of basic operations that a given computer can perform. One can distinguish the following types of instructions:

- *Data movement instructions*, which transfer data from one location to another. When there is a separate I/O address space, these instructions also include special I/O instructions. Stack manipulation instructions (e.g., **push**, **pop**) also fall into this category.
- *Integer arithmetic and logical instructions*, which can be one-operand (e.g., complement), two-operand, or three-operand instructions, with the latter offering a more compact program code. In some processors, different instructions are used for different data formats of their operands. For instance, there may be separate signed and unsigned multiply/divide instructions.
- *Shift and rotate instructions*, which perform either left or right shifts and rotations. There are two types of shifts,[1] logical and arithmetic, depending on what is transferred into the vacated positions.
- *Bit manipulation instructions*, which operate on specified fields of bits. The field is specified by its width and offset from the beginning of the word. Instructions usually include test (affecting certain flags), set, clear, and possibly others.
- *Multimedia instructions* can process multiple sets of small operands and obtain multiple results with a single instruction. The operations include packing and unpacking, arithmetic, comparisons, logic, and shifting of values that usually represent pixels. The pixels are represented as packed data types, such as eight bytes, four 16-bit words, or two 32-bit doublewords, all packed inside one 64-bit quadword.
- *Floating-point instructions*, which may include, depending on the system, floating-point data movement, arithmetic, comparison, square root, absolute value, transcendental functions, and others.
- *Control transfer instructions*, which consist primarily of jumps, branches, procedure calls, and procedure returns. We assume that jumps are uncon-

[1] Many modern microprocessors perform fast shifting by special hardware, such as barrel shifters.

ditional and branches are conditional. Some systems may also have return from exception instructions.

- *System control instructions*, which allow the user to influence directly the operation of the processor and other parts of the computer system.
- *Special function unit instructions*, which perform particular operations on special function units (e.g., graphic units). Another type of special instruction is the atomic instruction for controlling access to critical sections in multiprocessors.

Depending on how its operands are specified, an instruction can, in principle, be one of the following types: *register-register*, *memory-register*, *register-memory*, or *memory-memory*.

In a RISC ISA, all operations except load and store are register-register instructions (an ISA of this type is called a load/store ISA). As for addressing modes, the number of instructions is also reduced in a RISC ISA (e.g., up to 128).

Table 1.3. Program $\begin{array}{l} C=A+B \\ D=C-B \end{array}$ coded in four classes of ISA instruction formats

Register-Register	Machine Register-Memory	Accumulator	Stack
`load Reg1,A`	`load Reg1,A`	`load A`	`push B`
`load Reg2,B`	`add Reg1,B`	`add B`	`push A`
`add Reg3,Reg1,Reg2`	`store C,Reg1`	`store C`	`add`
`store C,Reg3`	`load Reg1,C`	`load C`	`pop C`
`load Reg1,C`	`sub Reg1,B`	`sub B`	`push B`
`load Reg2,B`	`store D,Reg1`	`store D`	`push C`
`sub Reg3,Reg1,Reg2`			`sub`
`store D,Reg3`			`pop D`

How are instructions encoded? The *instruction format* specifies the pieces of information needed to execute an instruction. Besides the instruction *opcode*, other addresses may be needed to specify *sources* (i.e., operands), *destination*, and the next instruction. The next instruction is explicitly specified in the case of control transfer instructions; for other types of instructions it is implicitly defined by the PC. With respect to arithmetic-logic instructions, we distinguish four classes of ISAs, each of them characterizing the corresponding machine (Table 1.3):

- 3-address instruction format consisting of |**opcode**|**Dest**|**Src1**|**Scr2**| and used by the *register-register* (also called load/store) machine.
- 2-address instruction format consisting of |**opcode**|**Dest/Src1**|**Scr2**|, often supported by *register-memory* machines.

- 1-address instruction format consisting of |`opcode`|`Src`| and supported by the *accumulator* machine.
- 0-address instruction format consisting only of |`opcode`| and supported by the *stack* machine.

ISA encoding can be fixed length with a 32-bit format or variable length. Most RISC ISAs use a 3-address instruction format where all instructions have a uniform length of 32 bits. CISC ISAs often use register-memory with variable instruction lengths. Variable instruction length can also be found in stack machines, exemplified today by the Java processors. Accumulator machines are today mostly found in microcontrollers.

1.3 Examples of RISC ISAs

In the following we give some examples of (super)scalar RISC instruction set architectures. These include SPARC, MIPS, ARM, HP PA-RISC, DEC Alpha, IBM POWER, and IBM/Motorola/Apple PowerPC. Many of them have appeared in several versions and found their implementation in several microprocessors. The primary objective of RISC ISAs was to be sufficiently simple so that implementations could have a very short cycle time, which would result in processors that could execute instructions at the fastest possible clock rate.

SPARC ISA. SPARC is the industry's only openly defined and evolved RISC architecture. Unlike other RISC designs, the SPARC Architecture Committee did not specify a hardware implementation, but an open ISA devoted to the community of SPARC vendors and users. After SPARC ISA version 7 (1986) there were two versions of the SPARC ISA: a 32-bit version 8 (1990) and a 64-bit version 9 (1992). Table 1.4 gives some vendors and their products complying with the SPARC ISA proposal.

Table 1.4. Some SPARC vendors and their processors

Vendor	Processor	Year	ISA Version
Fujitsu Microelectronics, Inc.	SPARClite	1992	8
Ross Technology	HyperSPARC	1993	8
Sun Microelectronics, Inc.	SuperSPARC	1993	8
SIDSA	SPARC	1994	8
HAL Computer Systems	SPARC 64	1995	9
Sun Microelectronics, Inc.	UltraSPARC	1995	9
T.sqware	SPARClet	1996	8
Fujitsu Microelectronics, Inc.	TurboSPARC	1996	8

MIPS ISA. MIPS Technologies (former MIPS Computer Systems) has defined several versions of ISA that were implemented in several CPU designs:

Table 1.5. MIPS II ISA

Data formats	byte, 16-bit halfword, 32-bit word, 64-bit doubleword big- or little-endian, ANSI/IEEE 754-1985
Register file	Integer (CPU) registers (64- or 32-bit): 32 registers `r0` to `r31`, program counter `PC`, two multiply and divide registers `HI` (remainder for divide) and `LO` (quotient for divide); `r0` is hardwired to a zero, `r31` is the link register for jumps and link instructions. Floating-point (FPU) registers: 32 floating-point registers `FGR0` to `FGR31`; can be configured as 16 64-bit registers; 32-bit implementation/revision register `FCR0` with implementation and revision number of the FPU, 32-bit control/status register `FCR31`.
Addressing modes	register, immediate, register indirect, displacement, PC-relative
Instruction set (163)	load/store (24), computational (51), jump and branch (22), special (2), exception (16), floating point (30), coprocessor (9), memory management (9)
Instruction formats	register-register, 3-address format Immediate (I-types): 6-bit opcode, 5-bit src register specifier, 5-bit dst register specifier or branch condition, 16-bit immediate value or branch displacement. Jump (J-types): 6-bit opcode, 26-bit jump target address. Register (R-type): 6-bit opcode, 5-bit src register specifier, 5-bit src register specifier, 5-bit dst register specifier, 5-bit shift amount, 6-bit function field

- MIPS I (1984, 32-bit, implemented in R2000, R3000),
- MIPS II (1990, 64-bit coprocessor, implemented in R6000, see Table 1.5),
- MIPS III (1991, 64-bit, implemented in R4000, R4400),
- MIPS IV (1994, 64-bit, in R5000, R7000, R8000, and R10000),
- MIPS V (1996, 64-bit, implemented in R12000).

In 1996, MIPS Technologies defined a multimedia extension to their ISA called[2] MDMX (MIPS Digital Media eXtensions) which is used by the R12000 processor (see Sect. 4.9.8). At the same time, a MIPS_{16} ISA was

[2] ...and pronounced "Mad Max"

defined to be used in a 16-bit TinyRISC processor. In Table 1.5 we give some basic features of MIPS II ISA.

Advanced RISC Machines ARM ISA. Since 1986, when the first ARM ISA appeared, four main ISAs of this family were defined. Table 1.6 presents several processor implementations based on the four ISAs. After the 64-bit ISA version 4, a low cost ISA version 4T was designed for the 16-bit Thumb instruction set (as was the case with the $MIPS_{16}$ ISA).

Table 1.6. ARM ISAs and implementations

ISA version	Implementations	Selected features
1	ARM1	
2	ARM2	Added multipliers, coprocessors
2a	ARM3	Added SWP, system coprocessor
3	ARM6, ARM7	Added 32-bit, more exception modes, separate processor state registers (PSR)
3G		Removed backward compatibility with 2a
3M	ARM7DM	Added long and signed multipliers
4	ARM8, StrongARM	Added system mode, signed-byte and halfword loads/stores
4T	ARM7T, ARM9T	Thumb instruction set

HP PA-RISC ISA. The PA-RISC 1.0 ISA was defined in the early 1980s (*Mahon et al.* [189], 1986) to be a single architecture that would efficiently span Hewlett-Packard's three computer lines: the HP3000 commercial minicomputers, the HP9000 technical workstations and servers, and the HP1000 real-time controllers. Before introduction, the project was named SPECTRUM. At introduction in 1986, it was known as HP's Precision Architecture, HP-PA, or just PA. Subsequently, the ISA was called PA-RISC. Since its introduction, the PA-RISC ISA has remained remarkably stable. Only minor changes were made over the next decade to facilitate higher performance in floating-point and system processing. In 1989, driven by the performance needs of the HP9000 technical workstation line, PA-RISC 1.1 ISA [131] was introduced with improved floating-point capabilities, and added architectural extensions to speed up the processing of performance-sensitive abnormal events, such as misses in the translation look-aside buffer (TLB, see p. 18). PA-RISC 1.1 also added support for both endian formats (previously, PA-RISC 1.0 was a consistently big-endian machine). The next ISA, PA-RISC 2.0, was the first to make user-visible changes to the core integer architecture (*Kane* [153], 1995). In addition to support for 64-bit integer data and 64-bit addresses, other user-visible changes have been added to enhance the performance of new user workloads.

Table 1.7. DEC Alpha ISA

Data formats	byte, 16-bit word, 32-bit longword, 64-bit quadword little-endian, ANSI/IEEE 754-1985, VAX floating-point
Register file	Integer registers: 32 64-bit registers `R0` to `R31`, program counter `PC`, `R30` is designated as a stack pointer (SP), `R31` is always equal zero (hardwired to a zero value). Floating-point registers: 32 64-bit floating-point registers `F0` to `F31`, `F31` is always equal zero (hardwired to a zero value).
Addressing modes	register, immediate, displacement, PC-relative
Instruction set (155)	integer load/store (12), integer control (14), integer arithmetic (20), logical and shift (17), byte manipulation (24), floating-point load/store (8), floating-point control (6), floating-point operate (47), miscellaneous (7)
Instruction formats	register-register, 3-address format Memory instructions: 6-bit opcode, 5-bit src register specifier, 5-bit src register specifier, 16-bit memory dst field, or function field (for miscellaneous instruction). Conditional branch instructions: 6-bit opcode, 5-bit branch condition, 21-bit branch displacement. Operate instructions: 6-bit opcode, 5-bit src register specifier, 5-bit src register specifier + 3-bit should be zero (if 12th bit is 0), or 8-bit literal (if 12th bit is 1), 7 bit function field, 5 bit dst register specifier. Floating-point operate instructions: 6-bit opcode, 5-bit src floating-point register specifier, 5-bit src floating-point specifier, 11-bit function field, 5-bit dst floating-point register destination. PALcode instructions: 6-bit opcode, 26-bit Privileged Architecture Library code.

For example, Multimedia Acceleration eXtension (MAX) and later MAX-2 have been added to speed up multimedia processing on the main processor, rather than on separate hardware. However, the principal aim of keeping the programming model stable has been carried forward as much as possible in the 64-bit version of the architecture.

DEC Alpha ISA. Alpha is a 64-bit load/store RISC ISA that was designed in 1992 with particular emphasis on the three elements that mostly affect performance: clock speed, multiple instruction issue, and multiple processors. Table 1.7 gives a brief description of Alpha ISA. In 1997 DEC announced Motion Video Instructions (MVI), an extension first implemented in the Alpha 21164PC processor.

IBM POWER ISA. The IBM's POWER (Performance Optimization With Enhanced RISC) Architecture ISA was defined in 1990 and incorporated characteristics common to most other RISC ISAs (*Levine and Thurber* [180], *Weiss and Smith* [323]). It was unique among the existing RISC architectures in that it was functionally partitioned, separating the functions of program flow control, integer computation, and floating-point computation. The architecture's partitioning facilitated the implementation of superscalar designs, in which multiple functional units concurrently executed independent instructions. In order to avoid the need for a separate address computation after each memory access during array manipulation, the update forms of most load/store instructions were included. These update forms perform the memory access and place the updated address (i.e., the address of the next location to be accessed) in the base address register. In addition, a common operation in many floating-point computational algorithms is to add a value to the product of two other values. This observation prompted the inclusion of a floating-point multiply-add instruction, which performs this operation.

IBM/Motorola/Apple PowerPC ISA. Early in 1991, IBM, Motorola, and Apple formed a partnership whose foundation was the use of a common ISA derived from the POWER architecture. The PowerPC ISA included most of the POWER instructions. Nearly all the excluded POWER instructions were those that executed infrequently and the compiler could replace by several other instructions that were in both POWER and PowerPC. The new PowerPC ISA was expected to permit a broad range of implementations from low-cost controllers to high-performance processors (with short cycle times, aggressive superscalar and multiprocessor features) and offer a 64-bit architecture that is a superset of the 32-bit architecture, thus providing binary compatibility for 32-bit applications. The PowerPC ISA achieved these goals and permits POWER customers to run their existing applications on new systems and to run new applications on their existing systems (see *Diefendorff et al.* [66, 67], *May et al.* [195], *Weiss and Smith* [323]).

1.4 Basic Structure of a RISC Processor and Basic Cache MMU Organization

A simple RISC CPU (as shown in Fig. 1.1) consists of an arithmetic logic unit (ALU) which is connected by two operand input buses and one result bus with the register file of thirty-two 32-bit general-purpose registers. Only load/store instructions load or store data from the D-cache to a register or vice versa. Instruction execution is controlled by the pipeline decode and control unit, which receives its instructions from the instruction fetch unit. The instruction fetch unit fetches a single instruction each cycle from the I-cache. The instruction address is provided by the program counter (PC) which is automatically incremented by four assuming the usual 32-bit instructions and a byte-addressable machine. In the case of previous control transfer instructions, the PC is loaded by a jump or branch target address.

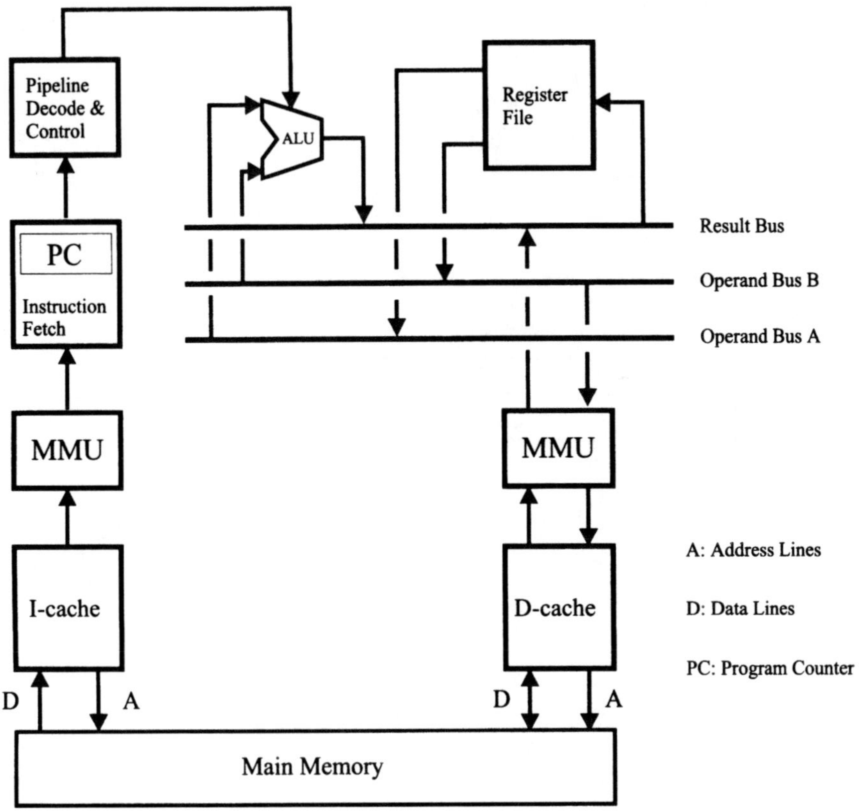

Fig. 1.1. Data path organization of a simple RISC processor

As semiconductor technology advanced towards yielding chips with higher component densities, the goal of RISC architecture has been to develop processors that can come close to issuing one instruction per clock cycle of the machine. The measures *clock cycles per instruction* (CPI) or *instructions per clock cycle* (IPC) are commonly used to characterize high performance processor architectures. The achievement of CPI $= 1$ has been made possible by two architectural features: *cache memories* and *instruction pipelining.* Today's multiple issue processors reach CPI < 1, or IPC > 1. Processors that can issue only one instruction at a time are said to be *scalar*; those that can issue more than one instruction simultaneously are said to be *superscalar.*

Caches. A salient feature of a modern RISC microprocessor is the use of on-chip L1 (level-one or primary) caches and associated MMUs. Both are described only briefly here. The primary cache is usually split into two cache memories, *I-cache* (instruction cache) and *D-cache* (data cache), located in the retrieval paths for instructions and data, respectively (see Fig. 1.1). Separate I-cache and D-cache eliminate the structural hazards that arise

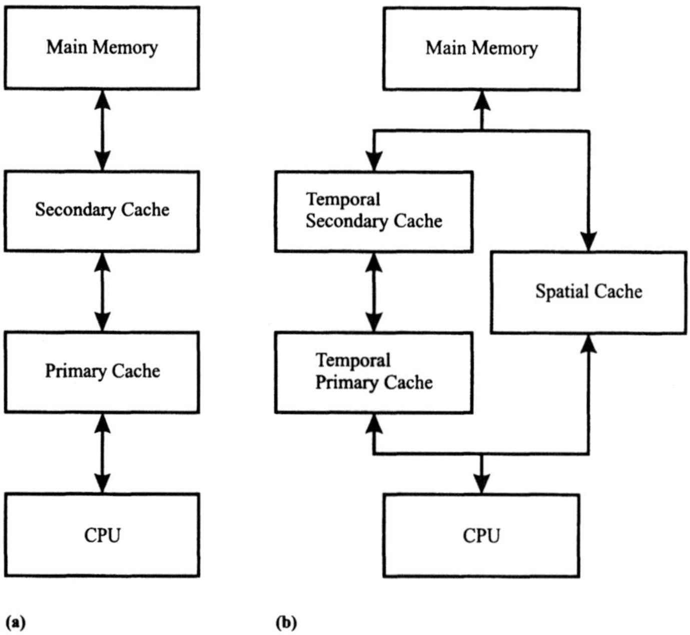

Fig. 1.2. Cache memory organization (a) conventional (b) split temporal/spatial

otherwise in a pipeline when the instruction fetch and a memory access occur in the same cycle. Each cache is organized into *sets* that are indexed by part of the effective address (instruction address for the I-cache and data address for the D-cache). Each set holds a single cache line (in the case of a *direct mapped* cache organization) or typically two or four cache lines (in the cases of so-called *2- or 4-way set-associative* cache organization). If there exists only a single set, the cache is called *fully associative.* A cache line typically consists of 32 bytes (denoting 8 instructions or an equivalent amount of data), a cache tag and state bits. During cache access the set is located by part of the address and then the tag of the cache line is compared with another part of the address. In the case of a n-way set-associative cache organization, all the tags within a set must be compared in parallel. If the tag and address part match and the cache line state bits signal a valid entry, a cache hit occurs and the least significant bits in the address locate the instruction or value to be loaded within the cache line. In the case of a cache miss, the missing cache line is fetched from main memory (or from a secondary, that is, L2 cache) resulting in several cycles of waiting time – bubbles in the pipeline. The newly arriving cache line replaces a cache line already in the denoted cache set (a pseudo LRU replacement strategy is often used in the case of a set-associative cache organization). Depending on the state bits, the replaced cache line has to be written back to memory in case of previous store accesses to that cache line. Note that the I-cache may be made read-only to reduce complexity, because instructions are only fetched from, but never stored in, the I-cache (at least when self-modifying code is disabled). If most memory accesses are cache hits in the on-chip cache memories, the average main memory access time and the number of pipeline bubbles are greatly reduced.

Cache memory architectures. Existing cache memory architectures are based on the application characteristics of *spatial locality* (i.e., items whose addresses are near one another tend to be referenced close together in time) and *temporal locality* (i.e., recently accessed items are likely to be accessed again in the near future). The existing cache architectures imply one cache organization (see Fig. 1.2a) which benefits from both localities. *Milutinović et al.* [204] describe the split temporal/spatial cache (see Fig. 1.2b), where they take special advantage of one locality type in one set of conditions, and the other locality type in another set of conditions. At compile-time, the data are classified as those exhibiting predominantly temporal locality or spatial locality. The "temporal" data need a cache hierarchy, with smaller cache capacity at each level satisfying the needs. The "spatial" data do not need any hierarchy, and a relatively small prefetch buffer is expected to satisfy the needs. This cache splitting makes the overall cache memory considerably smaller for approximately the same performance.

Memory Management Unit. The memory management is performed by the *memory management unit* (MMU), whose task is the translation of the virtual address into a physical address. Its primary functions are:

- Inclusion and management of a fast-access *translation lookaside buffer* (TLB) for virtual to physical page address translation. The TLB is organized like a fully associative cache and usually contains 32 to 256 entries. Accessing the TLB takes a single machine cycle or less.
- Support of a paging mechanism involved in the virtual memory organization, for the segmentation mechanism (if implemented) and for memory protection.

The MMU access is often overlapped with the set location during cache access – a cache organization that is called virtually indexed, physically tagged. Otherwise the MMU access can be done before the cache access (a so-called *physically addressed* cache) or after cache access in the case of a cache miss (a so-called *virtually addressed* cache). *Physically tagged* caches require the cache to be compared with the physical address from the MMU. In such environments cache miss detection may be a bottleneck in the MMU. A *virtually tagged* cache uses virtual addresses when attempting to find the required word in the cache. The least significant part of the virtual address is used to access one line of the cache (direct mapped) that may contain the required word. The most significant part of the virtual address is then compared with the tag address bits for a possible match, or cache hit. This scheme ensures that cache misses are quickly detected, and that addresses are translated only on a cache miss.

For more details on caches and MMU organization, see *Hennessy and Patterson* [134] and *Shriver and Smith* [258].

1.5 Basic Pipeline Stages

One of the major features of modern processors (especially RISC processors) is the use of a pipelined instruction execution to achieve an average CPI close to 1. Pipelining is an implementation technique whereby multiple instructions are overlapped in execution. It is not visible to the programmer. Each step is called a *pipe stage* or *pipe segment*. Pipeline stages are separated by clocked *pipeline registers* (also called *latches*). A *pipeline machine cycle* is the time required to move an instruction one step down the pipeline.

Ideally, in a *k-stage pipeline* an instruction is executed in k cycles by k stages. If instruction fetching into the pipeline continues, then at any time – assuming ideal conditions – k instructions will be handled simultaneously and it will take k cycles for each instruction to leave the pipeline. We define *latency* to be the total time needed for an instruction to pass through all k stages of the pipeline. The *throughput* of the pipeline is defined to be the number of instructions that can leave a pipeline per cycle. This rate reflects the computing power of a pipeline. In contrast to the $n * k$ cycles on a hypothetical non-pipelined processor, the execution of n instructions on a k-stage pipeline will take $k + n - 1$ cycles (assuming ideal conditions with latency k cycles and throughput 1). Hence, the resulting *speedup* is $n * k/(k + n - 1) = k/(k/n + 1 - 1/n)$. If the number of instructions that are issued to the pipeline is infinite, then the resulting speedup equals the number k of pipeline stages.

As an example of pipelined instruction execution we assume a simple instruction pipeline with the basic stages shown in Fig. 1.3. Overlapped execution of the five steps leads to a 5-stage pipeline. The pipeline execution

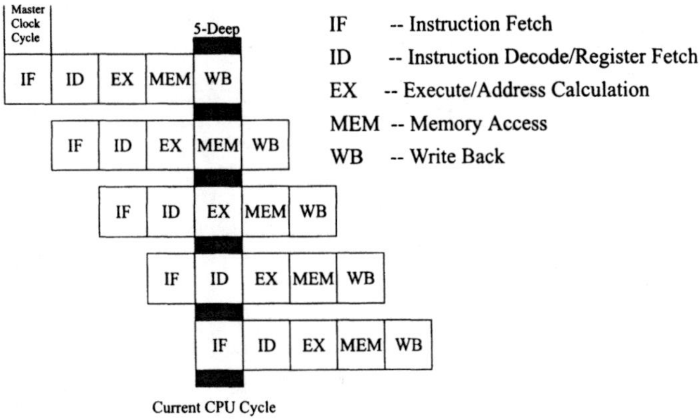

Fig. 1.3. Basic pipelining

proceeds in a smooth manner since each pipeline stage is accomplished in a single clock cycle. The RISC approach offers such a single cycle execution for most instructions. Such a pipeline can be found in the DLX[3] RISC of *Hennessy and Patterson* [134] and in the MIPS R3000 processor (Sect. 1.7.3). Today such simple RISC pipelines can be found as core pipelines in signal processors and in some multimedia processors.

Figure 1.4 shows the basic stages of the instruction pipeline in more detail. Pipeline stages are buffered by different pipeline registers:

- several *program counter registers* (PC) in the IF stage, between the IF/ID and between the ID/EX stages,
- the *instruction register* between the IF/ID stages,
- the *ALU input registers 1* and *2* and the *immediate register* between the ID/EX stages,
- the *conditional register*, the *ALU output register*, and the *store value register* between the EX/MEM stages, and
- the *load memory data register* and *ALU result register* between the MEM/WB stages.

During instruction execution the following sequence of steps is performed:

[3] DLX (pronounced "Deluxe") is a simple load/store architecture.

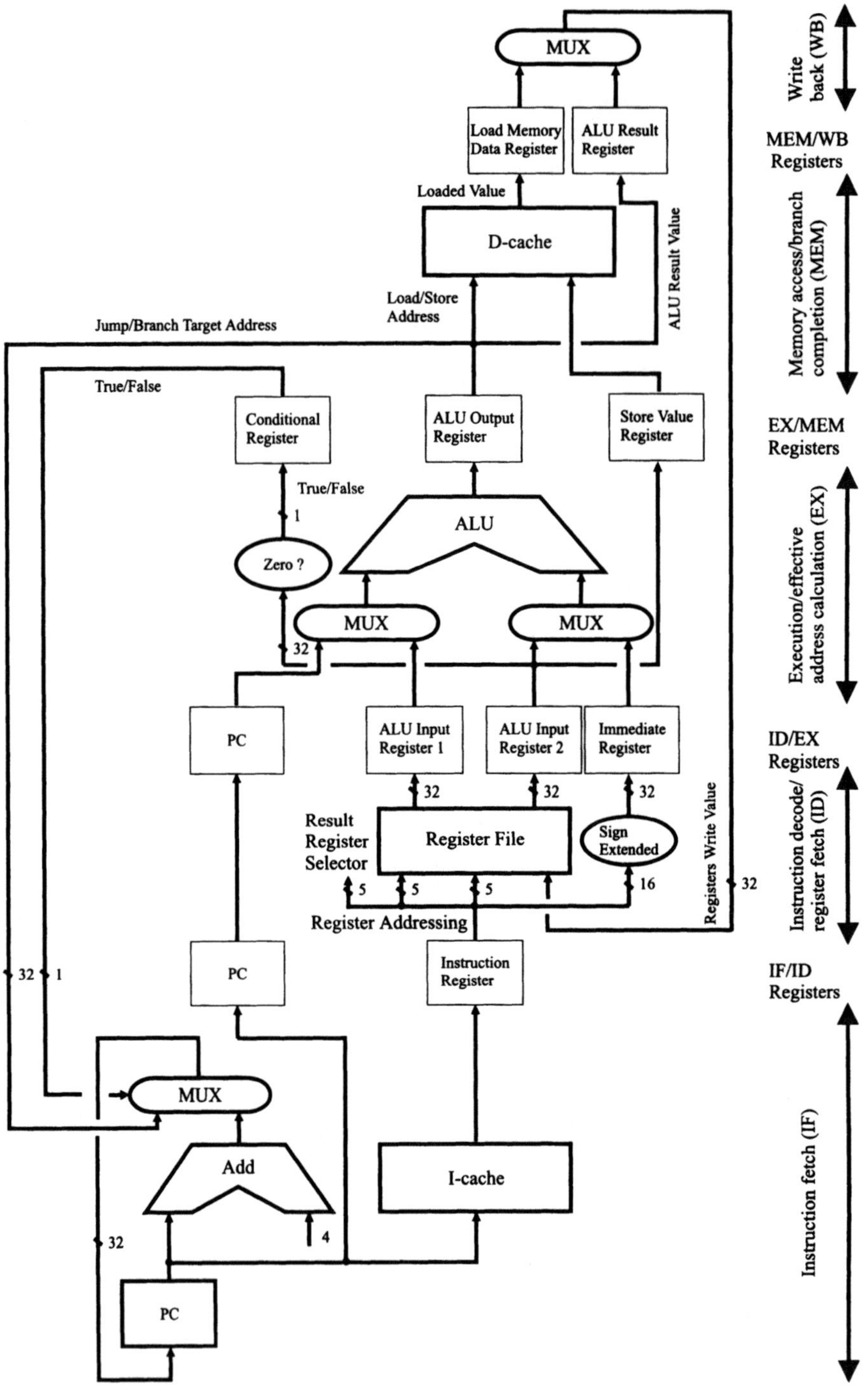

Fig. 1.4. The implementation of the DLX pipeline

1. In the *instruction fetch* (IF), the instruction pointed to by the PC is fetched from I-cache into the instruction register of the CPU, and the PC is incremented by four (assuming a fixed 32-bit instruction format and a byte-addressable processor) to point to the next instruction in memory. In the case of a previous control transfer instruction, the jump/branch target address from the MEM stage may be used to set the PC for the instruction to be fetched in the next cycle.
2. In the *instruction decode/register fetch* (ID), the instruction is decoded, and in the second half of the stage one of the following actions is performed depending on the instruction:
   a) *register-register* (e.g., arithmetic/logical), then the operands are transferred from the register file into the ALU input registers;
   b) *memory reference* (e.g., load/store), then part of the memory address is transferred from a register to ALU input register 1 and a displacement within the instruction is sign-extended and transferred to the immediate register (we assume displacement mode as the most complex addressing mode); in the case of a store instruction, the register value to be stored is transferred to the ALU input register 2;
   c) *control transfer* (e.g., branch on equal), then the displacement within the instruction is sign-extended and transferred to the immediate register (we allow PC-relative mode only) for computation of the jump/branch target address; in the case of a branch instruction, the register value that determines the branch direction is transferred to ALU input register 2 (we assume that a previous compare instruction produced a value that is stored in a general-purpose register).
3. During *execution/effective address calculation* (EX), the ALU operates on the operands from ALU input registers, from PC in ID/EX, or from the immediate register, and eventually puts the result into the ALU output register. The contents of this register depend on the type of instruction which selects the MUX inputs and determines the ALU operation. If the instruction is:
   a) *register-register* (e.g., arithmetic/logical), then the ALU outputs the result of the operation into the ALU output register;
   b) *memory reference* (e.g., load/store), then the ALU output register contains an effective memory address computed from ALU input register 1 and the immediate register; in the case of a store instruction, the ALU input register 2 (containing the register value to be stored) is transferred to the store value register;
   c) *control transfer*, then the ALU computes the jump/branch target address from the PC in ID/EX and the immediate register and stores it in the ALU output register and, at the same time, the branch direction (which determines whether the branch will be taken or not) is tested whether it is zero, and the Boolean result is stored in the condition register.

4. The *memory access/branch completion* (MEM) is performed only for load, store, and branch instructions. If the instruction is:
   a) *register-register*, then the ALU output register is transferred to the ALU result register;
   b) *load*, then the data is read from D-cache (as addressed by the ALU output register) and is placed in the load memory data register;
   c) *store*, then the data in the store value register is written into the D-cache (as addressed by the ALU output register);
   d) *control transfer*, then for jumps and taken branches, the PC in the IF stage is replaced by the ALU output register; for branches that are not taken, the PC remains unchanged (the MUX selection in the IF stage is done by the conditional register);
5. During *write back* (WB), the result of the instruction execution – *register-register* or *load* instruction – is stored into the register file in the first half of the phase. In particular, the load memory data register or the ALU result register is written into the register file (the flow of the register selector through the pipeline stages is not shown in the Fig. 1.4).

Only the data flow though the pipeline stages is shown in Fig. 1.4. Control information generated during the ID stage from the opcode flows through the subsequent pipeline stages and controls the multiplexers and the ALU operation.

Notice that all pipeline stages use different CPU resources. Thus, for example, after an instruction has been delivered to ID, resources used by IF become free and are used for fetching the next instruction. Ideally, each cycle another instruction is fetched and forwarded to the ID stage. The cycle time of the pipeline is dictated by the critical path, i.e., the slowest pipeline stage.

Ideal conditions mean that the pipeline has to be full. Unfortunately, there are several potential problems which may disrupt such a smooth instruction execution in the pipeline. For instance, if only one memory port exists and a load instruction is in the MEM stage, then memory read conflict appears between the IF and MEM stages. In this case, the pipeline has to stall one of the instructions until the required memory port is available. A stall is also called a pipeline *bubble*. All the various phenomena that can disrupt the smooth execution of a pipeline are referred to as pipeline *hazards*. In the next section, we will discuss pipeline hazards and the ways to eliminate them or, at least, minimize their effect.

1.6 Pipeline Hazards and Solutions

Three types of pipeline hazards can be distinguished:

- *Data hazards*, which arise because of the unavailability of an operand. For example, an instruction may require an operand that will be the result of a preceding, still uncompleted instruction.

- *Structural hazards*, which arise from some combinations of instructions that cannot be accommodated because of resource conflicts. For example, if the processor has only one register file write port and two instructions want to write to the register file at the same time.
- *Control hazards*, which arise from branch, jump, and other control flow change instructions. For example, if an instruction is a branch, and the branch is to be taken, then the flow of instructions into the pipeline has to be interrupted, and the branch target must be fetched before the pipeline can resume execution.

1.6.1 Data Hazards and Forwarding

Several types of dependence may exist between instructions $Inst_1$ and $Inst_2$, assuming that $Inst_1$ occurs before $Inst_2$:

- $Inst_2$ is *(true) data dependent* on $Inst_1$, if $Inst_1$ writes its output in a register *Reg* (or memory location) that $Inst_2$ reads as its input.
- $Inst_2$ is *antidependent* on $Inst_1$, if $Inst_1$ reads data from a register *Reg* (or memory location) which is subsequently overwritten by $Inst_2$.
- $Inst_2$ is *output dependent* on $Inst_1$ if both write in the same register *Reg* (or memory location) and $Inst_2$ writes its output after $Inst_1$.
- $Inst_2$ is *control dependent* on $Inst_1$, if $Inst_1$ must complete before a decision can be made whether or not to execute $Inst_2$.

A data dependence is sometimes also called *true* or *real* data dependence, while antidependences and output dependences are sometimes called *false* or *name* dependences. True dependences represent the flow of data through a program, while name dependences stem from the re-use of storage places (register or memory).

Data dependences between instructions may cause data hazards when $Inst_1$ and $Inst_2$ are so close that their overlapping within the pipeline would change their access order to *Reg*. The first three dependences generate the following three types of data hazards:

- *read after write* (RAW) hazard (caused by data dependence),
- *write after read* (WAR) hazard (caused by antidependence),
- *write after write* (WAW) hazard (caused by output dependence).

WAW hazards occur only in pipelines that write in more than one stage, or allow an instruction to proceed even when a previous instruction is stalled. WAR hazards may occur in a pipeline with a write stage preceding a read stage. Therefore, in the simple pipeline of Fig. 1.3, only a RAW hazard (demonstrated in Fig. 1.5) may appear, as described below.

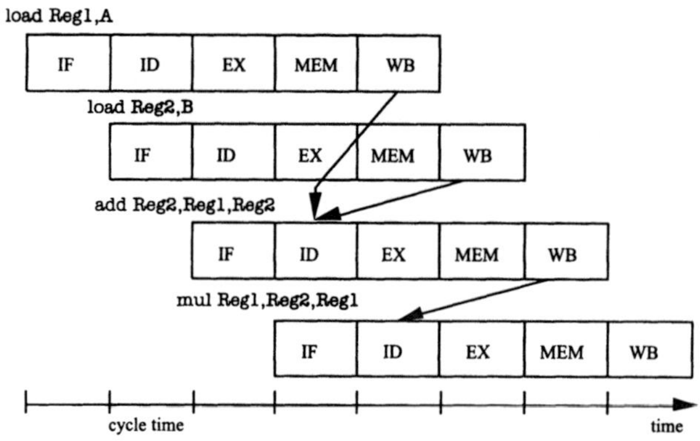

Fig. 1.5. Data hazard in an instruction pipeline

Problem (true data dependence). Consider a sequence of two register-register instructions, $Inst_1$ and $Inst_2$ with $Inst_2$ data dependent on $Inst_1$, and $Inst_1$ fetched before $Inst_2$. Suppose that the result of $Inst_1$ is to be transferred to $Inst_2$ via register *Reg*. No problem occurs if the two instructions are executed in a nonpipelined fashion. In a pipelined computation, $Inst_2$ reads *Reg* during its ID stage. If $Inst_2$ has been fetched immediately after $Inst_1$, then at that moment $Inst_1$ is still in its EX stage and will write the result into *Reg* during its WB stage two cycles later. Therefore, if no action is taken, $Inst_2$ reads the old value from *Reg* in its ID stage (demonstrated in Fig. 1.6).

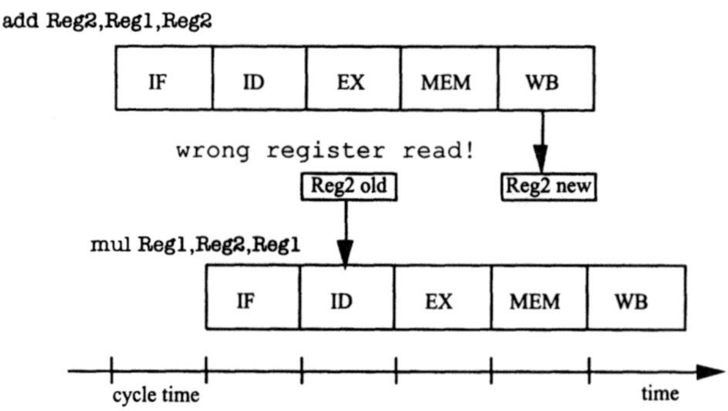

Fig. 1.6. Pipeline conflict due to a data hazard

Software solution. If the pipeline is not able to detect pipeline hazards by hardware, the compiler has to control pipeline execution. This can be done by putting no-op instructions after each instruction that causes or may cause (in the case of a branch) a pipeline hazard. Data hazards, and therefore pipeline stalls (or execution of no-ops), can be reduced by the compiler, which appropriately rearranges the program code (eliminating no-ops). In the example above, if possible, two instructions that do not generate new data hazards should be inserted between $Inst_1$ and $Inst_2$. This approach is called *instruction scheduling* or *pipeline scheduling*. Such a static approach was of major interest in the 1980s after pipelined processors became more widespread. For example, it was used in the MIPS family of microprocessors.

Hardware solutions. We distinguish the following three hardware solutions to the data hazard problem:

- *Interlocking*: The simplest way to deal with such a data hazard is to stall $Inst_2$ in the pipeline for two cycles. Hardware detection of pipeline hazards and stalling is called pipeline *interlocking*. This solution produces bubbles and considerably degrades speedup. In the above example, stalling produces two bubbles (see Fig. 1.7). Remember that in our pipeline we assume that the register write back is completed in the first half of the WB stage and that the same register value can be read again during the operand fetch in the second half of the ID stage.

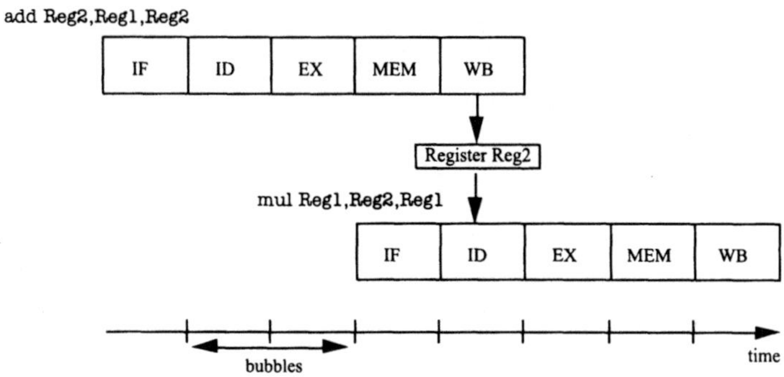

Fig. 1.7. Data hazard: hardware solution by interlocking

- *Forwarding*: There is a more sophisticated solution, requiring a hardware investment, which is called *forwarding*. The key insight is that $Inst_2$ need not wait until the result of $Inst_1$ is written in *Reg* during WB, if the result in the ALU output of $Inst_1$ in the EX stage can be immediately forwarded back to the ALU input of the EX stage as an operand for $Inst_2$. In our example, where both instructions are of the register-register type, forwarding removes all bubbles (see Fig. 1.8).
- *Forwarding with interlocking*: Unfortunately, forwarding does not resolve all types of data hazards. If $Inst_1$ is a load instruction, forwarding from EX is of no use, because the EX stage does not produce the value to be loaded, but the effective memory address in the ALU output register. Assuming that $Inst_2$ is data dependent on the load instruction $Inst_1$, then $Inst_2$ has to be stalled until the data loaded by $Inst_1$ becomes available in the

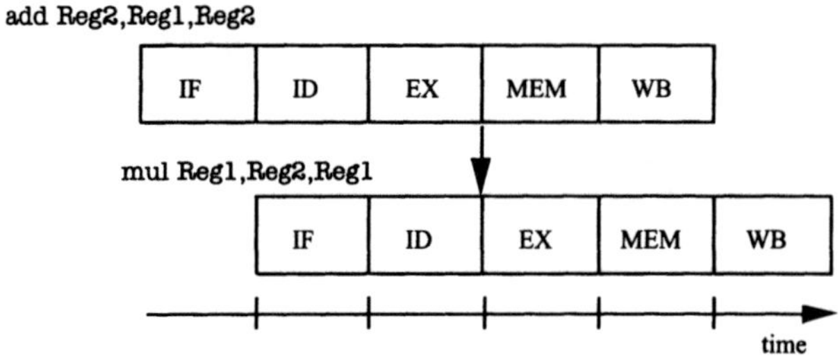

Fig. 1.8. Data hazard: hardware solution by forwarding

load memory data register in the MEM stage. Even when forwarding is implemented from MEM back to EX, one bubble occurs that cannot be removed (see Fig. 1.9 for the hazard and Fig. 1.10 for the resulting bubble).

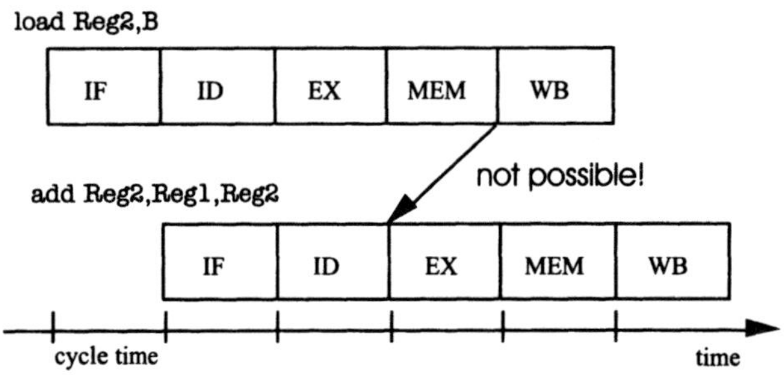

Fig. 1.9. Pipeline hazard due to data dependence unresolvable by forwarding

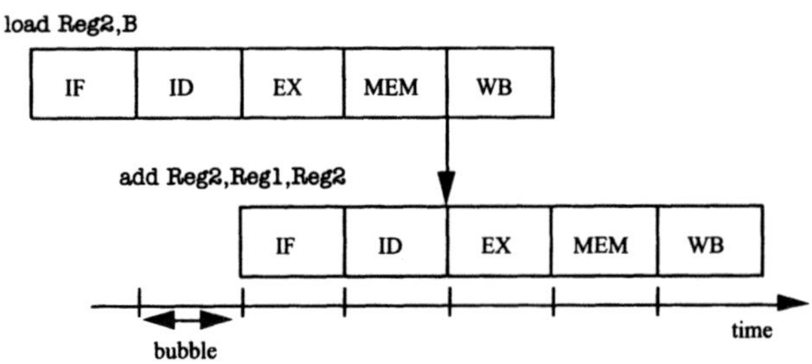

Fig. 1.10. Pipeline bubble due to data dependence

In contrast to static pipeline scheduling, where the compiler separates instructions causing data hazards, *dynamic* (i.e., run-time) pipeline scheduling is implemented in hardware. These solutions will be discussed in Sect. 3.2.

1.6.2 Structural Hazards

Problem (resource conflict). Structural hazards do not arise in the simple pipeline in Fig. 1.3. However, let us suppose that the MEM stage would be able to write back an ALU output in the case of a register-register instruction (from the ALU output register) into a register file with a single write port. Consider a sequence of two instructions, $Inst_1$ and $Inst_2$, with $Inst_1$ fetched before $Inst_2$, and assume that $Inst_1$ is a load, while $Inst_2$ is a data independent register-register instruction. Due to memory addressing, the data requested by $Inst_1$ arrives at the register file port at the same time as the result of $Inst_2$, causing a resource conflict (see Fig. 1.11).

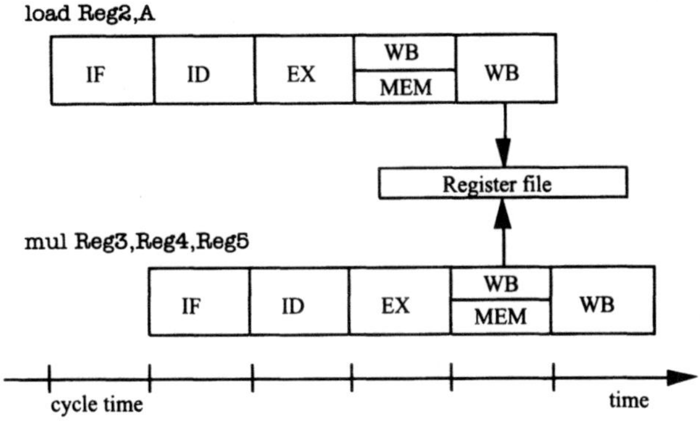

Fig. 1.11. A potential resource conflict due to a structural hazard

Solutions. When instructions cannot be simultaneously accommodated because of resource conflicts, two types of hardware based solutions exist:

- *Arbitration with interlocking*: Structural hazards can be resolved by hardware that performs resource conflict arbitration and interlocks the in program flow succeeding of the two competing instructions. Clearly, bubbles may result if this technique is used for resolving structural hazards. In the example above, the `load` instruction writes into the register file, while the `mul` instruction would be deferred, producing one bubble.
- *Resource replication*: The effects of structural hazards can be alleviated by the replication of hardware resources. No bubbles can appear in this way. In the example above, a register file with multiple write ports would enable simultaneous writes to different destination registers. In the case of the same destination register, either arbitration and interlocking is necessary, or (as in Fig. 1.11) the value produced by the `load` instruction is discarded and the ALU output register value of the `mul` instruction is used.

1.6.3 Control Hazards, Delayed Branch Technique, and Static Branch Prediction

Problem (control conflicts). Control hazards can be caused by jumps and by branches. Let $Inst_1$, $Inst_2$, $Inst_3$ ... be a sequence of instructions, fetched in this order, one immediately after the other.

Assume that $Inst_1$ is a jump. The jump address is computed in the EX stage and replaces the PC in the MEM stage, while $Inst_2$ is in the EX stage, $Inst_3$ is in the ID stage, and $Inst_4$ is in the IF stage. Assuming that the jump address does not point to $Inst_2$, $Inst_3$, or $Inst_4$, the prefetched instructions $Inst_2$, $Inst_3$, and $Inst_4$ should be canceled (i.e., nullified or flushed) and the instruction at the jump address fetched. The three instructions following $Inst_1$ form the so-called *delay slots.* A simple solution, which avoids cancellation, is to fill the delay slots with no-op instructions.

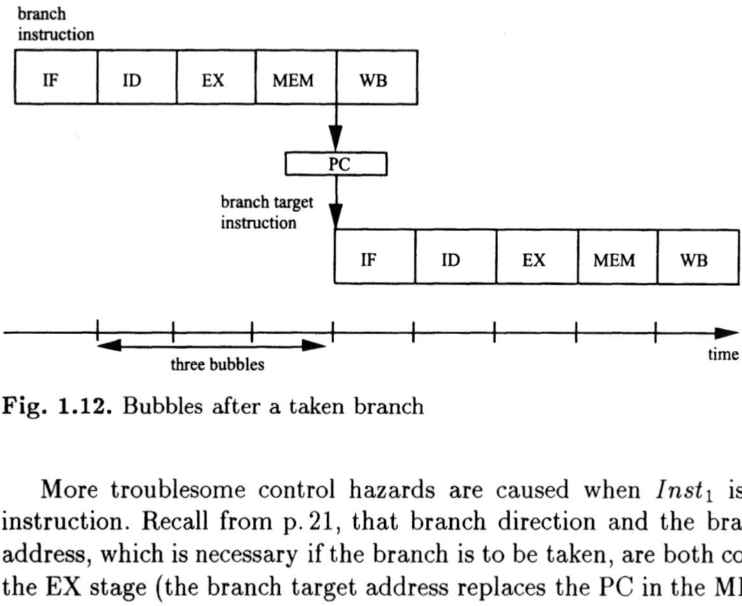

Fig. 1.12. Bubbles after a taken branch

More troublesome control hazards are caused when $Inst_1$ is a branch instruction. Recall from p. 21, that branch direction and the branch target address, which is necessary if the branch is to be taken, are both computed in the EX stage (the branch target address replaces the PC in the MEM stage). If the branch is taken, the correct instruction sequence can be started with a delay of three cycles since three instructions of the wrong branch path are already loaded in different stages of the pipeline (see Fig. 1.12).

In order to reduce the number of delay slots, the calculation of the branch direction and the branch target address should be done in the pipeline as early as possible. This could be in the ID stage after the instruction has become recognized as a branch instruction. However, then the ALU can no longer be used for calculating the branch target address, since it may still be occupied by the previous instruction. Recall, that this is a structural hazard, which can be avoided by an additional ALU for the branch target address calculation in ID stage. Assuming an additional ALU and a write-back of the branch target address to the PC already in the ID stage (if the branch is taken) only a single cycle delay slot arises, which can often be filled by instruction reordering at compile-time.

Although this reduces the branch delay to a single cycle, now a new pipeline hazard may arise. An ALU instruction followed by a branch on the result of the instruction will incur a data hazard stall even when the result value is forwarded from the EX to the ID stage (similar to the data hazard from a load with a succeeding ALU operation that needs the loaded value). The main problem with this pipeline reorganization is that the decode, the branch target address calculation, and the PC write-back must be done sequentially within a single pipeline stage. This may lead to a critical path in the decode stage that reduces the cycle rate of the whole pipeline.

Software solutions. Control hazards can be dealt with by several software-based techniques:

- *Delayed jump/branch technique*: The compiler fills the delay slot(s) with instructions that are in logical program order before the jump or branch. Of course, this is only possible if these instructions have no effect on the branch direction (we assume the branch target address does not point to one of the instructions in the delay slots). Notice that, in this case, the instructions moved within the slots are executed regardless of the branch outcome. This is the simplest solution viewed from hardware side. A cancellation of fetched instructions is not necessary, thus sparing hardware complexity (the simple pipeline in Fig. 1.4 is of that type). If there is lack of instructions that can be moved in the delay slots, no-op instruction(s) are used to fill up the slot(s). According to some program trace results, the probability of moving one instruction into the delay slot is greater than 60 %, that of moving two instructions is about 20 %, and that of moving three instructions is less than 10 %.
  Delayed branching was a popular technique in the first generations of scalar RISC processors, such as the IBM 801, Berkeley RISC I, and Stanford MIPS. In superscalar processors where more than one instruction can be fetched and processed simultaneously (see Sect. 4.1), the delayed branch technique complicates the instruction issue logic and the implementation of precise interrupts. However, due to compatibility reasons, it is often still in the ISA of some of today's microprocessors – for example, in SPARC- or MIPS-based processors.
- *Compiler-directed (static) branch prediction*: A bit in the opcode of the branch instruction allows the compiler to influence the prediction. Instructions are fetched from the predicted branch target address. If the prediction followed the wrong instruction path, then the instructions wrongly fetched are discarded from the pipeline. *Static branch prediction* means – in contrast to *dynamic branch prediction* – that the machine cannot dynamically alter the branch prediction. So static branch prediction comprises machine-fixed prediction (e.g., always predict taken) and compiler-driven prediction.

Hardware solutions. Control hazards can also be dealt with by hardware-based techniques:

- *Interlocking*: This is the simplest way to deal with control hazards. The hardware must detect that the $Inst_1$ is a branch and apply hardware interlocking to stall the next instruction $Inst_2$. For the pipeline in Fig. 1.3, this produces three bubbles in cases of jump or of (taken) branch instructions (since the effective branch target address is written back to the PC during the MEM stage, see p. 21).
- *Wired taken/not-taken prediction*: The static branch prediction can be wired into the processor by predicting that all branches will be taken (or all not taken).
- *Branch target address cache*: The *branch target address cache* (BTAC) is a small cache memory associated with the IF stage of the pipeline. The BTAC is a set of tuples each of which contains:

  Field 1: the address of a branch (or jump) instruction (which was executed in the past),

  Field 2: the most recent target address for that branch or jump,

  Field 3: information that permits a prediction as to whether or not the branch will be taken.

  The BTAC functions as follows: the IF stage compares the PC against the addresses of jump and branch instructions in BTAC (*Field 1*). Suppose that there is a match. If the instruction is a jump, then the target address is used as the new PC rather than incrementing the PC. All bubbles are removed. If the fetched instruction is a branch, a prediction is made based on information from BTAC (*Field 3*) as to whether the branch is to be taken or not. If it is to be taken, the most recent branch target address is read from BTAC (*Field 2*) and used to fetch the target instruction. Of course, a misprediction may occur. Therefore, when the branch direction is actually known in the MEM stage, the BTAC can be updated with the corrected prediction information and the branch target address.

  One of the critical issues is how large the BTAC should be and how it should be organized (e.g., associative memory, hashed). To keep the size of the BTAC as small as possible, only branches that are predicted to be taken are stored while those which turned out not to be taken may be removed. Since the hardware alters the prediction direction due to the "history" of the branch, this kind of branch prediction is an example of a simple *dynamic branch prediction.* The process described above implements a *one-bit predictor.* More dynamic branch prediction techniques follow in Sect. 4.3.

1.6.4 Multicycle Execution

Problem (multicycle execution). Consider a sequence of two instructions, $Inst_1$ and $Inst_2$, with $Inst_1$ fetched before $Inst_2$, and assume that

$Inst_1$ is a long-running (e.g., floating-point) instruction. Another example of a long-running instruction is the `load` instruction in our example pipeline processor. The `load` instruction needs two cycles to execute – one cycle to compute the effective address and a second cycle for MMU look-up and D-cache access.

To handle long-running $Inst_1$, it would be impractical to require that all instructions complete their EX stage in one clock cycle since that would mean accepting a slow clock, or using enormous amounts of logic, or both. Instead, the EX stage might be allowed to last as many cycles as needed to complete $Inst_1$. This, however, causes a structural hazard in the EX stage because the succeeding instruction $Inst_2$ cannot use the ALU in the next cycle.

Solutions. Several solutions to this problem can be identified:

- *Interlocking*: The simplest way to deal with such a structural hazard is to stall $Inst_2$ in the pipeline until $Inst_1$ leaves the EX stage.
- *Pipelining the EX stage*: If the EX stage is pipelined itself, the structural hazard is avoided, because the EX stage is able to accept another instruction in each cycle (throughput is 1).
- *Multiple functional units*: There may be multiple functional units so that $Inst_2$ may proceed to some other functional unit and overlap its EX stage with the EX stage of $Inst_1$.

Interlocking is inefficient as it produces bubbles and thus considerably degrades speedup.

Pipelining the EX stage is the solution that was chosen for `load` instruction execution in our example pipeline in Fig. 1.4 by providing separate EX and MEM stages (instead of a single combined EX/MEM stage). Striving for a simple hardware implementation of the two-cycle `load` instruction was the reason for deciding to forward the results of single-cycle arithmetic-logic instructions through the MEM stage, which delays write-back of the results by one cycle. Delaying write-back of results of simple instructions prevents WAW hazards in the pipeline.

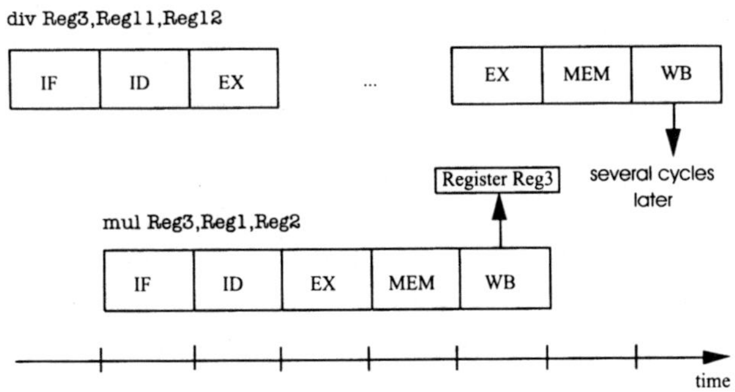

Fig. 1.13. Example of a WAW hazard

A more complex solution is the use of multiple FUs and simultaneous execution. This solution, however, implies that instructions may not complete in the original program order (see Fig. 1.13). Since the `div` instruction's EX stages may last from one to tens of cycles, the `mul` instruction proceeds to the WB stage before the `div` instruction. Unfortunately, such an *out-of-order* execution may cause a WAW hazard in the case where there is an output dependence between the two instructions. There are two solutions to solve the WAW hazard in Fig. 1.13:

- The `mul` instruction delays its write-back until the `div` instruction has written its result to the register which is subsequently overwritten.
- The more elegant solution is to write back the result of the `mul` instruction immediately and discard the result of the `div` instruction which is never used by another instruction in the example in Fig. 1.13. Unfortunately, the question arises as how to implement a precise interrupt in the case of, for example, a division by zero (solutions are given in Sect. 4.8.3).

Until now, we have looked at simple pipelined processors that use an *in-order execution* pipeline organization; that is, instructions are issued to the FUs and execution is initiated in exactly the same order as the dynamic sequence of instructions in the program. If multiple FUs are provided, an out-of-order execution is the next step. In the case of out-of-order execution, WAW hazards must be solved, and even an antidependence can cause a WAR hazard if a subsequent instruction starts execution and writes back its result before a previous instruction gets its operands. Solutions to this problem are delivered by the scoreboarding technique and by the Tomasulo's scheme, both described in Chap. 3.

1.7 RISC Processors

In this section some scalar RISC processors will be presented. We start with the first RISC processors – IBM 801, Berkeley RISC I and II, Stanford MIPS – which are the ancestors of all later commercial RISC processors. We also mention the GaAs microprocessor, based on Stanford MIPS, which was developed by RCA Corp. Next, we describe the SPARC family of processors, which originates from Berkeley's RISC project, and dates back to 1987 when Sun introduced the first SPARC-based computer Sun-4. After that, we describe the MIPS R3000 pipelined RISC processor that emerged in 1988 as one of the most well-known descendants of Stanford's MIPS project. Note that early RISC processors never contained a floating-point unit due to lack of chip space. There follows the MIPS R4400 RISC processor whose superpipelined architecture provides a good balance of integer and floating-point performance. Next we itemize several other pipelined RISC processors, including the Fairchild/Intergraph Clipper, ARM, Hewlett-Packard PA-RISC,

AMD 29000, and Motorola MC 88000. Finally, the pipelined processor picoJava I is also included in this section, although its stack-based ISA with too many instructions and variable-length instruction format violates the RISC philosophy.

1.7.1 Early Scalar RISC Processors

IBM 801 and ROMP (*Radin* [237]): The first system to formalize RISC principles was the IBM 801 project which began in 1975. The design goal was to speed up frequently used instructions while discarding complex instructions that slowed the overall implementation. Memory access was limited to load/store instructions (locking the register until complete, so most execution could continue). Branches were delayed, and instructions used a three operand format. Execution was pipelined, allowing CPI = 1. There were thirty-two 32-bit registers in the register file, but no floating point unit and no floating-point registers. IBM tried to commercialize the 801 design starting in 1977 but it was not successful.[4] Subsequently, in the mid-1980s, a commercial RISC-type processor, the Research Office products division Microprocessor (ROMP) was announced. Compared to the 801, it had a smaller percentage of instructions executing in a single cycle and 65 % of its instructions were 16-bit, while the others were 32-bit. The ROMP was used in the IBM RT 6150 and RT 6151 workstations.

Berkeley RISC I and II (*Patterson and Ditzel* [231], 1980): The term RISC came from Berkeley's project, which was the basis for the later SPARC processor. Because of this, their features are similar, including a windowed register file (10 global and 22 windowed, vs 8 and 24 for SPARC, see also Sect. 1.7.2) with `R0` wired to zero. Branches are delayed. All instructions have a bit to specify whether condition codes should be set, and execute in a 3-stage pipeline. In addition, next and current PC are visible to the user, and last PC is visible in supervisor mode. The Berkeley project also produced an I-cache with some innovative features, such as instruction line prefetch that identified jump instructions, frequently used instructions compacted in memory and expanded upon cache load, additional cache-chip support, and bits to map out defective cache lines.

Stanford MIPS (*Hennessy et al.* [133], 1981): The Stanford MIPS project was the basis for the MIPS R2000 and R3000. Stanford MIPS used the

[4] This was not the only innovative design developed by IBM which never saw daylight. Slightly earlier the Advanced Computer System pioneered superscalar design, speculative execution, delayed condition codes, multithreading, imprecise traps and instruction streamed interrupts, and load/store buffers, plus compiler optimization to support these features. It was expensive and incompatible with the System/360, so it was not pursued, but many ideas did find their ways into the expensive high-end mainframes.

compiler to eliminate register conflicts. Like the R2000, the Stanford MIPS had no condition code register, and a special `HI/LO` register pair was used in multiply and divide instructions. Unlike the R2000, the Stanford MIPS had only 16 registers, and two delay slots for load/store and branch instructions. The PC and the last three PC values were tracked for exception handling. Instructions were packed (like the Berkeley RISC), in that many instructions specified two operations that were issued in consecutive cycles (not decoded by the cache). In this way, it was a dual-issue VLIW, but executed sequentially. User assembly language was translated to packed format by the assembler.

RCA GaAs RISC/MIPS (*Milutinović et al.* [203], 1986): For its Star Wars program in mid-1980s, the US Department of Defense intended to develop a microprocessor chip having as much computing power as 100 DEC VAX 11/780 superminicomputers. One candidate for such a processor was a gallium arsenide (GaAs) version of the Stanford MIPS architecture. GaAs technology is, at the same power consumption, about a half order of magnitude faster than silicon technology, and several orders of magnitude more radiation hard. Unfortunately, GaAs is also characterized by some undesirable properties, such as high cost and low transistor count capability. The requirement specified a full 32-bit engine with a clock frequency of 200 MHz and computation rate 100 MIPS. In his book, *Milutinović* [202] takes the reader through all phases of the design of such a processor on a VLSI chip, which was very much ahead of its time.

1.7.2 Sun microSPARC-II

Since SPARC is a large family of RISC processors some of which will also appear as superscalar RISC processors in Chap. 4, we first briefly describe the SPARC's philosophy and an architectural concept, called register window, that is common to all of them. SPARC can be thought of as a branch stemming from the Berkeley RISC. It was designed by Sun Microsystems for their own use, but in keeping with the open philosophy, Sun licensed it to other companies (see Table 1.4). The history of the SPARC architecture dates back to 1987 when Sun introduced the Sun-4, the first SPARC-based computer. Over the years since its release it has become a vehicle for an array of chips from numerous vendors, and the foundation upon which many manufacturers are still basing new workstation products.

While most RISC CPUs at that time (e.g., AMD 29000, MIPS R2000, HP PA-RISC) were more conventional, the SPARC design was quite radical, omitting multiple cycle multiply/divide instructions[5] and using single-cycle "step" instructions instead. SPARC usually contains about 128 or 144 registers. At any time 32 registers are available – 8 of them are global, the rest

[5] Added in later versions.

are allocated in a so-called *window*, which is a subset of registers from the register file (see Fig. 1.14). The register window of the procedure currently running is called the *active* window (and is pointed to by the current window pointer CWP in the processor state register PSR). During a function call, the active window is moved 16 registers down the register file, so that 8 registers remain local while the upper and lower 8 registers are shared between functions for passing and returning values. On return, this window is moved up, so registers are loaded or saved only at the top or bottom of the window. This allows functions to be called in as little as one cycle with a parameter transfer of up to 8 register values.

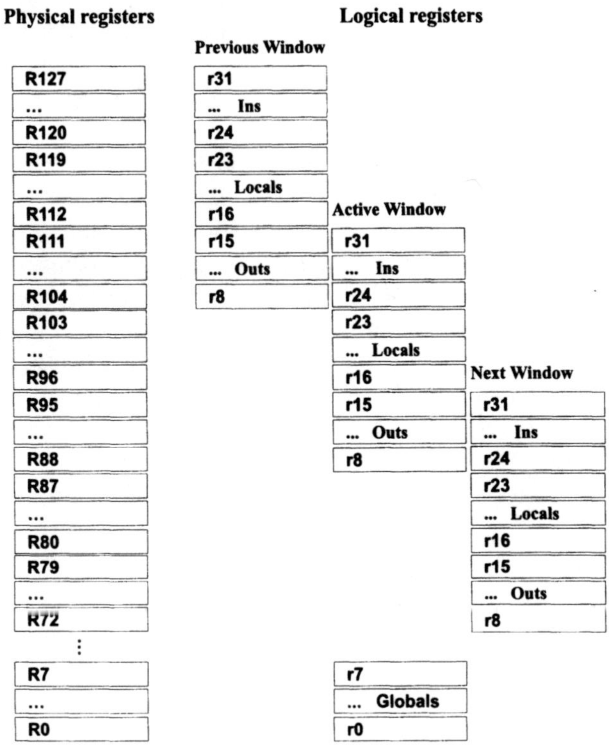

Fig. 1.14. Overlapping windows in SPARC

If the number of register windows is insufficient during program execution, a trap is signaled and a trap routine stores or loads register values to or from memory. To reduce loads and saves between functions, SPARC can be scaled up to 512 registers, thus allowing a circular stack of up to 32 overlapping windows. On the other hand, it can be scaled down to reduce context switch time, when the entire register set has to be saved. Function calls are usually much more frequent than context switches, so the large register file is usually an advantage, but compilers can now usually produce code that uses a fixed register set as efficiently as a windowed register file across function calls.

Like most RISC processors, global register `r0` is wired to zero to simplify instructions. SPARC is pipelined (though not as deeply as other RISC processors) and uses one branch delay slot. Like previous processors, a dedicated *condition code register* (CCR) holds comparison results.

SPARC is not a chip, but an architectural specification, so there are various designs of it (see Table 1.4). It has undergone many revisions, and now it supports multiply/divide instructions. Original versions were 32-bit, but 64-bit and superscalar versions were designed and implemented beginning with the Texas Instruments and Sun SuperSPARC in late 1992. Later its performance lagged behind other RISCs and even behind Intel x86 CISC processors until, in late 1995, the UltraSPARC-I and now UltraSPARC-II (see Sect. 4.9.5) and SPARC 64 multichip CPU have emerged. These superscalar processors will be covered in Chap. 4. Here we describe microSPARC-II, which is one of the latest scalar SPARC processors.

The microSPARC-II 32-bit microprocessor is a highly integrated, high-performance microprocessor. Implementing the SPARC ISA version 8 specification, it is well suited for low-cost uniprocessor applications. It is built with CMOS technology featuring 0.5 μm geometries, a 3-layer metal silicon process, with the core operating at low voltage for optimized power consumption. A block diagram of the microSPARC-II chip is shown in Fig. 1.15.

The microSPARC-II includes an *integer unit* (IU), an optimized *floating-point unit* (FPU), a *memory management unit* (MMU), I-cache and D-cache, programmable DRAM controller, SBus controller, graphics interface support, IEEE 1149.1 boundary scan test bus, power management and clock generation capabilities. Technical features of some of these components are given below:

- The IU executes SPARC integer instructions defined in the SPARC ISA version 8. The IU contains 136 registers organized in 8 windows and 8 global registers. It operates on prefetched instructions using a 5-stage pipeline. The throughput is improved by using *branch folding* and single cycle load/store instructions.
- The FPU (based on a Meiko design) fully executes all single and double precision floating-point instructions as defined in the SPARC ISA version 8. The FPU contains thirty-two 32-bit registers, a general-purpose execution unit, and a floating-point multiplier allowing in most cases the parallel execution of an **FPMUL** and another floating-point instruction. A 3-instruction deep queue of floating-point instructions is provided to increase the efficiency of concurrent floating-point and integer execution.
- The MMU translates 32-bit virtual addresses of each running process to 31-bit physical addresses in memory. The 3 high-order bits of the physical address are reserved to support memory mapping into 8 different address spaces. The unit also serves as an I/O MMU and controls the arbitra-

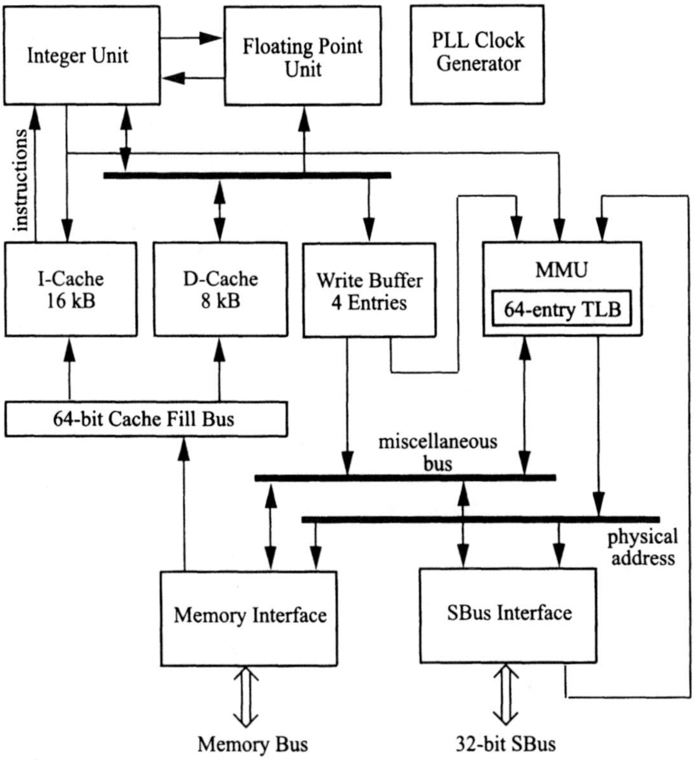

Fig. 1.15. Block diagram of the Sun microSPARC-II

tion between I/O, I-cache, D-cache, and TLB references to memory. The MMU contains a 64-entry fully associative TLB. It supports 256 contexts and protects memory so that a process can be prohibited from reading or writing to the address space of another process.

- The I-cache is a 16 kB direct-mapped, virtually indexed, virtually tagged cache. The I-cache data is organized as 512 lines of 32 bytes plus 32 tag bits. Cache refill is done two 32-bit words at a time. Cache streaming and bypass are supported.
- The D-cache is an 8 kB, direct-mapped, virtually indexed, virtually tagged, write-through cache with no write allocate. The data store is organized as 512 lines of 16 bytes plus 32 tag bits. Single-word integer and double-word floating-point read/write cache hits take one clock cycle. There is a four-deep store buffer to hold data being stored from the IU or FPU to memory or other physical devices. The store buffer is composed of 64-bit registers. Cache refill is done two 32-bit words at a time. Cache streaming and bypass are supported.
- The memory interface is provided by a complete DRAM controller which generates all the signals necessary to support up to 256 MB of system

memory. The DRAM bus is 64-bit wide with two parity bits, each covering 32 bits of data. The system DRAM is organized as eight banks, each of which may be 2 MB, 8 MB, or 32 MB depending upon the size of DRAM used.

- The SBus interface, as the principal I/O bus interface, performs all the functions necessary to connect the processor to the SBus, including dynamic bus sizing, cycle re-run control, burst cycle re-ordering, arbitration, and general SBus control. The SBus interface works with the MMU to arbitrate the system and memory resources and for I/O address translations.
- The chip has a five-wire *test access port* (TAP) interface to support internal scan, boundary scan and clock control. This interface is compatible with the IEEE 1149.1 specification for standard test access port and boundary scan architecture. This allows efficient access to any single chip in the daisy chain without board-level multiplexing. The TAP controller is a synchronous finite state machine (with 16 states) which controls the sequence of operations of the test circuitry, in response to changes at the bus. The TAP controller is asynchronous with respect to the system clock(s), and can therefore be used to control the clock control logic.

Taking advantage of optimized compiler technology and running with the internal CPU clock at 85 MHz, the throughput of the microSPARC-II has been measured above 64 SPECint92 and 54 SPECfp92.

1.7.3 MIPS R3000

The R3000 processor family (*Kane and Heinrich* [154]) stems from the Stanford MIPS and is most similar to the DLX. MIPS architecture recognizes four coprocessors, denoted CP0 through CP3. CP0 is the *system control coprocessor* that supports the virtual memory system and exception handling. It is always incorporated on the CPU chip. CP1 is the *floating-point coprocessor*, CP2 is reserved for future definition by MIPS, and CP3 provides some extensions to the MIPS ISA.

The R3000 family has been shipped in volume since 1988 as a second-generation MIPS RISC microprocessor (following up the R2000, the first commercially available RISC processor introduced in 1985 by MIPS Computer Systems, the predecessor of MIPS Technologies). One of the principal characteristics of the R3000 is its simplicity. The R3000 family consists of the R3000 CPU (with CP0 on-chip) fabricated at that time in 1.2 μm CMOS technology with an internal frequency of 40 MHz, the R3010 FPU chip (CP1), and a variety of other derivatives and support chips. CP1 includes thirty-two 32-bit registers and performs 32-bit (single precision) and 64-bit (double precision) ANSI/IEEE 754-1985 standard floating-point operations. The three operation units of CP1 (for adding/subtracting, multiplying, and dividing) can operate in parallel (*Rowen et al.* [248]). The derivatives and support chips were tailored for specific markets such as embedded controllers and

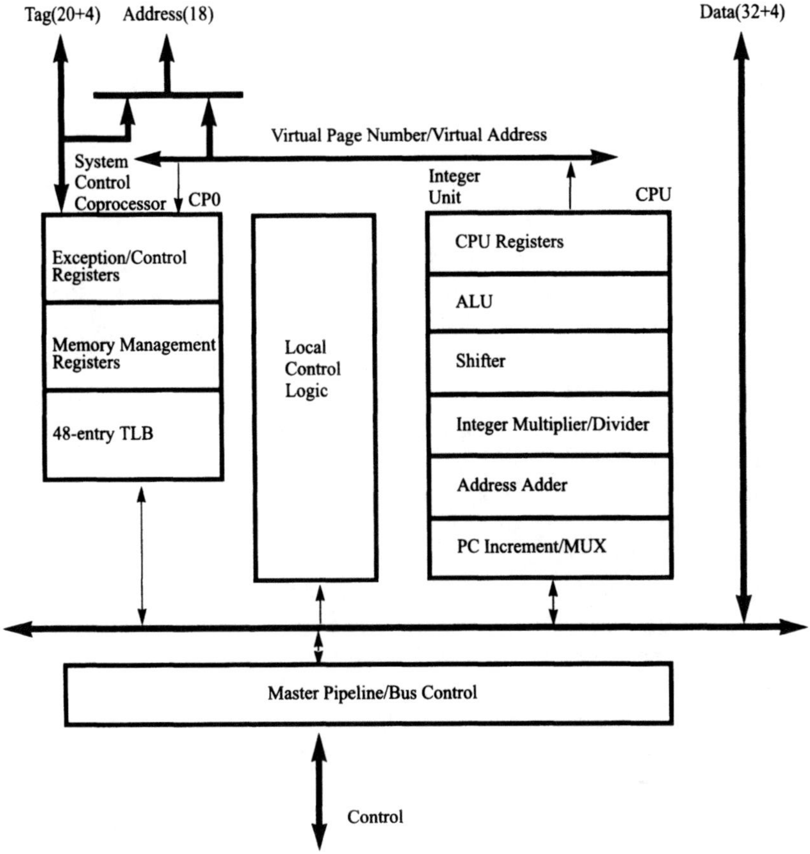

Fig. 1.16. Block diagram of the MIPS R3000

low-cost workstations. There is no CP3 coprocessor since the implementation of R3000 is based on MIPS I ISA. Instructions have fixed 32-bit format with three main format types: I-type (immediate), J-type (jump), and R-type (register). A block diagram of the R3000 chip is shown in Fig. 1.16.

The R3000 features a similar 5-stage pipeline as shown in Fig. 1.3:

1. In the IF stage the instruction physical address is read from the 48-entry TLB.
2. In the ID stage the instruction is fetched from external I-cache. Operands are read from the CPU register file while the instruction is being decoded.
3. During the EX stage the required operation is performed on the operand(s).

4. In the MEM stage external D-cache is accessed for load/store instructions.
5. During the WB stage the ALU result or the value loaded from the D-cache is stored into the register file.

One of the features of the MIPS architecture is the partial exposure of the pipeline to the programmer since the pipeline hardware is not able to recognize pipeline hazards.

1.7.4 MIPS R4400

The MIPS R4400 is a *superpipelined*[6] system with an external frequency of 75 MHz and the pipeline running at an internal frequency of 150 MHz. Although the R4400 is a fully-fledged 64-bit system, its instructions are uniformly 32-bit, as specified in Table 1.5. The processor has an on-chip *floating-point unit* (FPU) and an I-cache and D-cache with a capacity of 16 kB each. The processor is a 2.2 million transistor chip made in 0.6 μm CMOS technology (see *Kane and Heinrich* [154]).

A block diagram of the R4400 is shown in Fig. 1.17. There is an *integer unit* (CPU), a *floating-point unit* (FPU, also called CP1), a *system control coprocessor* (CP0), an I-cache and a D-cache. The caches are organized as direct mapped and are each 16 kB in size. The processor supports a secondary (L2) cache that can range in size from 128 kB to 4 MB and can either be split into I-cache and D-cache (Harvard approach) or unified (Princeton approach).[7] The R4400 caches allow the processor pipeline to execute at the rate of one clock cycle per instruction and also to minimize the load latency.

The CPU is capable of handling 64-bit operands and has thirty-two 64-bit general purpose registers. Two of them have assigned functions: `r0` is hardwired to 0, and `r31` is the link register used by jump and link instructions. There are three special purpose registers: `PC`, `HI` and `LO`. The latter two either hold the product of an integer multiply operation, or the quotient (`LO`) and remainder (`HI`) of an integer divide operation. Separate multiply and divide units allow multiplication and division to take place in parallel with other instructions.

The R4400 incorporates a CP0 on chip, responsible for translating virtual addresses to physical addresses, exception handling, as well as some

[6] In the original sense introduced in the context of the R4000 processor, *superpipelining* meant a long pipeline with an internal clock frequency that was twice as high as the I/O interface. Today, all processors are clocked two or three times as fast as the I/O. Therefore the term superpipeline is no longer used very often. Then it usually just means: a long pipeline.

[7] While the *Princeton architecture* has a single memory for instructions and data, the *Harvard architecture* has separate memories, so simultaneous data and instruction access do not conflict.

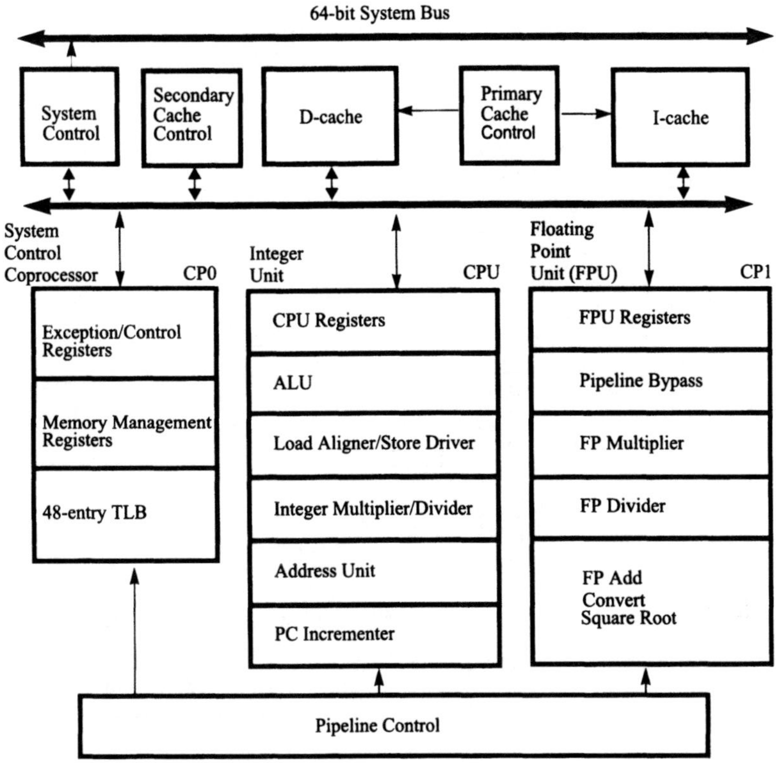

Fig. 1.17. Block diagram of the MIPS R4400

diagnostic capability. The CP0 contains a 48-entry TLB and several registers, some of which are used for memory management while the others are exception/control registers.

The CP1 is fully compliant to the ANSI/IEEE 754-1985 standard. CP1 has 32 registers which can be accessed as thirty-two 32-bit registers or sixteen 64-bit registers. There are separate add, multiply, and divide units to allow these operations to take place in parallel (with multiply/divide using the adder at the end of their operation).

The R4400 8-stage instruction pipeline is shown in Fig. 1.18. It applies superpipelining which effectively stretches R3000's 5-stage pipeline to the 8-stage pipeline in the R4400. In principle, the time-critical stages of the R3000 pipeline are split into two, or even three, stages. The splitting results in separate IF and IS stages instead of the single IF stage in R3000, and in separate DF, DS, and TC stages replacing R3000's MEM stage.

The R4400 pipeline stages are:

1. In the IF stage an instruction address is selected by the branch logic and the I-cache fetch begins. The TLB starts virtual to physical address translation.

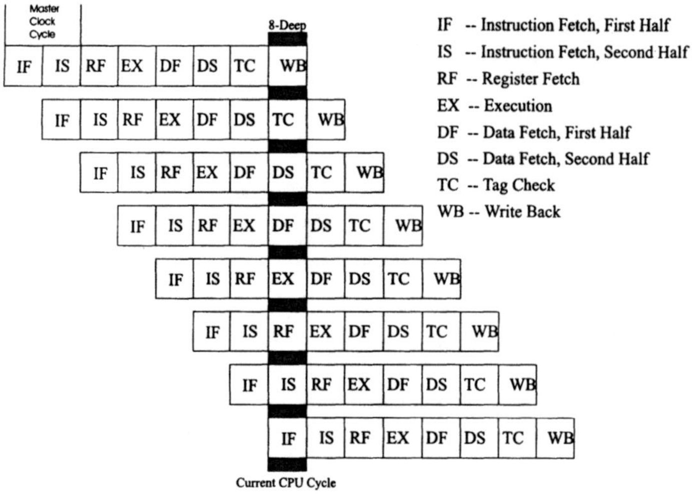

Fig. 1.18. Pipeline stages in the MIPS R4400

2. In the IS stage the I-cache fetch and the TLB translation are completed.
3. During the RF stage the instruction is decoded and checked for interlock conditions. The I-cache tag is checked against the page frame number obtained from the TLB. Any required operands are fetched from the register file.
4. The EX stage performs as follows:
   a) for register-register instructions, the ALU performs the arithmetic or logical operations;
   b) for load/store instructions, the ALU calculates the data virtual address;
   c) for branch instructions, the ALU determines whether the branch condition is true and calculates the virtual branch target address.
5. The DF stage performs as follows:
   a) register-register instructions perform no operations during the DF, DS, and TC stages;
   b) for load/store instructions, the D-cache fetch and the data address translation starts;
   c) for branch instructions, the target address translation and TLB update are initiated.
6. The role of the DS stage is:
   a) for load/store instructions, the D-cache fetch and data address translation are completed. The shifter aligns the data to the word or double word boundary;
   b) for branch instructions, the target address translation and TLB update are completed.
7. During the TC stage:

   a) for load/store instructions, the cache performs the tag check, i.e., the physical address from the TLB is checked against the cache tag to determine whether there is a hit or miss;
   b) for branch instructions no operations occur during the TC and WB stages.
8. In the WB stage:
   a) for register-register instructions, the result is written back to the register file;
   b) for load/store instructions the value is either loaded from D-cache to the register file or stored from register file to D-cache.

Superpipelining enables the MIPS R4400 to issue more than one instruction within a base clock cycle, but still one instruction after the other.

1.7.5 Other Scalar RISC Processors

Fairchild/Intergraph Clipper (1986): The Clipper C100 was developed by Fairchild and later sold to workstation-maker Intergraph which took over the chip development and in 1988 produced the C300 processor. Since it could not compete in processor technology, Intergraph decided to switch to Intel x86-based processors. The C100 was a three-chip set (like the Motorola MC88000 but predating it by two years), with a Harvard architecture CPU and separate cache chips for instruction and data. Instruction lengths were 16- and 32-bit. It had sixteen 32-bit user registers and eight 64-bit floating-point registers. There was a bank of sixteen supervisor registers which completely replaced the user registers. In addition, there were some microcoded instructions.

ARM (1986): Advanced RISC Machine (ARM, originally Acorn RISC Machine, from Cambridge, UK) was designed as a processor chip for the Archimedes home computer. ARM design was based partly on the Berkeley RISC and was fabricated by VLSI Technologies, Inc. It was a 3-stage pipeline, and operated in big-endian or little-endian mode. The first ARMs (ARM1, 2, and 3) were 32-bit CPU, but used 26-bit addressing. The next ARM6 series is a completely 32-bit CPU. It has user, supervisor, and various interrupt modes. The ARM architecture has sixteen registers (including user-visible PC as `R15`), though many registers are shadowed in interrupt modes so they need not be saved, for fast response (as in PA-RISC, see below). Features implemented for high code density include: barrel shifter (to perform arbitrary shifts within the same cycle, at no speed penalty), conditional execution on every instruction (to eliminate many branches), and load/store multiple instructions (for rapid context switching and memory transfer). A feature introduced by the ARM is that every instruction is *predicated* (see Sect. 4.3.4) using a 4-bit condition code. This idea was later used in the Texas Instruments TMS320C6x processors (Sect. 4.10.1), as well as in the forthcoming Hewlett-Packard and Intel processors based on IA-64 architecture (Sect. 4.10.2). ARM has also developed a low-cost 16-bit version called Thumb, which decodes a subset of ARM CPU instructions into 16 bits. Native ARM code can be mixed with Thumb code when the full instruction set is needed. The 16-bit instructions are exploded to native 32-bit ARM instructions without penalty – similar to the CISC decoders in Intel x86-compatible and Motorola MC 68060 processors, except they decode native instructions into a newer set, while Thumb does the reverse. Thumb programs can thus be 30 % to 40 % smaller than already dense ARM programs. The newer ARM7 series (from December 1994) increases performance by an optimized multiplier, and added DSP extensions including 32-bit and 64-bit multiply and multiply/accumulate instructions. The ARM7 is a small 32-bit microprocessor with very low power consumption. Its 3-stage pipeline features a combined shift ALU execution stage allowing a single instruction to specify one of its operands for shifting or rotation before it is passed to the ALU. It also doubles cache size to 8 kB. A full DSP coprocessor Piccolo (1997) adds an independent set of sixteen 32-bit registers (four of which can be used as 48-bit registers) and a complete DSP instruction set, using a RISC philosophy similar to the ARM itself. Piccolo has its own PC, interacting with the CPU which performs data load/store through I/O buffers connected to the coprocessor bus. Piccolo shares the main ARM bus, but uses a separate instruction buffer to reduce conflicts. Two 16-bit values packed in 32-bit registers can be computed in parallel, similar to the HP PA-RISC MAX-1 multimedia instructions. DEC has also licensed the architecture, and in 1996 developed the StrongARM SA-110 processor, running a 5-stage pipeline at 100 to 233 MHz and using only 1 W of power, with 5-port register file, faster multiplier, single cycle shift-add, and Harvard architecture (I-cache and D-cache are each 16 kB) (see *Santhanam* [255], *Jaggar* [149]). To fill the gap between ARM7 and StrongARM, ARM also developed the ARM8 series which includes many StrongARM features. ARM8 extends the ARM7 implementation by two additional pipeline stages and a new cache interface that allows instruction fetches in parallel with data accesses. Static branch prediction (backward branches are predicted taken, forward branches are predicted untaken) with a three-cycle misprediction penalty is applied.

HP PA-RISC (1986): The PA-RISC (Precision Architecture) was designed to replace Motorola MC 68000 processors in HP-3000 minicomputers and HP-9000 workstations. It has an unusually large instruction set for a RISC processor that includes, for example, a conditional skip instruction which is similar in concept to the condition bits in the ARM processor (see above). The instruction set is large because initial design of ISA took place before RISC philosophy became popular, and partly because careful analysis revealed that performance benefited from the instructions chosen. Despite this, PA-RISC has a simple design; for example, the entire original CPU had only 115 000 transistors, which is less than twice the number in the Motorola MC 68000. PA-RISC design is similar to most other mainstream RISC processors, like the Fairchild/Intergraph Clipper, and the Motorola MC 88000, in particular. It has a 5-stage pipeline with hardware interlocks for instructions taking more than one cycle, as well as result forwarding. It is a load/store architecture, originally with a Princeton architecture, later expanded to a Harvard architecture. It has thirty-two 32-bit integer registers (`GR0` wired to 0, `GR31` used as a link register for procedure calls), with seven shadow registers that preserve the contents of a subset of the `GR` set during fast interrupts (as in ARM). In addition, it has thirty-two 64-bit floating-point registers that can be also be used as sixty-four 32-bit or sixteen 128-bit registers.

AMD 29000 (1987): The AMD 29000 is another RISC processor descended from the Berkeley RISC design. Like the SPARC design that was introduced shortly after, the 29000 has a large set of registers split into local and global sets. The 29000 has 64 global registers. It allows variable-sized windows allocated from the 128-register stack cache. The current window or stack frame is indicated by a stack pointer. A pointer to the caller's frame is stored in the current frame, as in an ordinary stack. Spills and fills occur only at the ends of the cache, and registers are saved/loaded from the memory stack. This allows variable window sizes, from 1 to 128 registers. This flexibility, plus the large set of global registers, makes register management easier than in SPARC. There is no special condition code register – any general register is used instead, allowing several condition codes to be retained, though this sometimes makes code more complex. An instruction prefetch buffer (using burst mode) ensures a steady instruction stream. Branches to another stream can cause a delay, so the first four new instructions are cached – the next time a cached branch is taken, the cache supplies instructions during the initial memory access delay. Registers are not saved during interrupts, allowing the interrupt routine to determine whether the saving overhead is worthwhile. In addition, a form of register access control is provided. All registers can be protected, in blocks of four, from access. These features make the 29000 useful for embedded applications, which is where most of these processors are used. The 29000 also includes a memory management unit and support for the 29027 floating-point unit.

Motorola MC 88000 (1988): The Motorola MC 88000 is a 32-bit processor, one of the first RISC processors based on a Harvard architecture (*Alsup* [7], *Melear* [199]). It is similar to the HP PA-RISC in design although the MC 88000 is more modular, has a small and elegant instruction set, no special status register (condition codes may be stored in any general register), and lacks segmented addressing (thus limiting addressing to 32 bits). It has thirty-two 32-bit user registers and up to 8 distinct internal FUs. The processor is pipelined with interlocks and result forwarding. The MMU chip MC 88200 provides dual caches (including multiprocessor support) and support for the MC 88100, a version of MC 88000 processor. Multiple ALUs and floating-point units (with thirty-two 80-bit floating-point registers) and 2-issue instruction fetching were added to obtain the MC 88110 (*Diefendorff and Allen* [65], 1992), one of the first superscalar designs.

1.7.6 Sun picoJava-I

A non-RISC processor example of a small-scale microprocessor featuring a simple 4-stage pipeline is provided by Sun's picoJava-I microprocessor (*Wayner* [321], *O'Connor and Tremblay* [216], *McGhan and O'Connor* [197]) which is the first of a series of proposed JavaChips, potentially continued by the more complex microJava and UltraJava processors. Java processors are designed to execute Java bytecode instructions directly in hardware. Additionally, hardware support for stack manipulation and thread synchronization is provided. The picoJava core features a simple pipeline, but not a RISC instruction set. The picoJava core deviates from RISC principles by its stack-based ISA, a variable-length instruction format, a partly microcoded implementation, and too large an instruction set to count it as a RISC ISA.

Applications in the Java language are compiled to target the *Java Virtual Machine* (JVM). The JVM (see *Meyer and Downing* [200]) is the name of the (abstract) engine that actually executes a Java program compiled to Java bytecode. It details the instruction set, datatypes, operand stack, constant pool, method area, heap for run-time data, and the class file format. The JVM definition comprises a file format for the executable, called *class file*, and an instruction set that is called *Java bytecode*. Java bytecode can be interpreted, which is relatively slow, or compiled to native machine code by a *just-in-time* (JIT) compiler that speeds up execution on a normal microprocessor but expands the code size by a factor of 3 or more. The Java processors render JVM software interpretation or JIT compilers superfluous by direct execution of Java bytecode.

A main characteristic of the JVM instruction set is its *stack architecture* which means that all compute instructions (i.e., arithmetic/logical instructions) address their operands implicitly as top-of-stack and next-of-stack and write the result back to the top-of-stack (see p. 10). The lack of explicit operand specification leads to a zero-address addressing format for the compute instructions resulting in a very compact machine code by the use of different-length instruction formats. The general-purpose registers are replaced by a JVM stack that holds local variables, (frame local) operand stack and pointers to frames. Some special status information concerns the top-of-stack index, the thread status information, the current method, the method's class and constant pool, and the program counter.

Data types include Boolean, char, byte, short, reference, int, long, float (32 bit) and double (64 bit), both ANSI/IEEE 754-1985. The data format is big-endian as the network order.

The Java bytecode specifies 226 instructions. All opcodes have eight bits followed by a variable number of zero to four operand bytes. The most frequently executed instructions are just one byte long. One-byte instructions make up 62 % of bytecode instructions. There exist escape opcodes for instruction set extensions. The compact JVM instruction format yields an average instruction length of only 1.8 bytes, which is extremely small compared to the 4-byte instruction format that is standard for most of today's RISC processors.

Not all instructions of the JVM are implemented in hardware in the picoJava-I[8] instruction set. Most of the hardware-implemented instructions execute in one to three cycles. Of the instructions not implemented directly in hardware, those deemed critical for system performance are implemented in microcode, for example, method invocation. This group of moderately complicated instructions contains about 30 bytecode instructions. A small microcode ROM contains the microcode; the picoJava core uses two approximately 2 kB ROMs, one in the integer unit and the other in the optional floating-point unit. The remaining group of about 30 instructions are either very complicated or require services from the underlying operating system, or both. These instructions are emulated by core traps – an example would be creating a new object, which is a less frequently used instruction.

The Java security model does not allow direct access of memory. There are no Java bytecode instructions that allow access to arbitrary memory locations. Java bytecode instructions only operate upon object references, the physical storage location of the objects are not known to the programmer. The JVM relies on library calls to the underlying operating system of its host system.

Consequently, for a Java processor additional instructions are necessary to implement low-level hardware management. In picoJava-I core, 115 instructions additional to the JVM are defined as extended instructions in reserved opcode space with 2-byte opcodes, the first byte being one of the reserved JVM opcode bytes. The extended bytecodes implement arbitrary load/store, cache management, internal register access, and other miscellaneous instructions necessary to allow the programmer to write system-level code. Hence, the picoJava-I ISA is no longer a JVM ISA, because the JVM cannot execute extended bytecode instructions. Moreover, no program containing extended bytecode instructions defined for the picoJava-I can be regarded as safe in the sense of the Java security model.

The picoJava-I processor consists of a core that includes a RISC-style 4-stage pipeline (see Fig. 1.19), an integer FU, and an optional floating-point

[8] The description in this section is given for the picoJava-I core [216]; some features are also mentioned that are only in the picoJava-II core [197].

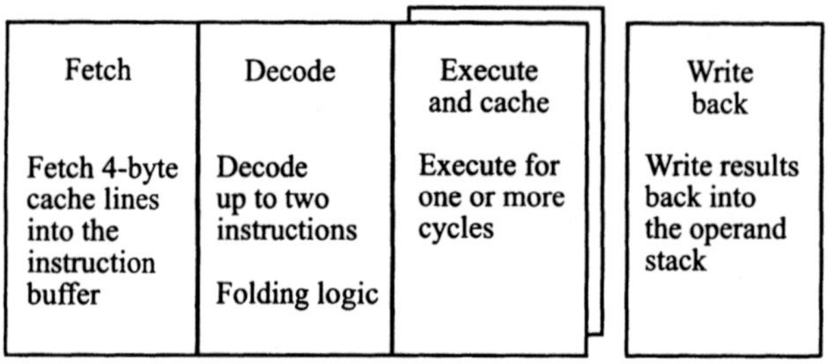

Fig. 1.19. picoJava-I pipeline

FU. The core is extended by an optional direct-mapped I-cache with zero to 16 kB and 8-byte line size, and an optional 2-way set-associative write-back D-cache with zero to 16 kB and 32-bit line size (see Fig. 1.20).

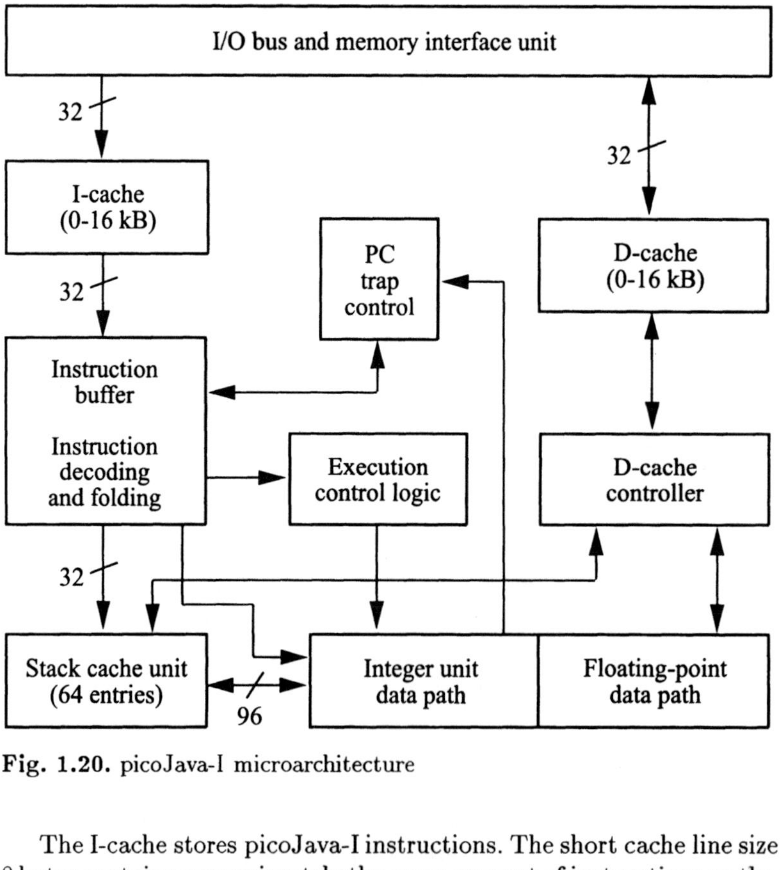

Fig. 1.20. picoJava-I microarchitecture

The I-cache stores picoJava-I instructions. The short cache line size of only 8 bytes contains approximately the same amount of instructions as the normal 16-byte line used by other RISC processors due to the compact instruction format of the Java bytecode.

A 12-byte instruction buffer decouples the I-cache from the rest of the core pipeline. The instruction buffer write in during the instruction fetch stage is four bytes and the read out by the decode stage is five bytes. The picoJava-I decodes each cycle up to five bytes at the head of the instruction buffer and sends the decoded instructions to the execution stage.

The branch prediction always predicts not taken. The 4-stage core pipeline yields a two-cycle penalty when the branch is taken. Branch delay slots can only be used by microcode and are not available to the compiler that compiles to the bytecode of the JVM.

Instructions stay for one or more cycles in the execution stage. The optional floating-point unit follows the ANSI/IEEE 754-1985 single and double precision standard. It is not internally pipelined. Compute instructions only operate on stack data and never on memory data. The D-cache can be accessed by load or store instructions during the execution stage. Execution result values and values loaded from the D-cache or memory are written back into the stack cache in the fourth pipeline stage.

JVM's stack architecture is implemented in the picoJava-I processor by a 64-entry on-chip hardware stack – called *stack cache* or *stack register file* – that is used instead of a general-purpose register file. The stack register file is managed as circular buffer (see Fig. 1.21). The top-of-stack pointer wraps around. The stack cache contains both integer and floating-point data which facilitates the (rather infrequent) passing of operands between integer and floating-point units. Data from the constant pool or from local variables that are arranged deeper in the stack have to be pushed on top of the stack before being used for execution. Access to these areas is provided by local-variable load and store instructions defined by the JVM. Each method invocation creates a call frame on the stack at run-time. The frame contains the parameters for the method, its local variables, the frame-state-like return address and the monitor entered. The hardware stack allows direct parameter passing without requiring any copying of the parameters – an option important for object-oriented languages that typically rely upon plenty of small method invocations (*O'Connor and Tremblay* [216]).

A hardware mechanism called a *dribbler* maintains an automatic spill and refill of the hardware stack concurrently to the instruction execution in the pipeline. When the hardware stack is almost full, the dribbler writes the oldest stack cache entries to the D-cache. Similarly, when the number of valid entries gets too low, the dribbler transfers JVM stack entries back in the hardware stack. The points at which the dribbler decides to spill or refill stack cache entries depends on *high* and *low water marks* set in a control register (*O'Connor and Tremblay* [216]).

Since access of computing instructions is limited to the top portion of the stack, a frequently occurring instruction combination consists of an local-variable load instruction with a succeeding compute instruction that con-

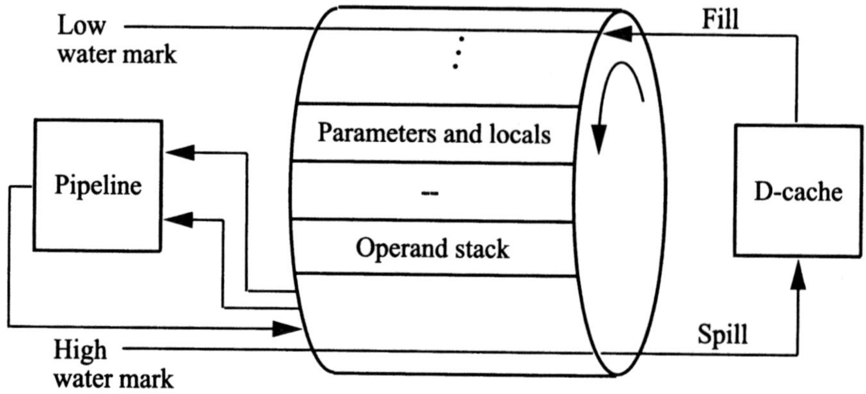

Fig. 1.21. picoJava I stack architecture

sumes the loaded value. Elimination of this extra load step during execution improves the performance of the stack architecture.

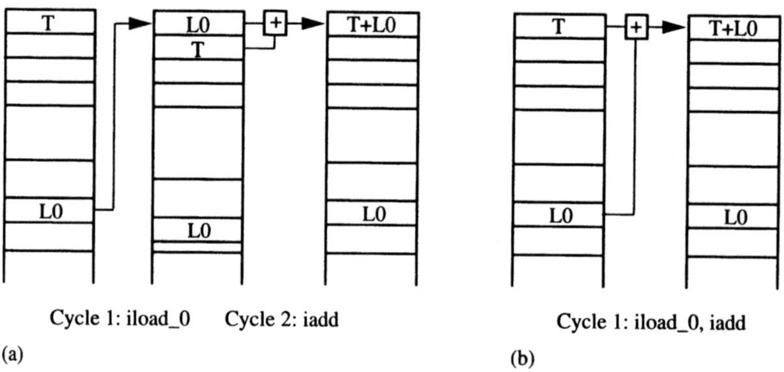

Fig. 1.22. Example of folding **(a)** without folding, the processor executes `iload_0` during the first cycle and `iadd` during the second cycle **(b)** with folding, `iload_0` and `iadd` execute in the same cycle

picoJava-I relies on a *folding* operation that executes a local-variable load instruction in the same cycle as the succeeding compute instruction (Fig. 1.22). The instruction decoder detects this situation and folds the instructions together in the execution stage. Simulations show that folding eliminates approximately 15 % of the total dynamic instruction count (Table 1.8, [216]).

The picoJava-II core which will be implemented in the microJava-701 processor goes one step further. Up to four instructions can be folded, if moves of local data to the top of stack are immediately followed by compute instructions that consume the data just moved, and/or compute operations

Table 1.8. JVM instruction frequencies without and with folding

Instruction class	Dynamic frequency before folding %	Dynamic frequency after folding %	Instructions folded %
Local-variable loads	34.5	24.4	10.1
Local-variable stores	7.0	7.0	0.0
Loads from memory	20.2	20.2	0.0
Stores to memory	4.0	4.0	0.0
Compute (integer/floating-point)	9.2	9.2	0.0
Branches	7.9	7.9	0.0
Calls/returns	7.3	7.3	0.0
Push constant	6.8	2.0	4.8
Miscellaneous stack operations	2.1	2.1	0.0
New objects	0.4	0.4	0.0
All others	0.6	0.6	0.0
Total	**100.0**	**85.1**	**14.9**

are immediately followed by local stores of the result just computed. Based on a set of grouping rules, the picoJava core scans the incoming stream of bytecodes looking for sequences of instructions that can be folded together dynamically (*McGhan and O'Connor* [197]).

A 4-way instruction folding is demonstrated in Fig. 1.23. Two local load instructions `iload_0` and `iload_1` which load the values from local integer variables L0 and L1 to the top of stack, are followed by an integer addition `iadd`, and by an `istore_2` instruction that removes the result from the top op stack and stores it in local integer variable `L2`. All four instructions can be folded into a single operation, which is equivalent to a 3-address operation `iadd L2, L0, L1` of a RISC ISA where `L0`, `L1`, and `L2` are register specifiers.

Most of the stack architecture overhead can be removed by such extensive folding techniques. Measurements indicate that between 23 and 37 % of all instructions executed become folded (*McGhan and O'Connor* [197]).

The picoJava-I core contains a two-entry cache of the two most recent monitors that the current thread has entered. Associated with each entry is a counter that indicates how many times the thread has entered the monitor, since re-entry of a monitor by the same thread is possible by the Java language specification. The monitor entry count is incremented when the monitor is entered and decremented when exited. If the counter is zero, the monitor is exited completely. On each monitor entry, the monitor cache is associatively examined, incremented, and decremented by hardware which speeds up performance compared to a pure software solution. However, replacements of monitor cache entries are raised by core traps and managed in software.

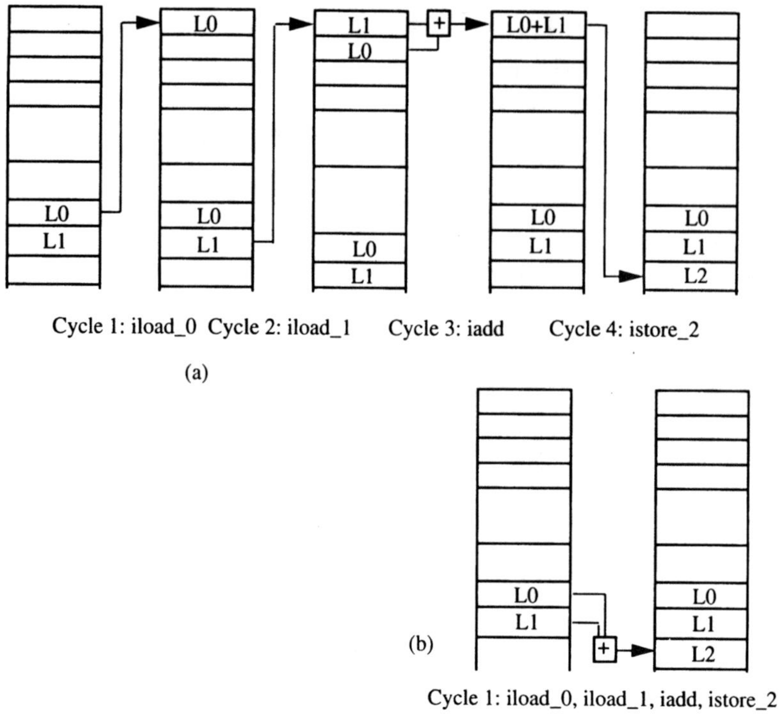

Fig. 1.23. Example of 4-way folding (a) without folding (b) with folding

Simulations show that picoJava-I features excellent performance compared to an Intel Pentium or an Intel 80486 processor running bytecode. Benchmarks run 15 to 20 times faster than a 80486 with interpreter and still 5 times faster than a Pentium with a JIT compiler and equal clock rate (see *O'Connor and Tremblay* [216]).

Second in the series of Sun's Java chips is the microJava-701 [283, 284] which implements a 32-bit picoJava-II core. Besides Java bytecode C-compiled code is also supported by the instruction set. The microJava-701 features a 6-stage pipeline, and an extensive folding that allows up to four instructions executed per cycle. Further system features are integrated on-chip: a memory controller, and a I/O bus controller. Volume production of microJava-701 was scheduled for the end of 1998 in 0.25 μm CMOS technology with 2.8 million transistors and a 200 MHz clock cycle rate.

The picoJava-I targets the market of embedded appliances. Stated by Sun Microelectronics Whitepapers [282], Java's simple, secure and small object-oriented code promotes clean interfaces and software re-use, while its distributed nature makes it a natural choice for network applications. Java is quickly establishing itself as a standard for the market of low-cost, embedded network computers. By the end of the decade, the average home will contain between 50 and 100 microcontrollers. There will be millions of cellular phones, set-top boxes, personal digital assistants, low-cost network terminals, and other Internet appliances operating in a networked environment and highly optimized for small applications running at top speed.

The direct execution of Java bytecode, the short pipeline, the hardware stack in combination with the dribbler, the garbage collection, and the monitor support provide excellent performance of Java-based code. Java bytecode is extremely dense by its stack architecture. Embedded market requirements are supported by making caches and floating-point units optional. Hard real-time requirements often require an exact cycle count for a running service routine. Cache misses, the dribbler hardware, and garbage collection may dynamically enhance the number of executed cycles, thereby potentially missing hard real-time deadlines. However, by an appropriate setting of the water marks in the control register the dribbler hardware can be disabled.

The main drawback for Java processors caused by the JVM stack architecture is, however, that the stack architecture disables most instruction-level parallelism (except for folding) that is exploited by multiple-issue processors. It will be very difficult to design a superscalar Java processor as is proposed by the more complex of Sun's Java processors. One solution might be a dynamic translation of stack register accesses to accesses to a directly addressed general-purpose register, set by a modified register-renaming stage. This step would allow the elimination of the stack architecture register set and thereby open the way to exploit more instruction-level parallelism.

1.8 Lessons learned from RISC

The goal of RISC architecture was to develop designs that can come close to issuing one instruction on each clock cycle. This has been made possible by:

- using hardwired instead of microcoded control,
- supporting a small set of equal-length instructions most of which are of the register-register type,
- relying on the optimizing compilers,
- relying on the high-performance memory hierarchy,
- and, especially, using instruction pipelining.

Since the pipelined and superpipelined RISC processors regarded in this chapter issue only one instruction at a time, they are scalar.

Recall from Sect. 1.7.2 that, besides microSPARC, there is a spectrum of other SPARC microprocessors. What differentiates them from the microSPARC is that they can simultaneously issue more than one instruction per cycle. For example, HyperSPARC issues two instructions per cycle, SuperSPARC three instructions, and UltraSPARC four instructions per cycle. This so-called superscalar or multiple-issue approach offers CPI < 1. Since instructions can be evaluated in parallel, we call such a potential overlap among instructions *instruction-level parallelism* (ILP). *Rau and Fisher* [239] distinguish the following main types of processor architectures capable of utilizing ILP by combining processor and compiler design techniques:

- *Sequential architectures* where the program provides no explicit information regarding instruction parallelism,
- *Dependence architectures* where the program explicitly indicates the dependences between instructions, and
- *Independence architectures* where the program conveys information as to which instructions are independent of one another.

Sequential architectures are represented by superscalar processors which still retain result serialization as required by von Neumann architecture. The basic superscalar principles will be discussed in Sect. 4.1, a detailed study follows in Sects. 4.2–4.8. In Sect. 4.9 we will describe several state-of-the-art superscalar processors. Chapter 2 is devoted to dataflow processors, which belong to dependence architectures. Independence architectures are represented by VLIW processors that are covered in Sect. 4.10 and EPIC processors described in Sect. 4.10.2.

2. Dataflow Processors

Dataflow stands apart as being the most radical of all approaches to parallelism and the one which has been the least successful ...
... If any practical machine based on dataflow ideas and offering real power emerges, it will be very different from what the originators of the concept had in mind.

Maurice V. Wilkes
Computing Perspectives
(Morgan Kaufmann Publishers, 1995)

... these instructions [of the Intel Pentium Pro] *are ... executed in dataflow order (when operands are ready) ...*

Robert P. Colwell and Randy L. Steck
A 0.6μm BiCMOS Processor with Dynamic Execution
(Int'l Solid State Circuits Conference, February 1995)

2.1 Dataflow Versus Control-Flow

Control-Flow. The most common computing model (i.e., a description of how a program is to be evaluated) is the von Neumann control-flow computing model. This model assumes that a program is a series of addressable instructions, each of which either specifies an operation along with memory locations of the operands or it specifies (un)conditional transfer of control to some other instruction. A control-flow computing model essentially specifies the next instruction to be executed depending on what happened during the execution of the current instruction. The next instruction to be executed is pointed to and triggered by the program counter PC. This instruction is executed even if some of its operands are not available yet (e.g., uninitialized).

Dataflow. The dataflow model represents a radical alternative to the von Neumann computing model since the execution is driven only by the availability of operands. It has no PC and global updatable store, i.e., the two features of the von Neumann model that become bottlenecks in exploiting parallelism.[1]

[1] The serialization of the von Neumann computing model is a serious limitation for exploiting more parallelism in today's microprocessors – e.g., superscalars.

In the context of parallel computing, the earliest use of the word "dataflow" dates back to 1960s when *Karp and Miller* [155] studied a dataflow-like model of computation. The history of dataflow computers and languages began in the late 1960s with two separate strands of research whose similarities were not to be generally recognized until the end of the following decade (*Glauert et al.* [101], 1985). The earliest work is mostly concerned with languages and notations for parallel computation. The language work led to the development of the so-called *single-assignment languages*, and eventually to the design of a *single-assignment machine* (see the LAU System on p. 61). The notational research formed the foundations of the later design of dataflow computer systems. The first architecture to embody the dataflow computing model was developed in the mid-1970s[2] by *Dennis and Misunas* [64]. By the late 1970s it was clear that dataflow and single-assignment were synonymous, and subsequent dataflow machines have all been designed in conjunction with a single-assignment language. These languages are characterized by the *single-assignment rule* which means that a variable may appear on the left side of an assignment only once within the area of the program in which it is active. High-level single-assignment programming languages include, for example, VAL (*Ackerman and Dennis* [1], 1979), Id (*Heller and Traub* [132], 1985), and LUCID (*Wadge and Ashcroft* [315], 1985).

A program, which is written in a single-assignment language, is compiled into a *dataflow graph* which is a directed graph consisting of named *nodes*, which represent instructions, and *arcs*, which represent data dependences among instructions. During the execution of the program, data propagate along the arcs in data packets, called *tokens*. This flow of tokens enables some of the nodes (instructions) and fires them.[3] Figure 2.1 shows two dataflow graphs. The acyclic graph in Fig. 2.1a represents the function `Stats` which is defined in VAL below and computes the mean and standard deviation of three input values. The function returns two real values, `Mean` and `StDev`, which are defined in the `let` part of the program.

```
function Stats (x,y,z: real returns real, real);
 let
    Mean := (x + y + z)/3;
    StDev := SQRT((x**2 + y**2 + z**2)/3 - Mean**2);
 in
    Mean, StDev
```

[2] In 1967, however, *Tomasulo* applied a limited dataflow design to the floating-point unit for the IBM System 360 Model 91 (see Sect. 3.3.2).

[3] There is also a diametrically opposite method of evaluating dataflow graphs called *demand-driven* execution where an enabled node is fired if there is a demand for the result. The demand-driven execution model leads to so-called *reduction machines* (*Treleaven et al.* [299]). Reduction machines came in two flavors, graph reduction and string reduction. In data-driven (e.g., dataflow) computers, the availability of operands triggers the execution of the operation to be performed on them, whereas in demand-driven (e.g., reduction) computers the requirement for a result triggers the operation that will generate it.

```
    endlet
  endfun
```

The cyclic graph in Fig. 2.1b represents the following Id program segment

```
( initial j <- n; k <- 1
  while j > 1 do
    new j <- j - 1;
    new k <- k * j;
  return k )
```

which computes factorial $n!$ of integer n. The **CHOOSE** node outputs either the token from F-input or the token from T-input, depending on the token on the control input. The token on the control input of the **SWITCH** node selects either the T-output or F-output, to which the input token will be sent.

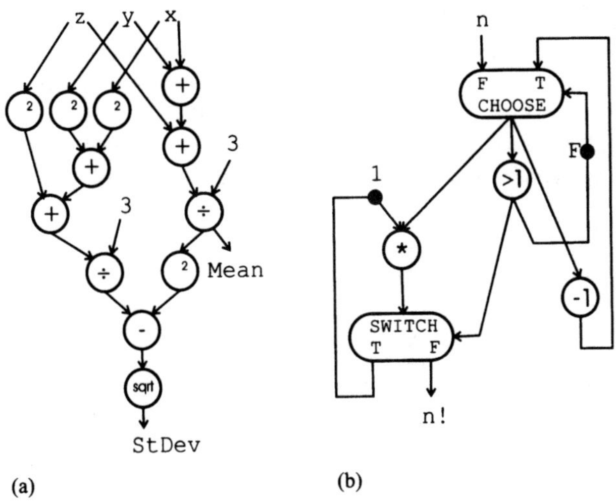

Fig. 2.1. Dataflow graphs (a) simple statistic function (b) factorial

Two important characteristics of dataflow graphs are *functionality* and *composability*. Functionality means that evaluation of a graph is equivalent to evaluation of the corresponding mathematical function on the same input data. Composability means that graphs can be combined to form new graphs.

The execution of a dataflow graph proceeds according to the instruction enabling and instruction firing rules. The instruction (node) *enabling rule* is:

An instruction is enabled (i.e., executable) if all operands are available to it.

Note that in the von Neumann model, an instruction is enabled if it is pointed to by the PC. The computational rule of the dataflow model, also known as the instruction (node) *firing rule*, specifies when an enabled instruction is actually executed. The basic instruction firing rule is:

An instruction is fired when it is enabled
(and when the resources are available).

The effect of firing an instruction is the consumption of its input data (operands), the execution of the instruction, and the generation of output data (results).

Because of the single-assignment rule, parallelism is not constrained by antidependences and output dependences as it is in conventional imperative languages. Control dependences are transformed into data dependences. Structural hazards are mostly ignored in dataflow literature, where unbounded hardware resources are assumed. We have enhanced the usually idealistic firing rule by demanding resource availability in the rule stated above. Dataflow is said to be *self-scheduling* since instruction sequencing is constrained only by data dependences among instructions. Thus, the flow of control is the same as the flow of data among various instructions.

There are computer architectures that support the *pure* dataflow computation model as described above, such as the static, the dynamic, and the explicit token store architecture. Advanced architectures, however, support *augmenting* the dataflow computation model with traditional control-flow mechanisms, such as multithreading, large-grain computation, complex machine operations, RISC approach, etc.

2.2 Pure Dataflow

Let us first explain the basic principles of pure dataflow computer architectures. A dataflow computer executes a program by receiving, processing, and sending out tokens, each containing some data and a tag. Dependences between instructions are translated into tag matching and tag transformation. Processing starts when a set of matched tokens arrives at the execution unit. The instruction which has to be fetched from the instruction store (according to the tag information) contains information about what to do with data and how to transform the tags. The matching unit and the execution unit are connected by an asynchronous pipeline, with queues added between the stages to smooth out workload variations. Some form of associative memory is required to support token matching. It can be a real memory with associative access, a simulated memory based on hashing, or a direct matched memory. Each of these three solutions has its proponent but none is absolutely suitable.

Due to its elegance and simplicity, the *pure dataflow* model has been the subject of many research efforts. Since the early 1970s, a number of dataflow computer prototypes have been built and evaluated, and different designs and compiling techniques have been simulated (see *Treleaven et al.* [299], *Srini* [278], *Gaudiot and Bic* [98], *Šilc et al.* [263]).

Clearly, an architecture supporting the execution of dataflow graphs should support the flow of data. Depending on the way the data are handled, several types of dataflow architectures have emerged in the past:

- *static*,
- *dynamic*, and
- *explicit token store.*

We describe them in Sects. 2.2.1–2.2.3, respectively.

2.2.1 Static Dataflow

The *static* (also called *single-token-per-arc*) dataflow architecture was first proposed by *Dennis and Misunas* [64], 1975. At the machine level, a dataflow graph is represented as a collection of *activity templates*, each containing the *opcode* of the represented instruction, *operand slots* for holding operand values, and *destination address fields*, referring to the operand slots in subsequent activity templates that need to receive the result value. Each token only consists of a value and a destination address. Note that different tokens flowing to the same destination, one after the other, could not be distinguished. Therefore, the static dataflow approach allows *at most one token* to reside on any one arc. This is accomplished by extending the basic firing rule as follows:

An enabled node is fired if there is no token on any of its output arcs (and when the resources are available).

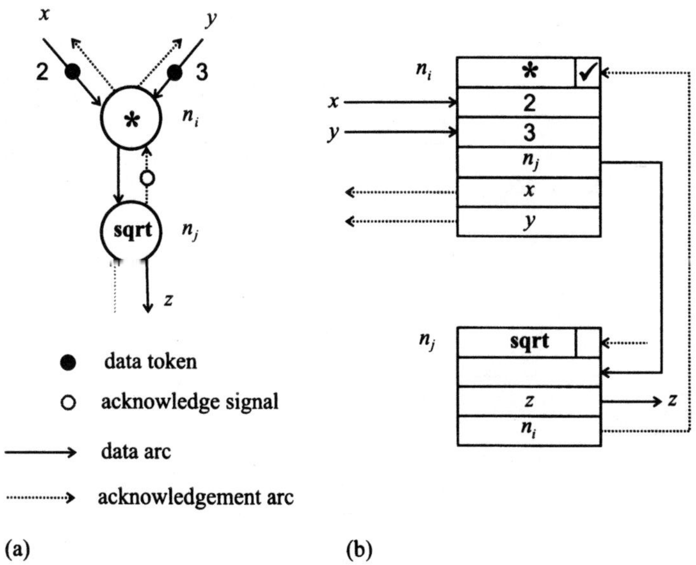

Fig. 2.2. Acknowledge signals in **(a)** a dataflow graph **(b)** activity templates

To implement the restriction of at most one token per arc, *acknowledge signals*, traveling along additional arcs from consuming to producing nodes, are used as additional tokens (Fig. 2.2). Thus, the firing rule can be changed to its original form:

A node is fired at the moment when it is enabled (and when the resources are available).

The major advantage of the single-token-per-arc dataflow model is its simplified mechanism for detecting enabled nodes (see Fig. 2.2).

Unfortunately, this model of dataflow has a number of serious deficiencies. Consecutive iterations of a loop can only be pipelined, which limits the amount of parallelism that can be exploited. Another undesirable effect is that, due to acknowledgment tokens, the token traffic is doubled. In addition, there is a lack of support for programming constructs that are essential to any modern programming language (e.g., procedure calls, recursion). Despite these shortcomings, several static machines were constructed and served as the theoretical and practical basis for subsequent dataflow computers.

MIT Static Dataflow Machine

This machine was designed by *Dennis and Misunas* [64] at the Massachusetts Institute of Technology (MIT) (Cambridge, MA) as a direct implementation of a single-token-per-arc dataflow model. It comprises a set of *processing*

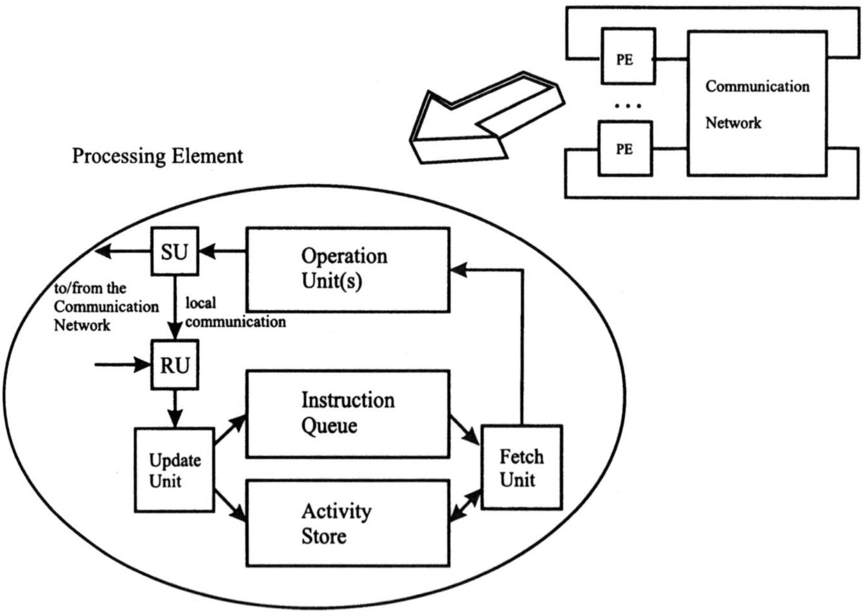

Fig. 2.3. Communication structure and processing element of the MIT Static Dataflow Architecture

elements (PEs) interconnected through a *communication network* (CN) as shown in Fig. 2.3.

Activity templates reside in an *activity store* (AS), and addresses of fired instructions (activity templates) reside in an *instruction queue* (IQ). This queue is accessed by the *fetch unit* (FU) that removes the first entry in the IQ, uses it for fetching the corresponding opcode, data, and destination list from AS, packs them into an operation token, forwards this token to an available *operation unit* (OU), and finally clears the operand slot in the template. The OU executes the operation specified by the opcode using the corresponding operands, generates result tokens for each destination, and sends them to the *send unit* (SU). The SU decides whether the token's destination is in a local or a remote PE. If the destination is local, the token is sent to the local *receive unit* (RU) that, in turn, passes it on to the *update unit* (UU). If the destination is not local, however, the result token is routed to the destination PE through the CN. All units operate concurrently, so instructions are processed in a pipeline fashion. Note that large delays in the CN do not affect the performance as long as enough fired instructions are present in each PE.

A prototype of the MIT Static Dataflow Machine was built in the early 80s with eight PEs (each emulated by a microprogrammable microprocessor) and an equidistant packet routing network using 2×2 routing elements. The machine was used primarily as an engineering model.

Another implementation of the MIT Static Dataflow Machine was the cell block version of that machine which was not a multiprocessor of dataflow PEs, but a highly parallel machine consisting of cell blocks and FUs interconnected by a distribution and an arbitration network.

Other projects

LAU System (*Plas et al.* [234], 1976): Researchers at the CERT (Toulouse, France) carried out the first project to go through the entire process of graph, language, and machine design. Their system, called LAU[4], was built in 1979. The LAU system used static graphs and had its own dataflow language based on single-assignment rule. LAU's architecture contained five major processing sections connected in a ring. Four of the five sections managed I/O, task and job supervision, and backing memory. The fifth section contained up to 32 PEs, each of which had its own local memory for graph and data values. Each PE was a 16-bit microprogrammed processor built around the AMD 2900 bit-slice microprocessor. The PC was replaced by two memories: the *instruction control memory* (ICM) and the *data control memory* (DCM). ICM handled control bits associated with each instruction while DCM managed the bit associated with each data operand. The LAU prototype system demonstrated that a dataflow computer could be built.

[4] "Language à assignation unique" (LAU) is a French acronym for single-assignment computation.

DDM1 (*Davis* [60], 1978): The Utah Data Driven Machine (DDM1) was designed at the University of Utah (Salt Lake City, UT) and was completed at the Burroughs Interactive Research Center (La Jolla, CA). The machine organization was based on the concept of recursion. It was tree-structured, with each PE connected to up to eight descendant PEs. Each PE consisted of an agenda queue with firable instructions, a local program memory, and an "atomic" processor. Another significant aspect of this work was the high-level graphical language designed to be used on the DDM1 (see *Boekelheide* [30]).

TI's DDP (*Cornish et al.* [54], 1979): The Distributed Data Processor (DDP) was a system, designed at Texas Instruments, which aimed to investigate the potential of dataflow as the basis of a high-performance computer. The DDP was not commercially exploited, but a follow-on design with Ada as the programming language had its application in military systems.

NEC Image Pipelined Processor (*Temma et al.* [290], 1985): NEC Electronics (Japan) developed the first dataflow VLSI microprocessor chip μPD7281. Its architecture was a pipeline with several blocks (i.e., working areas), organized in a loop. The program was stored in two tables, called a link table and a function table, while data memory was used for temporarily storing the tokens to be matched. An address generator and a flow controller were responsible for matching two tokens and temporarily storing them in the queue before sending them to the processing unit. The μPD7281 had a very powerful instruction set designed specifically for digital image processing algorithms such as restoration, enhancement, compression, and pattern recognition.

HDFM (*Vedder et al.* [311], *Gaudiot et al.* [99], both 1985): The Hughes Dataflow Multiprocessor (HDFM) project began in 1981 at Hughes Aircraft Co. (USA), prompted by the need for high performance, reliable, and easily programmable processors for embedded systems. The HDFM consisted of many, relatively simple, identical PEs connected by a global packet-switching network. The interconnection (3D cube) network was integrated with PEs for ease of expansion and minimization of VLSI chip types. The communication network was designed to be reliable, with automatic retry on garble messages, distributed bus arbitration, alternate path packet routing, and a failed PE translation table to allow rapid switch-in and use of spare PEs. The communication network was able to accommodate up to 512 PEs in an $8 \times 8 \times 8$ configuration. The following candidates proposed for the HDFM network were evaluated by *Exum and Gaudiot* [82]: a 3D cube network, a multistage network, a hypercube network, a mesh network, and a ring network. The processor engine was a 3-stage pipelined processor with three operations overlapped: instruction fetch and dataflow firing rule check, instruction execution, and result and destination address combining to form a packet.

2.2.2 Dynamic Dataflow

The performance of a dataflow machine significantly increases when loop iterations and subprogram invocations can proceed in parallel. To achieve this, each loop iteration or subprogram invocation should be able to execute as a separate instance of a re-entrant subgraph. This replication, however, is only conceptual. In a real implementation, only one copy of any dataflow graph is actually kept in memory. Each token has a *tag* consisting of the address of the instruction for which the particular data value is destined and other information defining the computational context in which that data is to be used. This context is sometimes called the value's *color*. Each arc can be viewed as a bag that may contain an *arbitrary* number of tokens with different tags. The enabling and firing rule is now:

A node is enabled and fired as soon as tokens with identical tags are present on all input arcs (and when the resources are available).

U-interpreter. A method for assigning tags to each execution of an instruction was called the *U-interpreter* (for unraveling interpreter). Each token consists of a *tag* and *data*. The tag comprises the *context field* c that uniquely identifies the context in which the instruction is to be invoked, the *initiation number* i that identifies the loop iteration in which this activity occurs, and the *instruction address* n. Note that c can itself be a tag. Since the destination instruction may require more than one input, each token also carries the number of its destination *port* p. We represent a token by $\langle c.i.n, data\rangle_p$.

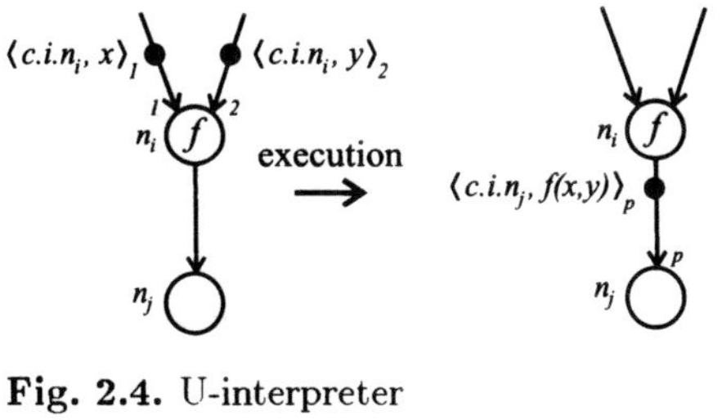

Fig. 2.4. U-interpreter

In the following, we describe the tag generation mechanism as it was proposed by *Gostelow and Arvind* ([105], 1982). Here, if the node n_i performs a dyadic function f, and if the port p of n_j is the destination of n_i, then we have

$$in : \{\langle c.i.n_i, x\rangle_1, \langle c.i.n_i, y\rangle_2\} \quad out : \{\langle c.i.n_j, f(x,y)\rangle_p\}$$

as can be seen in Fig. 2.4.

Figure 2.5a,b shows two basic nodes, referred to as the **MERGE** and **SWITCH**, which are used to represent branches and loops.

The branch construct (Fig. 2.6a) is described by

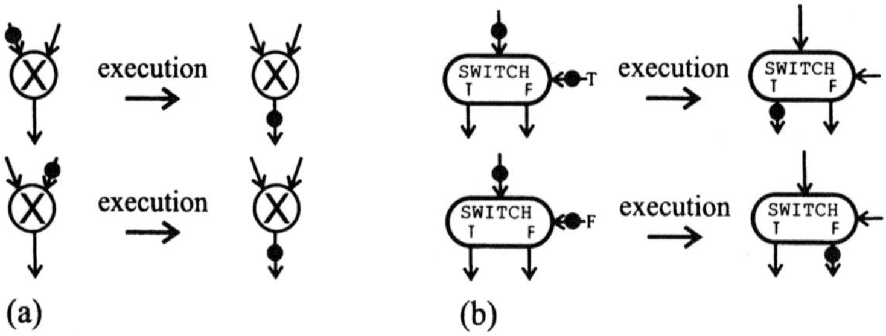

Fig. 2.5. U-interpreter (a) MERGE node (b) SWITCH node

$$in : \{\langle c.i.n_i, x\rangle_{data}, \langle c.i.n_i, b\rangle_{control}\} \quad out : \begin{cases} \{\langle c.i.n_j, x\rangle\} \text{ if } b = \mathrm{T} \\ \{\langle c.i.n_k, x\rangle\} \text{ if } b = \mathrm{F} \end{cases}$$

Merging of f and g outputs with a MERGE node causes no tag duplication. It is surprising that no one ever considered evaluating branches speculatively, although this is easy due to the single-assignment rule and obviously realized with the graph in Fig. 2.6b. In Fig. 2.6b, f and g represent the *then*-path and the *else*-path, respectively. Both paths can be executed concurrently because the single-assignment rule guarantees data independence. The SWITCH-node is replaced by a CHOOSE-node (see Fig. 2.6b)[5] that transports the token either from the subgraph f or from the subgraph g to the output line, depending on the predicate token b. The loop construct (Fig. 2.6c) uses, besides MERGE and SWITCH, additional operators L, $\mathtt{L}^{-1}$, D, and $\mathtt{D}^{-1}$. L is described by

$$in : \{\langle c.i.n_i, x\rangle\} \quad out : \{\langle c'.1.n_k, x\rangle\},$$

where $c' = \langle c.i.n_i\rangle$. The D operator was introduced because the initiation number of a token in a loop must be incremented every time a token goes around a loop. This is accomplished by D as follows:

$$in : \{\langle c'.j.n_j, x\rangle\} \quad out : \{\langle c'.j+1.n_k, x\rangle\}.$$

If after $k-1$ iterations the loop predicate P becomes false, SWITCH sends the last token with initiation number k to the $\mathtt{D}^{-1}$ operator, which resets the initiation number to 1:

$$in : \{\langle c'.k.n_l, x\rangle\} \quad out : \{\langle c'.1.n_m, x\rangle\}.$$

The $\mathtt{L}^{-1}$ sends its input token to an activity whose context and initiation number are identical to those of the activity that initiated this loop:

$$in : \{\langle c'.1.n_m, x\rangle\} \quad out : \{\langle c.i.n_n, x\rangle\}.$$

The function application (Fig. 2.6d) uses operators A, $\mathtt{A}^{-1}$, BEGIN, and END. The A operator creates a new context c' within which the function to be

[5] The CHOOSE-node is not part of the U-interpreter.

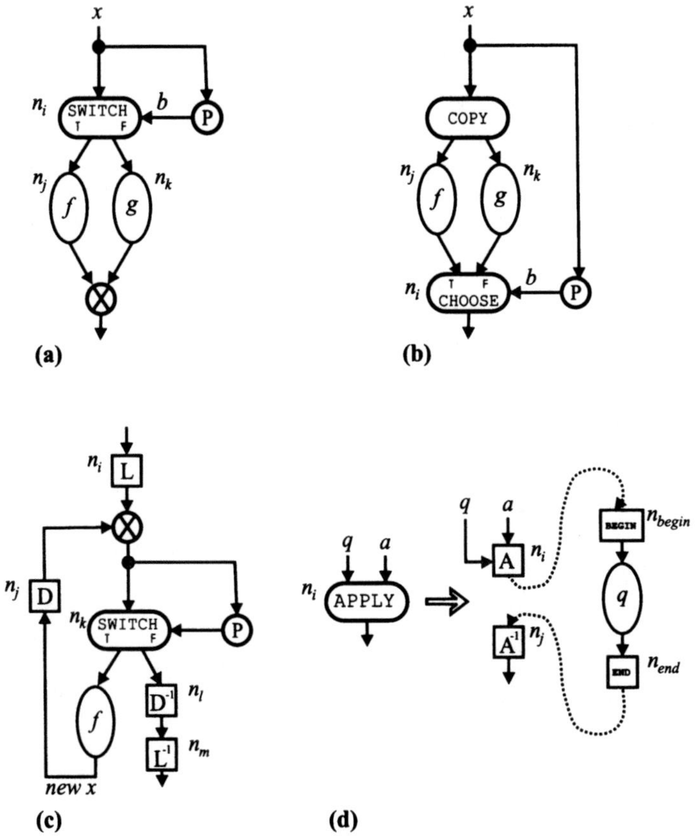

Fig. 2.6. U-interpreter **(a)** branch **(b)** speculative branch evaluation (CHOOSE-node) **(c)** loop **(d)** function application

applied arrives on arc q, while the argument on which this function is to be applied is passed on arc a:

$$in : \{\langle c.i.n_i, q\rangle_{func}, \langle c.i.n_i, a\rangle_{arg}\} \quad out : \{\langle c'.1.n_{begin}, a\rangle\},$$

where $c' = \langle c.i.n_j\rangle$ and n_j is the address of the $\mathtt{A}^{-1}$ operator. The **BEGIN** operator simply replicates tokens for each fork in its output arc. The **END** operator returns the result to the caller by unstacking the return address:

$$in : \{\langle c'.1.n_{end}, q(a)\rangle\} \quad out : \{\langle c.i.n_j, q(a)\rangle\}.$$

The $\mathtt{A}^{-1}$ operator replicates its output for its successors.

I-structure. The single-assignment rule in conjunction with a complex data structure means that each update of a data structure consumes the structure and the value producing a new data structure. However, this is awkward or even impossible to implement. To solve the problem of complex data structures, the concept of *I-structures* (for incremental structures) has been proposed by *Arvind and Thomas* ([17], 1981). An I-structure may be viewed as a data repository obeying the single-assignment rule. That is, each element of the I-structure may be written only once but it may be read any number of times. The basic idea is to associate with each element *status bits* and a *queue of deferred reads*. The status of each element of the I-structure can be:

- PRESENT, meaning that the element can be read but not written,
- ABSENT, meaning that a read request has to be deferred but a write operation into this element is allowed,
- WAITING, which means that at least one read request of the element has been deferred (since nothing has been written yet).

The state transition diagram for the I-structure operations is given in Fig. 2.7.

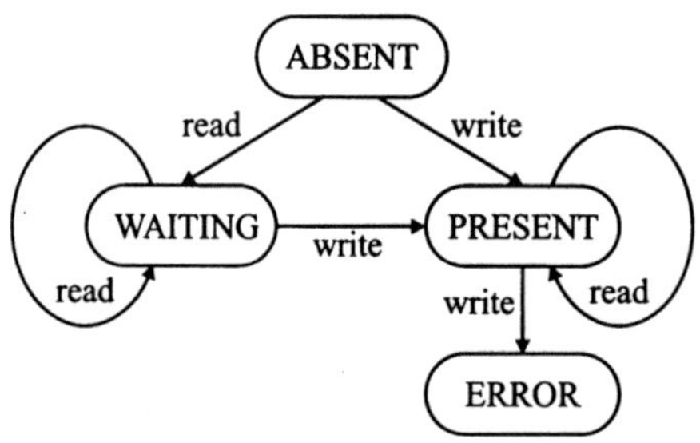

Fig. 2.7. State transition diagram for I-structure operations

After an element of the data structure has become defined (recall that that can happen exactly once), all deferred reads, which are kept in the associated queue, immediately become satisfied. Thus, the I-structure makes it possible to use a data structure before it is fully defined. In addition, it allows defining complex data structures from existing, though partially defined, data structures.

The following three elementary operations are defined on I-structures:

- *Allocate*, which reserves a specified number of elements for a new I-structure,
- *I-fetch*, which retrieves the contents of the specified I-structure element (if the element has not yet been written, this operation is automatically deferred),
- *I-store*, which writes a value into the specified I-structure element (if that element is not empty, an error condition is reported).

These elementary operations are used to construct nodes **SELECT** and **ASSIGN** as described in Fig. 2.8. The operation $x = A[j]$ is performed by the **SELECT**

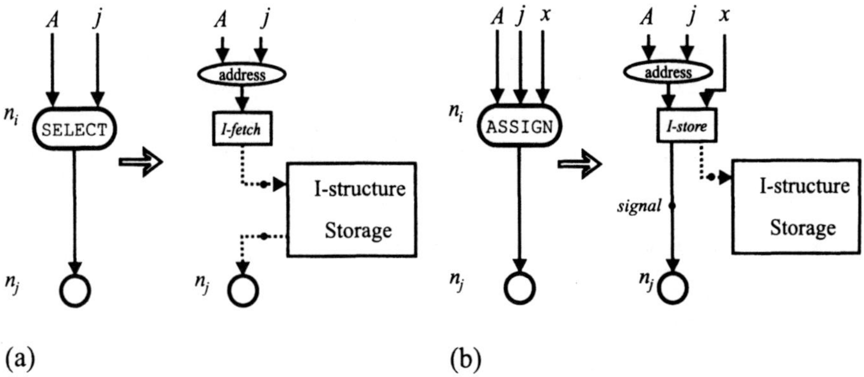

Fig. 2.8. I-structure (a) selection (b) assignment

node. First, the address a of the j-th element in the I-structure A is computed. The operation *I-fetch* then sends a *read*-token $\langle c.i.n_j, a,$ "*read*"$\rangle$ to the I-structure storage, thus asking for the contents of the element at the address a. If the status of the storage location a is PRESENT, the element contains a value, say x, so the value is read and a token $\langle c.i.n_j, x\rangle$ is sent to the node n_j. However, if the status is WAITING, the read request is deferred. Similarly, an ABSENT status is changed to WAITING and the request is enqueued in the associated queue as in the previous case. The *I-fetch* instruction is implemented as a *split-phase* memory operation, meaning that a read request issued to an I-structure is independent in time from the response received and thus does not cause a wait by the issuing PE. The operation $A[j] = x$ is performed by the **ASSIGN** node. After the address a is computed, *I-store* sends a *write*-token $\langle a,$ "*write*"$, x\rangle$ to the I-structure storage. The *I-store* instruction also generates a *signal*-token $\langle c.i.n_j\rangle$. In spite of firing the **ASSIGN** node, x may be written into the I-structure later. If the status of $A[j]$ is ABSENT, then x is written and the status is set to PRESENT. If the status is WAITING, the same action as in the previous case is performed first, activating all deferred read requests for that I-structure element. If a write-token arrives at a non-empty location (status is PRESENT), it is treated as a run-time error due to the single-assignment rule.

Dataflow architectures that use the model of execution whereby tags are attached to tokens are called *dynamic* (or sometimes *tagged-token*) dataflow architectures. This model was proposed by *Watson and Gurd* [320] at the University of Manchester (England), and simultaneously by *Arvind et al.* [14] at MIT.

The major advantage of the tagged-token dataflow model is better performance (compared with static predecessors) as it allows multiple tokens on each arc, thereby unfolding more parallelism. One of the main problems of the tagged-token dataflow model was efficient implementation of the unit that collects tokens with matching colors. For the sake of efficiency, an associative memory would be ideal. Unfortunately, it would not be cost-effective since the amount of memory needed to store tokens waiting for a match tends to be very large. As a result, all existing machines use some form of hashing technique which is typically not as fast as associative memory. In the following, we list the most important tagged-token dataflow projects.

MIT Tagged-Token Dataflow Architecture

This project originated at the University of California at Irvine (CA) and continued at MIT. The Irvine dataflow machine, proposed by *Arvind et al.* [14], 1978, implemented a version of the U-interpreter and array handling mechanisms to support I-structures. The machine was proposed to consist of multiple clusters interconnected in a *token ring*. Each cluster consisted of four PEs sharing local memory through a local bus and memory controller.

The MIT Tagged-Token Dataflow Architecture (TTDA) was a modified Irvine machine, but was still based on the Id language (*Arvind and Nikhil* [16], 1987). Instead of using a token ring, a *n-cube packet network* for inter-processor communication was used (Fig. 2.9). The I-structure storages were addressed uniformly and collectively implemented a global address space. A single *processing element* (PE) and a single *I-structure storage* suffice to constitute a complete dataflow computer.

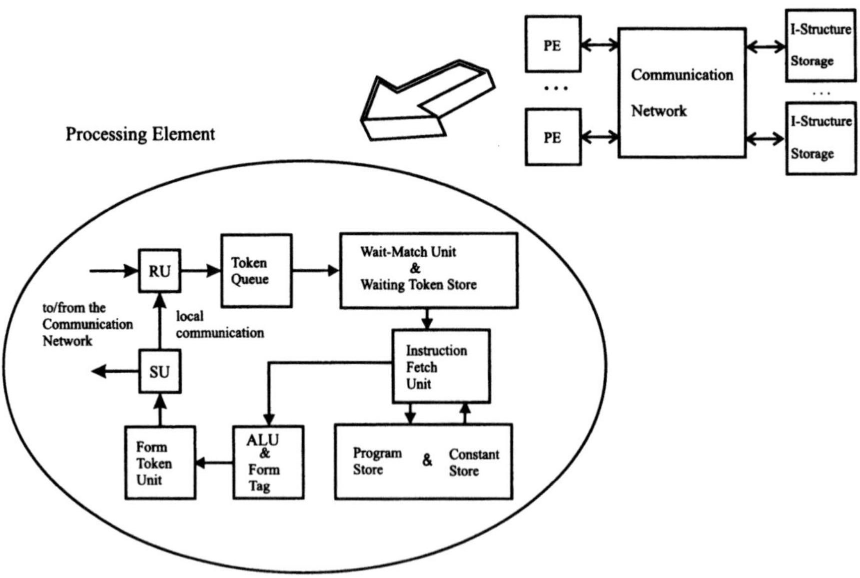

Fig. 2.9. Communication structure and processing element of the MIT TTDA

A program under execution is distributed over the *program storages* (PSs) of different PEs. Tokens entering the PE go through the *token queue* (TQ) and the following sections. They are first removed from the TQ by the *wait-match unit* (WMU). The WMU is a memory containing a pool of waiting tokens. If the entering token is destined for a monadic operator, it goes straight to the *instruction fetch unit* (IFU). Otherwise, a matching phase is initiated. This involves comparing the token's tag with the tags of all tokens currently held in the WMU. If a match occurs, the token is extracted from the WMU, and the two matching tokens are passed on to the IFU. However, if the WMU does not contain the partner, the token is left in the WMU to wait for its partner. The WMU is thus the rendezvous point for pairs of arguments for dyadic operators. It behaves as an *associative* memory, but is implemented with hashing methods. The tag on the operand token entering the IFU identifies the instruction to be fetched from the PS. The fetch instruction may also include a literal or a reference to a constant to be used as an operand. In the latter case, the constant is fetched immediately from the *constant store*. When a complete executable packet is assembled, it is passed to the ALU for execution. The ALU computes the result and, in parallel, derives new tags. The *form token unit* (FTU) takes the result and these tags from the ALU and combines them into the result token. The result token is forwarded to the local TQ, to another PE if the destination address is non-local, or to the I-structure storage if the operation is an access to a data structure.

The implementation of the token-matching stage is critical for the TTDA's performance. The TTDA requires that a token is checked against all other waiting tokens for a possible match. That means that the TTDA requires a moderately large associative memory. The dataflow graph is dyadic (i.e., nodes may need two input tokens). The experiences gained by the simulated TTDA (never built) were used in building its successor, the Monsoon, which is described in Sect. 2.2.3.

Manchester Dataflow Machine

The researchers led by *Watson and Gurd* ([320], 1979) at the University of Manchester focused on the construction of a prototype dataflow computer. The graph structure in this machine was static with token labels to keep different procedure calls separate.

The Manchester Dataflow Machine (MDM) was first simulated in 1977/78, and implemented as a hardware prototype in 1981. It consisted of one *processing element* (PE) and two *structure storage modules* connected with a simple 2×2 switch. The MDM prototype is shown in Fig. 2.10.

Two structure storages can hold a total of 1 M data values with access rates of 750 k reads per second and half as many writes per second. The structure storage provides reference counts to support garbage collection of structures that are no longer needed by the program.

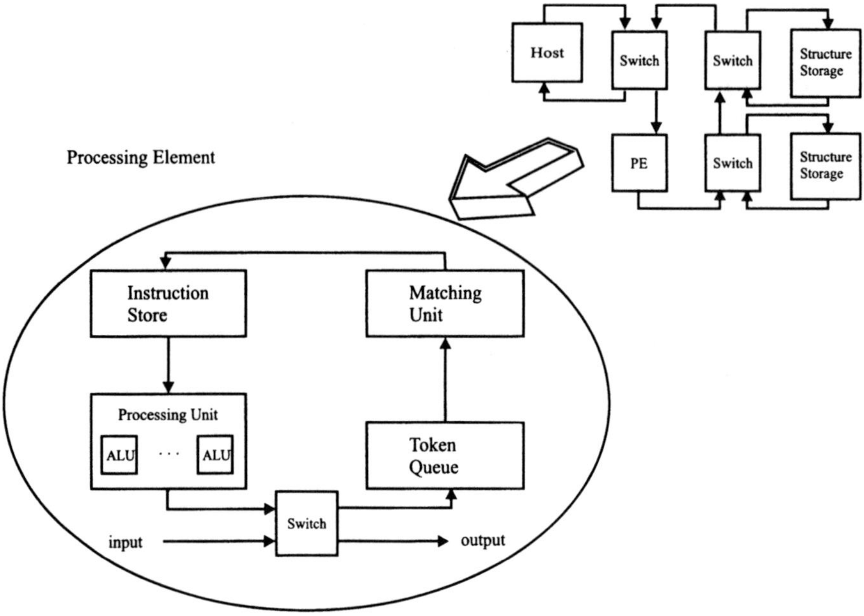

Fig. 2.10. Communication structure and processing element of the MDM

Each PE had a pipelined internal structure, with tokens passing through a *token queue* (TQ) module, a *matching unit* (MU) and an *instruction store unit* (ISU) before being processed by one of 20 ALUs in a parallel *processing unit* (PU). The ALUs were microcoded, and different instructions took quite different times to execute. Microinstructions are 48 bits wide, and the clock period is 67 ns. If all 20 ALUs can be utilized fully, this gives approximately a 6 MIPS rate for the whole computer. A hardware hashing mechanism is used instead of an associative memory in the MU that can hold up to 1.25 M unmatched tokens waiting for their partners. The MU has a 155 ns clock period and a 150 ns memory cycle time, so it gives rates of 1.29 M matches per second for dyadic operators and 6.45 M bypasses per second for monadic operators. The ISU is a buffered RAM with a capacity of 64 k instructions. The clock period is 40 ns, and the memory access time is 150 ns, resulting in a maximum processing rate of 2 M instruction fetches per second.

To overcome the restriction in the MDM that only two successor instructions could be specified in one instruction, the TUPlicate operator (iterative instruction) was introduced by *Böhm et al.* [31]. The TUPlicate operator reduced the size of the code and led to significant reductions in execution time, especially for large programs.

The Extended Manchester Dataflow Machine (EXMAN) from the Indian Institute of Science (Bangalore, India) (*Patnaik et al.* [228]) incorporated three major extensions to the basic MDM:

- a multiple MUs scheme,
- an efficient implementation of the array data structure, and
- a facility to execute re-entrant routines concurrently.

To allow all storage functions to be performed concurrently, a prototype parallel structure store was developed by *Kawakami and Gurd* [159].

Other projects

NTT's DPAS (*Takahashi et al.* [287], 1983): The Dataflow Processors Array System (DPAS), developed at Nippon Telephone and Telegraph (Japan), was a dynamic tagged-token machine intended for large scientific calculations. A hardware experimental system Eddy, consisting of 4×4 PEs, was built and used to test some applications.

DDDP (*Kishi et al.* [162], 1983): The Distributed Data Driven Processor (DDDP) from OKI Electric Ind. (Japan) had a centralized tag manager and performed token matching by a hardware hashing mechanism similar to that of the Manchester Dataflow Machine. A prototype consisting of four PEs and one structure store connected by a ring bus achieved 0.7 MIPS.

SIGMA-1 (*Hiraki et al.* [137], 1984): The SIGMA-1 system is a supercomputer for large-scale numerical computation and has been operational since early 1988 at the Electrotechnical Laboratory (Tsukuba, Japan). It consists of 128 PEs and 128 structure elements interconnected by 32 local networks (10×10 crossbar packet switches) and one global two-stage Omega network. Sixteen maintenance processors are also connected with the structure elements and with a host computer for I/O operations, system monitoring, performance measurements, and maintenance operations.

PIM-D (*Ito et al.* [147], 1986): The Parallel Inference Machine PIM-D was proposed to be one of the candidates for a parallel inference machine in the Fifth Generation Computer System and was a joint venture of ICOT and OKI Electric Ind. (Japan). This machine was constructed from multiple PEs and multiple structure memories interconnected by a hierarchical network and exploited three types of parallelism: OR parallelism, AND parallelism, and parallelism in unification.

Q-p (*Asada et al.* [18], *Nishikawa et al.* [215], both 1987): The one-chip data-driven processor Q-p was specifically designed to be a one-chip functional element that was easy to program to form various dedicated processing functions. Particular design decisions were taken to achieve high flow-rate data-stream processing capabilities. In the Q-p, a novel bi-directional elastic pipeline processing concept was introduced to implement token matching. The Q-p was developed jointly by Osaka University, Sharp, Matsushita Electric Ind., Sanyo Electric, and Mitsubishi Electric (Japan).

DDA (*Koren et al.* [166], 1988): A Data-Driven VLSI Array (DDA) was designed at Technion (Haifa, Israel) consisting of a set of PEs with each PE connected to six neighbors. Before the program starts, the corresponding dataflow graph is mapped into the DDA by assigning PEs to nodes and links to arcs. During execution, the computation front propagates through the DDA, as through the dataflow graph. The DDA is capable of executing any arbitrary algorithm. Performance of the DDA was enhanced by improving the architecture (*Weiss et al.* [322]) and the mapping algorithm (*Robič and Vilfan* [244]).

PATTSY (*Narashimhan and Downs* [210], 1989): The Processor Array Tagged-Token System (PATTSY) was an experimental system that was developed at the University of Queensland (Brisbane, Australia), which supported the dynamic model of dataflow execution. PATTSY had a host computer that provided the user-interface to the machine, accepted user programs and converted them into dataflow graphs. These graphs were mapped onto the PEs, but the actual scheduling of operations was carried out at run-time. A prototype with 18 PEs was operational. It used an IBM-PC as host while the PEs were built from Intel 8085-based single-board microcomputers.

CSIRAC II (*Egan* [75], 1990): The origins of the CSIRAC II dataflow computer date from 1978. It was built at the Swinburne Institute of Technology (Hawthorn, Australia). The architecture is unusual in that the temporal order of tokens with the same color on the same graph arc is maintained.

SDFA (*Snelling* [272], 1993): The Stateless Data-Flow Architecture (SDFA) was designed at the University of Manchester and inspired by the Manchester Dataflow Machine. As its name implies, the SDFA system had no notion of states. There were no structure stores, and only extract-wait functionality was provided in the matching stores. All the instructions in the instruction-set were simple and based on RISC principles. There were no iterative or vector-style instructions producing more than two tokens per execution.

2.2.3 Explicit Token Store Approach

One of the main problems of tagged-token dataflow architectures is efficient implementation of token matching. To eliminate the need for costly associative memory, the concept of the *explicit token store* (ETS) has been proposed by *Papadopoulos* ([224], 1988). The basic idea is to allocate a separate frame in the *frame memory* (Fig. 2.11) for each active loop iteration or subprogram invocation. A frame consists of *slots* where each slot holds an operand that is used in the corresponding activity. Since access to slots is direct (i.e., through offsets relative to the frame pointer), no associative search is needed.

In the middle of Fig. 2.11 a (part of a) program graph is depicted, with input token $\langle \mathsf{FP}, \mathsf{IP}, 3.01 \rangle$ entering the $*$ node. The token consists of a pointer

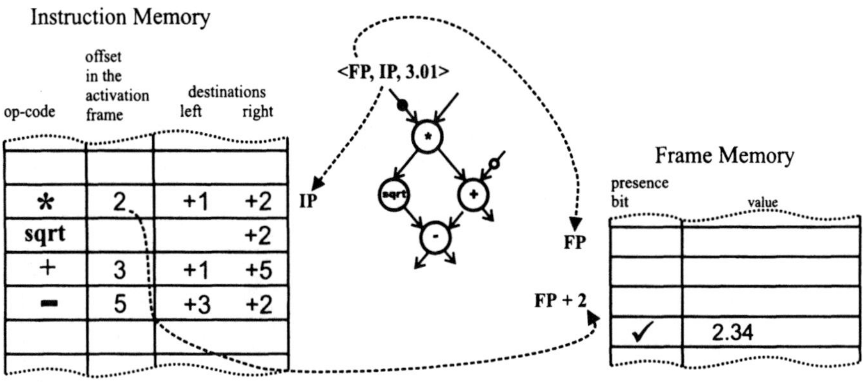

Fig. 2.11. Explicit token store

FP to the corresponding frame, a pointer IP to the instruction in the *instruction memory*, and one *operand*:

$$\langle \mathsf{FP}, \mathsf{IP}, operand \rangle.$$

FP and IP form the token's *tag*. The instruction fetched from the location IP specifies the *opcode* (e.g., $*$), the *offset* in the associated frame where the match between the corresponding input operands will take place (e.g., $\mathsf{FP}+2$), and IP-relative displacement(s) to the destination instruction(s) in the instruction memory (e.g., instructions at $\mathsf{IP}+2$ and $\mathsf{IP}+1$ with opcodes $+$ and *sqrt*, respectively).

Figure 2.12 illustrates how matching is performed in the ETS model. An unusual characteristic of ETS frames is that each slot has associated a *presence bit* specifying the disposition of the slot. The dynamic dataflow firing rule is implemented by a simple state transition of these presence bits. For example (see Fig. 2.12), at the moment $t = 0$, the token $\langle \mathsf{FP}, \mathsf{IP}, 3.01 \rangle$ arrives and, since it is the first of the input tokens, it is treated as follows. The slot $\mathsf{FP} + 2$ is found empty, so the operand 3.01 is deposited in that slot and the presence bit is set on. At the moment $t = 1$ the second token $\langle \mathsf{FP}, \mathsf{IP}, 2.0 \rangle$ arrives. Since the slot $\mathsf{FP}+2$ is found to be full (presence bit is on), the waiting operand 3.01 is extracted from it (leaving the slot empty). The presence of both operands causes the instruction at IP to be fired, producing at $t = 2$ two result tokens, $\langle \mathsf{FP}, \mathsf{IP}+2, 6.02 \rangle$ and $\langle \mathsf{FP}, \mathsf{IP}+1, 6.02 \rangle$. Notice, that at $t = 1$ another input token $\langle \mathsf{FP}', \mathsf{IP}, 7.4 \rangle$ arrived. This token belongs to some other part of computation, for example, to the next loop iteration.

Bounded loops. The ETS concept can be used for resource control in dataflow computers. Most of the critical resources, in particular, concerning the size of the token storage, will be consumed by an unlimited number of concurrently active loop iterations. Hence, the *k-bounded loops* constraint

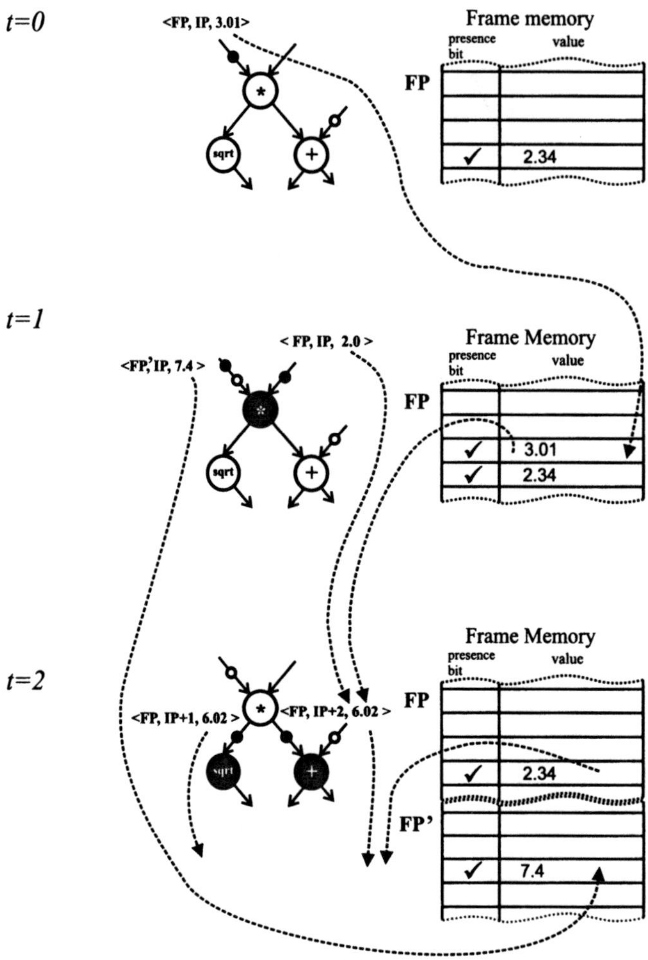

Fig. 2.12. Explicit token store matching scheme

was devised by *Culler* [56], allowing at most $k-1$ consecutive loop iterations to be concurrently active in k ETS frames (Fig. 2.13).

The implementation of *Culler*'s idea uses a new *gate operator* G, a *synchronization tree*, and $\mathtt{D}_k$ and $\mathtt{D}_k^{-2}$ operators. The G operator has two inputs (for control and data) and functions as a loop throttle by passing one token from its data input to the output for each token on its control input. At the end of each iteration, a new control token is generated by combining the output of all $\mathtt{D}_k$ operators into a single value using the synchronization tree.

The $\mathtt{D}_k$ operator (with modified semantics of the D operator from p. 64) sets the iteration number i of tokens to $i+1 \bmod k$. When a control token with iteration number $i+1 \bmod k$ has passed the synchronization tree, it is guaranteed that the iteration executing in ETS frame i has terminated and

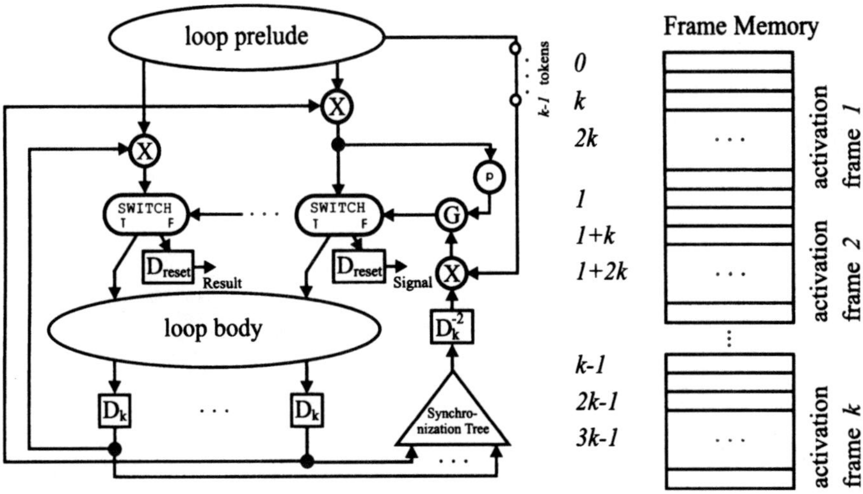

(a) (b)

Fig. 2.13. Explicit token store **(a)** bounded loop schema **(b)** frames are re-used by iterations separated by k

no token with iteration number i is still circulating. Next the control token with iteration number $i + 1 \bmod k$ in its tag passes the $\mathtt{D}_k^{-2}$ operator which decrements the iteration number by $-2 \bmod k$ yielding the new iteration number $i - 1 \bmod k$ in the tag.

The *k-bounded loops* scheme is initialized (during compile-time or during load-time of the program) by $k-1$ control tokens (with loop iteration numbers $0, \ldots, k-2$ in their tags) on the arc from the loop prelude to the gate operator `G`. $k - 1$ loop iterations may be activated in ETS frames $0, \ldots, k - 2$. Hence, there are at most $k - 1$ consecutive loop iterations active at the same time; however, k frames are needed. The iteration with iteration number $i-1 \bmod k$ can be started, as soon as the iteration with iteration number $i \bmod k$ has terminated. Thus tokens from different iterations but with the same iteration number $i-1$ are prevented from meeting on the arcs before the **SWITCH** nodes.

The ETS principle was developed in the Monsoon project, but is used in most recent dataflow architectures, for example, as the so-called direct matching in EM-4 and Epsilon-2 machines (see Sect. 2.3.1 below).

Monsoon, an Explicit Token Store Machine

The Monsoon dataflow multiprocessor was built jointly by MIT and Motorola (USA) (*Papadopoulos and Culler* [225], 1990). In Monsoon, dataflow PEs are coupled with each other and with I-structure storage units by a multistage packet-switching network (Fig. 2.14).

The main objective of the Monsoon dataflow processor architecture was to alleviate the waiting/matching problem by using an ETS. Each frame resides entirely on a single PE. The FP and IP are conventionally segmented by the PE as follows, $tag = \mathsf{N}_{PE} : (\mathsf{FP.IP})$, where N_{PE} is the PE's number.

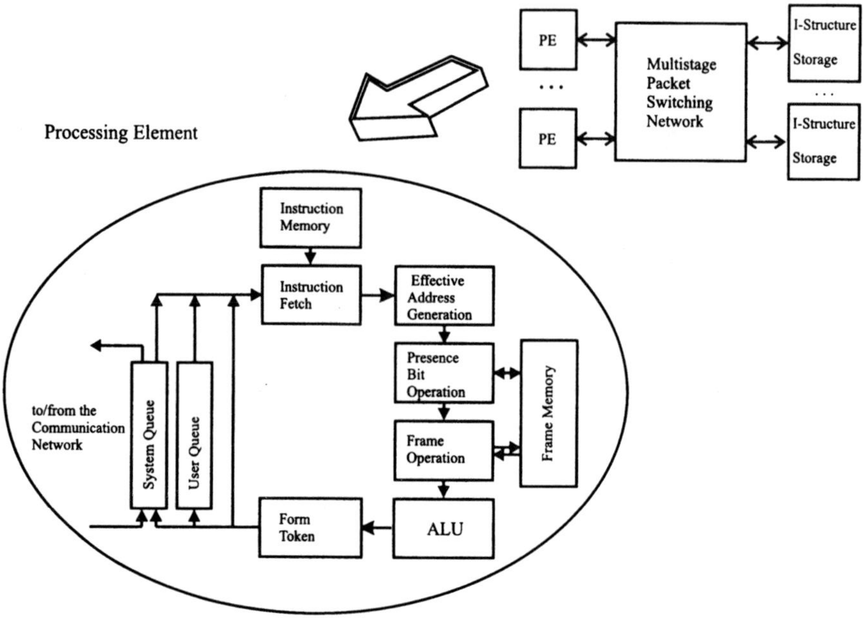

Fig. 2.14. Communication structure and processing element of the Monsoon

Each PE uses an 8-stage pipeline (Fig. 2.14). The first stage is the *instruction fetch* stage which precedes token matching (in contrast to dynamic dataflow processors with associative matching units). Such a new arrangement is necessary since the operand fields in an instruction denote the offset in the memory frame that itself is addressed by the tag of a token. The explicit token address is computed from the frame address and operand offset. This is done in the second stage (*effective address generation*), which is the first of three pipeline stages that perform the token matching. In the third stage, called *presence bit operation*, a presence bit is accessed to find out whether the first operand of a dyadic operation has already arrived. If not, the presence bit is set and the current token is stored in the frame slot of the frame memory. Otherwise, the presence bit is reset and the operand is retrieved from the slot. Operand storing or retrieving is the task of the fourth pipeline stage – the *frame operation* stage. The next three stages are execution stages, in the course of which the next tag is also computed concurrently. The eighth (*form-token*) stage forms one or two new tokens that are sent to the network, stored in a user token queue, a system token queue, or directly recirculated to the instruction fetch stage of the pipeline. In the following, we give some details of the Monsoon system:

- Processing element:
  - 10 MHz clock
  - 256 kW Instruction Memory (32-bit wide)
  - 256 kW Frame Memory (word[6] + 3 presence bits)
  - Two 32 k-token queues (system, user)
- I-structure storage:
  - 4 MW (word + 3 presence bits)
  - 5 M requests/s
- Network
  - Multistage, pipelined
  - Packet Routing Chips (PaRC, 4 x 4 crossbar)
  - 4 M tokens/s/link (100 MB/s)

Since September 1990, a 1 PE $\times$ 1 I-structure memory configuration (also referred to as the two-node system) has been operational while the first 8 $\times$ 8 configuration (16-node system) was delivered in fall 1991. In total, sixteen two-node Monsoon systems were constructed and delivered to universities across the USA and two 16-node systems were delivered to MIT and Los Alamos National Laboratories (USA).

Until the mid-1990s, dataflow architectures did not permit the use of traditional storage models. In order to bring the dataflow computational model closer to the control-flow model, *Kavi et al.* ([157], 1995), studied cache design issues applicable to dataflow architecture, in particular, the design of I-cache, D-cache, and I-structure cache memories in the ETS model.

2.3 Augmenting Dataflow with Control-Flow

Pure dataflow computers based on the single-token-per-arc or tagged-token dataflow model usually perform quite poorly with sequential code. This is due to the fact that an instruction of the same thread can only be issued to the dataflow pipeline after the completion of its predecessor instruction. In the case of an 8-stage dataflow pipeline, instructions of the same thread can be issued at most every eight cycles. If the load is low, for instance, for a single sequential thread, the utilization of the dataflow processor drops to one-eighth of its maximum performance. Another drawback is the overhead associated with token matching. For example, before a dyadic instruction is issued to the execution stage, two result tokens have to be present. The first token is stored in the waiting-matching store, thereby introducing a *bubble* in the execution stage(s) of the dataflow processor pipeline. Only when the second token arrives can the instruction be issued. Clearly, this may affect the

[6] Word size: 64-bit data + 8 bits type tag

system's performance, so bubbles should not be neglected. For example, the pipeline bubbles sum up to 28.75 % when executing the Traveling Salesman program on the Monsoon machine (*Papadopoulos and Culler* [225]).

Since a context switch occurs in fine-grain dataflow after each instruction execution, no use of registers is possible for optimizing the access time to data, avoiding pipeline bubbles caused by dyadic instructions, and reducing the total number of tokens during program execution.

One solution to these problems is to combine dataflow with control-flow mechanisms. The possible symbiosis between dataflow and von Neumann architectures was investigated by a number of research projects developing von Neumann/dataflow hybrids (see Table 2.1). The spectrum of such hybrids is quite broad, ranging from simple extensions of a von Neumann processor with a few additional instructions to specialized dataflow systems attempting to reduce overhead by increasing the execution grain size and employing various scheduling, allocation, and resource management techniques developed for von Neumann computers. These developments show that dataflow and von Neumann computers do not necessarily represent two entirely disjoint worlds, but rather two extreme ends of a spectrum of possible computer systems.

Several techniques for combining control-flow and dataflow emerged, such as:

- *threaded dataflow*,
- *large-grain dataflow*,
- *dataflow with complex machine operations*,
- *RISC dataflow*,
- *hybrid dataflow*.

We will describe these in the following five subsections. For a comparison of the techniques, see also *Beck et al.* [24].

2.3.1 Threaded Dataflow

By the term *threaded dataflow* we understand a technique where the dataflow principle is modified so that instructions of certain instruction streams are processed in succeeding machine cycles. In particular, in a dataflow graph (program) each subgraph that exhibits a low degree of parallelism is identified and transformed into a sequential thread. Such a thread of instructions is issued consecutively by the matching unit without matching further tokens except for the first instruction of the thread.

Threaded dataflow covers the repeat-on-input technique used in Epsilon-1 and Epsilon-2 processors, the strongly connected arc model of EM-4, and the direct recycling of tokens in Monsoon. Data passed between instructions of the same thread is stored in registers instead of written back to memory. These registers may be referenced by any succeeding instruction in the thread. Single-thread performance is thereby improved. The total number

Table 2.1. Augmenting dataflow with control-flow

Principle Architecture	Instruction execution parallelism		Execution mode	Synchronization of threads	Matching unit	Registers	Samples
von Neumann	one control thread			----	no	yes	standard microprocessors
Multithreaded (von Neumann)	several active control threads	von Neumann		full/empty bits locked	no	yes	HEP APRIL Tera MTA
Large-Grain Dataflow	several active control threads			matching operations	yes	yes	LDF 100
Large-Grain Dataflow with Complex Machine Operations	several active control threads with complex machine operations		vector pipelining	matching operations	yes	yes	DGC Stollman
Dataflow with Complex Machine Operations	complex machine operations	dataflow	vector pipelining	matching operations	yes	yes	ASTOR
Threaded Dataflow	several active control threads		repeat-on-input	matching operations	yes	yes	Monsoon Epsilon- 1 & 2 EM-4
Fine-Grain Dataflow	single instructions			matching operations	yes	no	MADAME MIT TTDA SIGMA-1

of tokens needed to schedule program instructions is reduced, which in turn saves hardware resources. Pipeline bubbles are avoided for dyadic instructions within a thread. Two threaded dataflow execution techniques can be distinguished:

- *direct token recycling*,
- *consecutive execution* of the instructions of a single thread.

Direct token recycling is used in the Monsoon dataflow computer. It allows a particular thread to occupy only a pipeline frame slot in the 8-stage pipeline. This implies that at least eight threads must be active for full pipeline utilization to be achieved. This cycle-by-cycle instruction interleaving of threads is used in a similar fashion by some multithreaded von Neumann computers (Sect. 6.3.1).

To optimize single-thread performance, Epsilon and EM-4 execute instructions from a thread consecutively. The circular pipeline of fine-grain dataflow is retained. However, the matching unit has to be enhanced with a mechanism that, after firing the first instruction of a thread, delays matching of further tokens in favor of consecutive issuing of all instructions of the thread which has been started. For example, in a *strongly connected arc* model, each arc of the dataflow graph is classified as either a normal arc or a strongly connected arc. The set of nodes that are connected by strongly connected arcs is called the *strongly connected block*. Recall that the standard firing rule is that a node is fired when all input arcs have matching tokens. The enhancement for strongly connected blocks is that such a block is fired if its "source" nodes are enabled and the execution of the whole block is conducted as a unit, that is, without applying the standard dataflow firing rule for other nodes in that block. Figure 2.15 shows an example of strongly connected blocks (threads). There are two strongly connected blocks, A and B. When node 5 or node 6 is fired, block A or block B is executed exclusively. Cycling of tokens that would normally flow through the other block is suppressed in the pipeline. For example, if node 5 becomes enabled before node 6, nodes 5, 8, and 10 will be fired before any node in block B.

In all threaded dataflow machines, the central design problem is the implementation of an efficient synchronization mechanism (see *Sakai* [252]). The *direct matching* is a synchronization mechanism that needs no associative mechanisms. As in ETS (see Sect. 2.2.3), a matching area in operand memory is exclusively reserved for a single function instance. This area is called an *operand segment* (OS). The code block in instruction memory corresponding to the OS is called a *template segment* (TS) (see Fig. 2.16). The top address of TS is called the *template segment number* (TSN). An *operand segment number* (OSN) points to the top of the OS (as IP in ETS). A token comprises an *operand*, OSN, a *displacement* (DPL), and a *synchronization flag* (SF):

$$\langle \mathsf{SF}, \mathsf{OSN}, \mathsf{DPL}, \mathit{operand} \rangle .$$

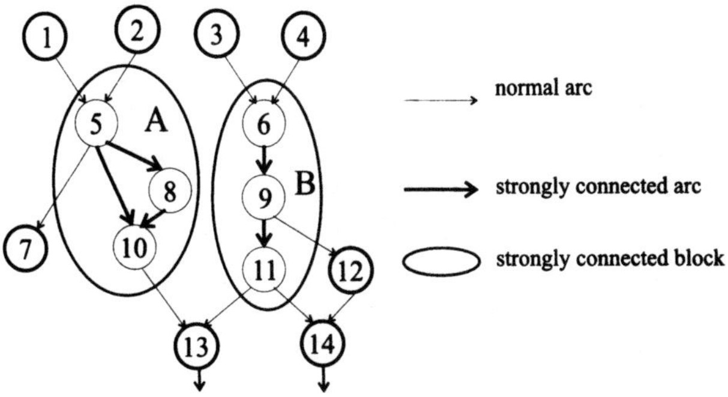

Fig. 2.15. Strongly connected blocks (threads)

DPL serves both as a displacement of the destination instruction in the instruction memory and a displacement of the matching operand in the operand memory. SF indicates the type of synchronization which can be either monadic, left dyadic, right dyadic, or immediate. The matching address is produced by concatenating OSN and DPL. The instruction address is derived by concatenating TSN and the DPL. Each slot in an OS also has a *presence bit*. A dyadic matching is performed by a test-and-set[7] of the presence bit.

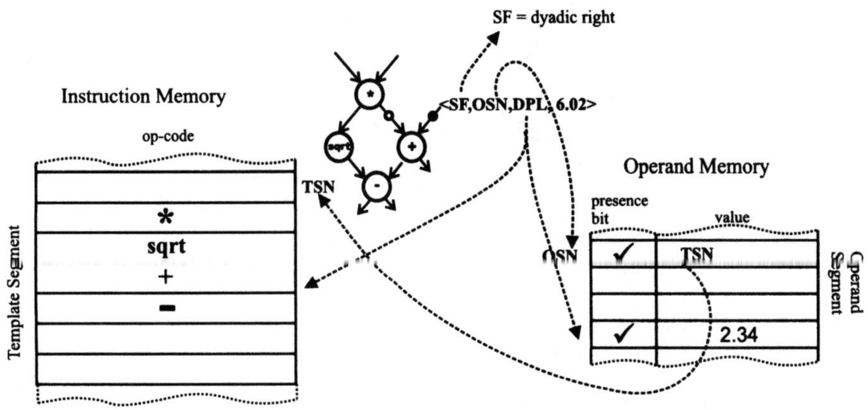

Fig. 2.16. Direct matching: Instruction Memory and Operand Memory

Some highly influential threaded dataflow projects, EM-4 with its successor EM-X, Monsoon, Epsilon-2, and RWC-1 are described below.

[7] If the presence bit has already been set, the partner data will be read, the bit will be cleared, and the instruction will be executed. Otherwise the arriving data will be stored there and the presence bit will be set.

EM-4 and EM-X

In the EM-4 project (*Sakai et al.* [254], 1989) at Electrotechnical Laboratory (Tsukuba, Japan), the essential elements of a dynamic dataflow architecture using frame storage for local variables are incorporated into a single chip processor. In this design each *strongly connected* subgraph of a function body is implemented as a thread that uses registers for intermediate results. The EM-4 was designed for 1 024 PEs. The EM-4 prototype with 80 PEs has been operational since May 1990. Each PE consists of a processor and I-structure memory.

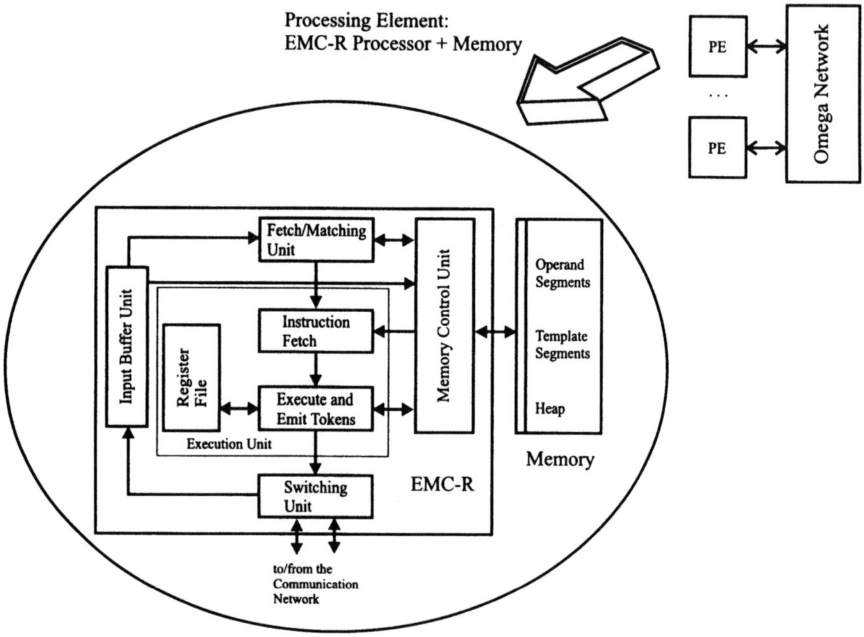

Fig. 2.17. Communication structure and processing element of the EM-4

The organization of the EM-4 is shown in Fig. 2.17. Each PE consists of a single-chip gate-array processor EMC-R (without floating-point hardware but including a switching unit for the network) and I-structure memory (1.25 MB SRAM). The EMC-R consists of a *switching unit* (SU), an *input buffer unit* (IBU), a *fetch/matching unit* (FMU), an *execution unit* (EU), and a *memory control unit* (MCU). The EMC-R communicates with the network through a 3×3 crossbar SU. The processor and its memory (containing OSs and TSs) are interfaced with the MCU. The IBU is used as a token store. A 32-word FIFO type buffer is implemented using a dual-port RAM on chip. If this buffer is full, a part of the off-chip memory is used as secondary buffer. The

FMU is used for matching tokens and fetching instructions. It performs direct matching for packets and instruction sequencing for a strongly connected block (thread). The heart of the EMC-R is the EU, which fetches instructions until the end of the thread (if the next instruction is strongly connected with the current instruction, instruction fetch and data load of the next instruction are overlapped with the execution). Instructions with matching tokens are executed. Instructions can emit tokens or write to a *register file*.

In 1993, an upgrade to EM-4, called EM-X, was developed by *Kodama et al.* [164]. It was designed to support:

- latency reduction by fusing the communication pipeline with the execution pipeline,
- latency hiding via multithreading, and
- run-time latency minimization for remote memory access.

The EM-4 can access remote memory by invoking packet system handlers on the destination PE. Clearly, when the destination PE is busy, remote memory requests are blocked by the current thread execution. To remedy this, the EM-X supports a direct remote memory read/write mechanism, which can access the memory independently of thread execution. For these reasons, the EMC-Y single-chip processor was used in EM-X (instead of the EMC-R that was used in EM-4). Some characteristics of the EM-X machine are listed below:

- EMC-Y processor:
  - 1.0 µm CMOS gate-array technology
  - Total 80 593 gates
  - 20 MHz clock
  - 40 MW/s[8]
  - Performances: 20 MIPS and 40 MFLOPS
  - Network throughput: 10 Mpackets/s/port
- Memory:
  - 1 MW SRAM (20 ns)
- Network:
  - Circular Omega
  - Average 5.06 hops and max. 8 hops on 80 PEs system
  - Deadlock prevention with 3 banked buffers

Some performance parameters of the 80 PEs EM-X system (operational since 1994):

- 1.6 GIPS and 3.2 GFLOPS (peak)
- 1.26 µs remote memory read latency
- 2.86 µs remote function call latency
- 37.2 MB/s point-to-point throughput
- 7.8 µs barrier synchronization

[8] Word size: 32-bit data + 6-bit data tag + 2-bit parity

- Linpack ($N = 5000$): 600 MFLOPS
- Radix Sort (16 M $\times$ 32 bits): 1.3 s

Other projects

Monsoon (*Papadopoulos and Traub* [226], 1991): The Monsoon dataflow processor (see also Sect. 2.2.3) can be viewed as a cycle-by-cycle interleaving multithreaded computer due to its capability of direct token recycling. Using this technique a successor token is directly fed back in the 8-stage pipeline, bypassing the token store. Another instruction of the same thread is executed every eighth processor cycle. Monsoon allows the use of registers (eight register sets are provided) to store intermediate results within a thread, thereby digressing from the pure dataflow execution model.

Epsilon-2 (*Grafe and Hoch* [106], 1990): The Epsilon-2 machine developed at Sandia National Laboratories (Livermore, CA), supports a fully dynamic memory model, allowing single-cycle context switches and dynamic parallelization. The system is built around a module consisting of a processor and structure unit, connected via a 4 $\times$ 4 crossbar to each other, an I/O port, and the global interconnection network. The structure unit is used for storing data structures such as arrays, lists, and I-structures. The Epsilon-2 processor retains the high performance features of the Epsilon-1 prototype, including direct matching, pipelined processing, and a local feedback path. The ability to execute sequential code as a grain provides RISC-like execution efficiency.

RWC-1 (*Sakai et al.* [253], 1993): The Massively Parallel Architecture Laboratory at Real World Computing Partnership (Tsukuba, Japan) developed a massively parallel computer RWC-1 which is a descendant of EM-4 (as is EM-X). A *multidimensional directed cycles ensemble* (MDCE) network connects up to 1 024 PEs. Two small-scale systems, Testbed-I with 64 PEs and Testbed-II with 128 PEs are used for testing and software development. The PE is based on *reduced interprocessor-communication architecture* (RICA) which employs superscalar execution (2-issue), a floating-point multiplier/adder module, and offers:

- a fast and simple message handling mechanism,
- a hard-wired queuing and scheduling mechanism,
- a hard-wired micro-synchronization mechanism,
- integration of communication, scheduling and execution, and
- simplification of the integrated structure (*Matsuoka et al.* [194]).

2.3.2 Large-Grain Dataflow

Another technique for combining dataflow with control-flow is referred to as the *coarse-grain* (also large-grain) dataflow model. It advocates activating macro dataflow actors in the dataflow manner while executing instruction sequences, represented by actors, in the von Neumann style. Large-grain dataflow machines typically decouple the matching stage (sometimes called signal stage, synchronization stage, etc.) from the execution stage by use of FIFO buffers. Pipeline bubbles are avoided by decoupling. Off-the-shelf microprocessors can be used to support the execution stage. Most of the more recent dataflow architectures fall into this category and are listed below. Note that they are often called *multithreaded machines* by their authors.

StarT

The StarT (sometimes also written as *T), started by *Nikhil et al.* ([214], 1992), is a direct descendant of dataflow architectures, especially of the Monsoon, and unifies them with von Neumann architectures. StarT has a scalable computer architecture designed to support a broad variety of parallel programming styles including those which use multithreading based on non-blocking threads (see Sect. 6.3.1). A StarT node consists of the *data processor* (dP), which executes threads, the *synchronization coprocessor* (sP), which handles returning load responses and join operations, and the *remote-memory request processor* (RMem) for incoming remote load/store requests. The three components share local *node memory* (see Fig. 2.18). The node is coupled with a high performance network having a fat-tree topology with high cross-section bandwidth.

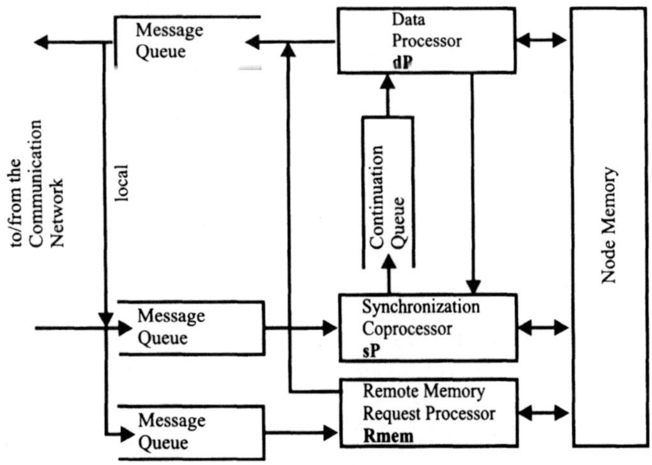

Fig. 2.18. MIT StarT node architecture (concept)

In mid-1991 the StarT project was launched by MIT and Motorola. Due to its on-chip *special-function unit* (SFU), the Motorola 50 MHz 2-issue superscalar RISC microprocessor 88110 was chosen as the basis for the node implementation. However, in order to keep the communication latency to a minimum, a number of logic modules were added to the 88110 chip so that it acted as a tightly-coupled network interface. The resulting chip was called the 88110MP (MP for multiprocessor) with a 10–20 machine cycles overhead for sending and receiving data between the node and the network. Two 88110MP microprocessors were used to implement the StarT node (see Fig. 2.19). The first one operated as dP, with its SFU serving as sP. dP and sP were optimized for long and short threads, respectively. The second 88110MP was tailored to act as RMem to handle remote memory requests from other nodes to the local node memory (64 MB).

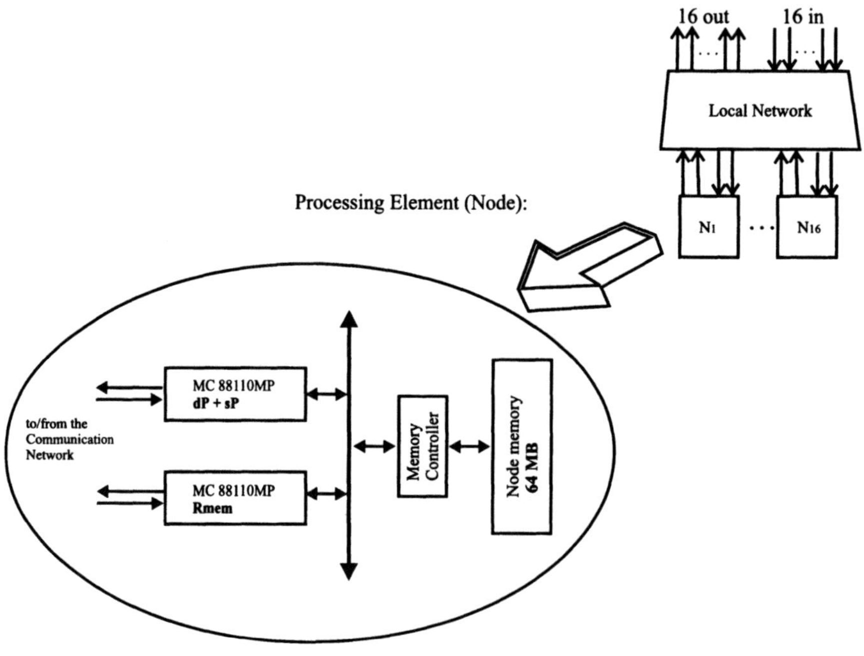

Fig. 2.19. Communication structure and processing element of the StarT

The fat-tree network was based on the MIT Arctic packed routing chip (*Boughton* [35]) that was twice as fast as Monsoon's PaRC (see p. 77) and was expected to drive the interconnection network at 1.6 GB/s/link in each direction with packet sizes ranging from 16 to 96 bytes. Sixteen nodes were packaged into a "brick" (approximately 23 cm cube) with 3.2 GFLOPS and 3 200 MIPS peak performance. Sixteen bricks can be interconnected into a

256-node machine (1.5 m cube) with the potential to achieve 50 GFLOPS and 50 000 MIPS.

As reported by *Arvind et al.* [15], MIT decided to go back to the drawing board and to start afresh on PowerPC-based StarT machines after Motorola and IBM started manufacturing the PowerPC family of RISC microprocessors. Thus, the PowerPC 620 was planned in a StarT-ng machine (*Ang et al.* [12]) but the architecture was redesigned once again, this time around a 32-bit PowerPC 604, and was called the StarT-Voyager machine (*Ang et al.* [13]). This machine, however, bears little resemblance to the original StarT architecture and no similarity to Monsoon.

Other projects

TAM (*Culler et al.* [57], 1991): The Threaded Abstract Machine (TAM) from the University of California at Berkeley is an execution model for fine-grain interleaving of multiple threads, that is supported by an appropriate compiler strategy and program representation, instead of elaborate hardware. TAM's key features are placing all synchronization, scheduling, and storage management under explicit compiler control.

ADARC (*Strohschneider et al.* [280], 1994): In the Associative Dataflow Architecture (ADARC) the processing units are connected via an associative communication network. The processors are equipped with private memories that contain instruction sequences generated at compile-time. The retrieval of executable instructions is replaced by the retrieval of input operands for the current instructions from the network. The structure of the associative switching network enables full parallel access to all previously generated results by all processors. A processor executes its current instruction (or instruction sequence) as soon as all requested input operands have been received. ADARC was developed at the J.W.Goethe University (Frankfurt, Germany).

Pebbles (*Roh and Najjar* [245], 1995): The Pebbles architecture from Colorado State University (Fort Collins, CO) is a large-grain dataflow architecture with a decoupling of the synchronization unit and the execution unit within the PEs. The PEs are coupled via a high-speed network. The local memory of each node consists of an instruction memory, which is read by the execution unit, and a data memory (or frame store), which is accessed by the synchronization unit. A ready queue contains the continuations representing those threads that are ready to execute. The frame store is designed as a storage hierarchy where a frame cache holds the frames of threads that will be executed soon. The execution unit is a 4-way superscalar microprocessor.

MTA (*Hum et al.* [140], 1994) and *EARTH* (*Maquelin et al.* [192], 1995): The EARTH (Efficient Architecture of Running Threads), developed at McGill University and Concordia University (Montréal, Canada), is based on the MTA (Multithreaded Architecture) and dates back to the Argument Fetch Dataflow Processor. An MTA node consists of an *execution unit* (EU) that may be an off-the-shelf RISC microprocessor and a *synchronization unit* (SU) to support dataflow-like thread synchronization. The SU determines which threads are ready to be executed. The EU and SU share the processor local memory which is cached. Accessing data in a remote processor requires explicit request and send messages. The EU and SU communicate via FIFO queues: a ready queue containing ready thread identifiers links the SU with the EU, and an event queue holding local and remote synchronization signals connects the EU with the SU, but also receives signals from the network. A register-use cache keeps track of which register set is assigned to which function activation. MTA and EARTH rely on non-blocking threads. The EARTH architecture is implemented on top of the experimental (but rather conventional) MANNA multiprocessor.

2.3.3 Dataflow with Complex Machine Operations

Another technique to reduce the overhead of the instruction-level synchronization is the *use of complex machine instructions*, for instance, vector instructions. These instructions can be implemented by pipeline techniques as in vector computers. Structured data is referenced in block- rather than element-wise fashion, and can be supplied in a burst mode. This deviates from the I-structure scheme where each data element within a complex data structure is fetched individually from a structure store.

Another advantage of complex machine operations is the ability to exploit parallelism at the subinstruction level. Therefore, such a machine has to partition a complex machine operation into suboperations that can be executed in parallel. The use of a complex machine operation may spare several nested loops. The use of FIFO buffers allows the machine to decouple the firing and the execution stages to bridge the different execution times within a mixed stream of simple and complex instructions issued to the execution stage. As a major difference to conventional dataflow architectures, tokens do not carry data (except for the values `true` or `false`). Data is only moved and transformed within the execution stage. This technique is used in the Decoupled Graph/Computation Architecture, the Sto*ll*mann Dataflow Machine, and the ASTOR architecture. These architectures combine complex machine instructions with large-grain dataflow, described above. The structure-flow technique proposed for the SIGMA-1 enhances these fine-grain dataflow computers by structure load/store instructions that can move, for instance, whole vectors to/from structure store. Arithmetic operations are executed by the cyclic pipeline within a PE.

ASTOR

The Augsburg Structure-Oriented Architecture (ASTOR) was developed by *Zehendner and Ungerer* ([334], 1987) at the University of Augsburg (Germany). It can be viewed as:

- a dataflow architecture that utilizes task level parallelism by the architectural structure of a distributed memory multiprocessor,
- instruction-level parallelism by a token-passing computation scheme, and
- subinstruction-level parallelism by SIMD evaluation of complex machine instructions.

Sequential threads of data instructions are compiled to dataflow *macro actors* and executed consecutively using registers. A dependence construct describes the partial order in the execution of instructions. It can be visualized by a dependence graph. The nodes in a dependence graph represent *control constructs* or *data instructions*; the directed arcs denote control dependences between the nodes. Tokens are propagated along the arcs of the dependence graph. To distinguish different activations of a dependence graph, a tag is assigned to each token. The firing rule of dynamic dataflow is applied, that is, a node is enabled as soon as tokens with identical tags are present on all its input arcs. However, in the ASTOR architecture tokens do not carry data.

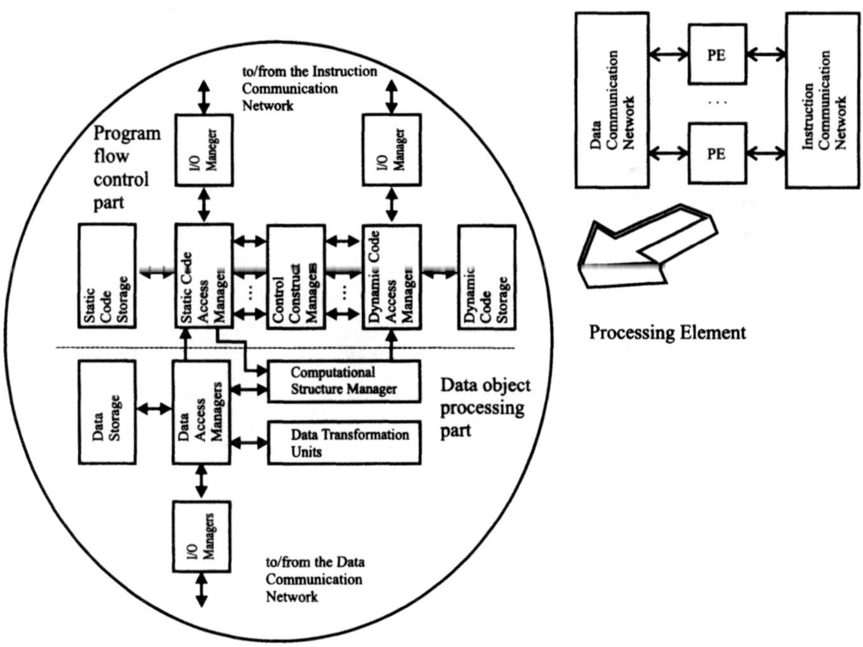

Fig. 2.20. Communication structure and processing element of the ASTOR

The ASTOR architecture consists of PEs connected by an *instruction communication network* to transfer procedure calls and a *data communication network* for parameter passing (Fig. 2.20). No global storage is used. Due to the separation of code and data objects, each PE consists of two loosely coupled parts:

- The *program flow control part* consists of static and dynamic code storage, the static and the dynamic code access manager[9], the I/O managers, and the control construct managers (individually named call, loop, choice, and dependency managers).
- The *data object processing part* consists of data storage, several data access managers, an I/O manager, some data transformation units, and the computational structure manager.

All managers in a PE work in parallel to each other. Asynchronous processing and decoupling of the managers is achieved by buffering the links between them.

Other projects

Stollman Dataflow Machine (*Glück-Hiltrop et al.* [102], 1989): The Stollman dataflow machine from Sto*ll*man GmbH (Hamburg, Germany) is a coarse-grain dataflow architecture directed towards database applications. The dataflow mechanism is emulated on a shared-memory multiprocessor. The query tree of a relational query language (such as SQL) is viewed as dataflow graph. Complex database query instructions are implemented as coarse-grain dataflow instructions and (micro-)coded as a traditional sequential program running on the emulator hardware.

DGC (*Evripidou and Gaudiot* [81], 1991): In the Decoupled Graph/Computation (DGC) architecture, developed at the University of Southern California (Los Angeles, CA), the token matching and token formatting and routing are reduced to a single graph operation called *determine executability*. The decoupled graph/computation model separates the graph portion of the program from the computational portion. The two basic units of the decoupled model (*computational* unit and *graph* unit) operate in an asynchronous manner. The graph unit is responsible for determining executability by updating the dataflow graph, while the computation unit performs all the computational operations (fetch and execute).

2.3.4 RISC Dataflow

Another stimulus for dataflow/von Neumann hybrids was the development of *RISC dataflow* architectures that support the execution of existing software

[9] The term *manager* is used to characterize an abstract component of a PE. The manager is an autonomously acting unit that solves a specific task.

written for conventional processors. Using such a machine as a bridge between existing systems and new dataflow supercomputers should have made the transition from imperative von Neumann languages to dataflow languages easier for the programmer. The basic philosophy underlying the development of the RISC dataflow architecture can be summarized as follows:

- use a RISC-like instruction set,
- change the architecture to support multithreaded computation,
- add fork and join instructions to manage multiple threads (see Fig. 2.21),
- implement all global storage as I-structure storage, and
- implement load/store instructions to execute in split-phase mode.

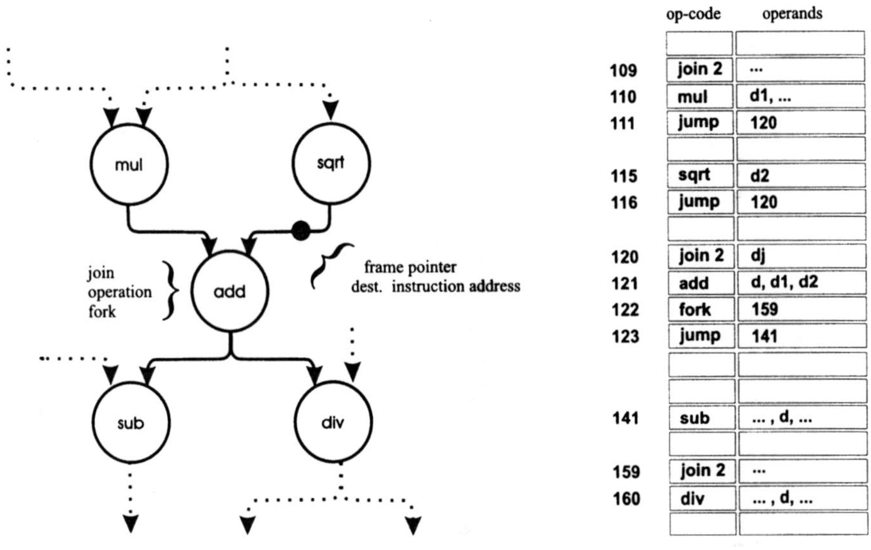

(a) (b)

Fig. 2.21. "RISCifying" dataflow **(a)** conceptual **(b)** encoding of graph

P-RISC Architecture

The Parallel RISC (P-RISC) architecture based on the above principles was developed at MIT by *Nikhil and Arvind* ([213], 1989). It consists of a collection of PEs (with local memory) and *global memory* (GM), interconnected through a packet-switching communication network (Fig. 2.22).

Following the principles underlying all RISC architectures, the ALU of P-RISC PEs distinguishes between *load/store* instructions, which are the only instructions accessing GM (implemented as I-structure storage), and *arithmetic/logical* instructions, which operate on local memory (registers). Fixed instruction length and one-cycle instruction execution (except for load/store instructions) are characteristics of this processor. In addition, P-RISC lacks any explicit matching unit. Instead, all operands associated with a sequential thread of computation are kept in a frame in local *program memory* (PM). Each execution step makes use of an ⟨IP, FP⟩ pair (similar to Monsoon), where IP serves to fetch the next instruction while FP serves as the base for fetching and storing operands. The pair is called *continuation* and corresponds to the tagged part of a token in a tagged-token dataflow machine. To make P-RISC multithreaded, the stack of frames must be changed to a tree of frames, and a separate continuation must be associated with each thread. The frame tree allows different threads of instructions to access different branches of the tree concurrently while the separate continuation extends the concept of a single PC and a single operand base register to multiple instances. Continuations

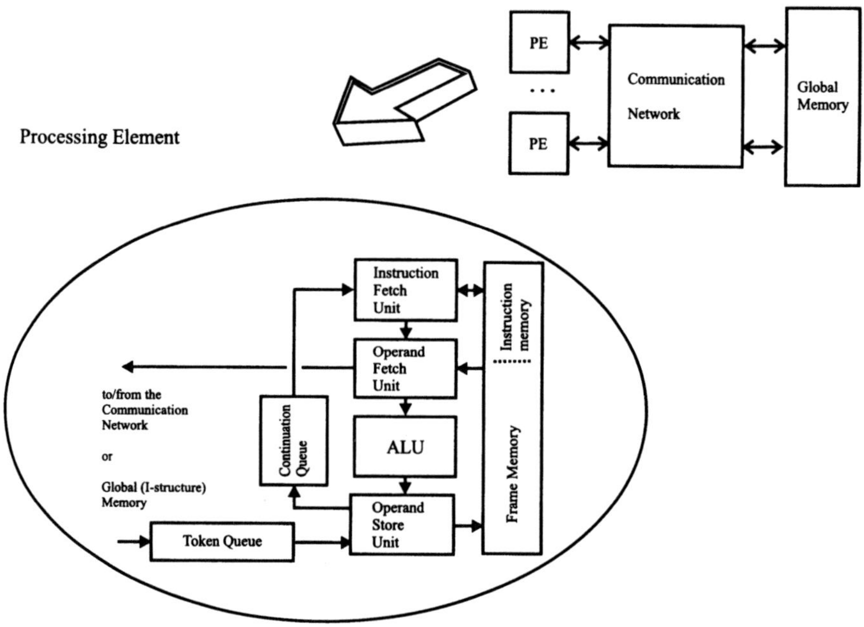

Fig. 2.22. Communication structure and processing element of the P-RISC

of all active threads are held in the *continuation queue* (CQ). At each clock cycle, a continuation (also called a token) is dequeued and inserted into the pipeline (Fig. 2.22). It is first processed by the *instruction fetch unit* (IFU), which fetches from *instruction memory* (IM) the instruction pointed to by IP. Next, operands are fetched from PM by the *operand fetch unit* (OFU). The OFU uses operand offsets (specified in the instruction) relative to the FP. The executable token is passed to the ALU or, in the case of a load/store instruction, to the GM[10]. The execution of an ALU instruction produces result tokens and new continuations. Result tokens are stored in the appropriate frame in (local) *frame memory* (FM) by the *operand store unit* (OSU). Continuations are new ⟨FP, IP⟩-pairs, generated by incrementing the current IP value or, in the case of a branch instruction, replacing it by the target pointer. They are enqueued in the CQ of the local PE.

2.3.5 Hybrid Dataflow

The first attempts towards hybrid dataflow computing were made in the early 1980s in quite diverse directions. Some of them are listed below.

JUMBO (*Treleaven et al.* [299], 1982): The Newcastle Data-Control Flow Computer JUMBO was built at the University of Newcastle-upon-Tyne (UK) to study the effects of the integration of dataflow and control-flow computation. It has a packet communication organization with token matching. There are three principal units (matching unit, memory unit, processing unit) interconnected in a ring by FIFO buffers.

PDF (*Requa and McGraw* [240], 1983): The Piecewise Dataflow Architecture (PDF) was designed to address the Lawrence Livermore National Laboratory's (Livermore, CA) needs for supercomputing power. This architecture blended the strengths found in SIMD, MIMD, and dataflow architectures. The PDF contained a SIMD processing unit and a scalar processing unit that managed a group of processors (MIMD). Programs that ran on the PDF were broken into basic blocks and scheduled for execution using dataflow techniques. Two levels of scheduling divided responsibility between the software and hardware. The time-consuming analysis of when blocks can overlap was done at the compile-time. Individual instruction scheduling was done in hardware.

MUSE (*Brailsford and Duckworth* [36], 1985): The Multiple Stream Evaluator (MUSE) from University of Nottingham (UK) was a structured architecture supporting both serial and parallel processing that allowed the abstract structure of a program to be mapped onto the machine in a logical way. The MUSE machine had many features of dataflow architecture but in its detailed operation it steered a middle course between a pure dataflow machine, a tree machine, and the *graph reduction* approach (see p. 56).

RAMPS (*Barkhordarian* [22], 1987): The Real Time Acquisition and Multiprocessing System (RAMPS) was a parallel processing system developed at EIP Inc. (USA) for high-performance data acquisition and processing

[10] To solve the memory latency problem, the load/store instructions are implemented to operate in a split-phase manner (see the `I-fetch` instruction on p. 67).

applications. RAMPS used dataflow at a macro level while tasks were executed using a sequential model of computation.

MADAME (*Šilc and Robič* [261], 1989): The Macro Dataflow Machine (MADAME) from Jožef Stefan Institute (Ljubljana, Slovenia) was suitable for execution of acyclic dataflow (sub)graphs. While the pure dataflow scheme defines an asynchronous computation, MADAME defines a synchronous operation principle whereby more efficient run-time code is generated due to compile-time instruction scheduling (*Šilc and Robič* [262], 1993). This approach is in a certain way opposite to the strongly connected arc approach used in the EM-4 machine (see p. 80). The organization of the MADAME is circular and consists of five units: *instruction store*, *multiple functional units*, *append unit*, *insert unit*, and *I/O unit*. For a detailed account of this organization, see *Šilc and Robič* [261]. For each instruction the most appropriate fire time is computed at compile-time, as well as a functional unit identifier indicating where the instruction will be executed. Then, at run-time, ready instructions are not executed in the same order as they enter the ready instruction pool. Instead, the insert unit stores each incoming ready instruction into the ready instruction pool at the proper place, depending on the instruction's precomputed fire time and functional unit identifier. This makes it possible to schedule the ready instructions in a way that improves the machine's performance (*Robič et al.* [243], 1987). Note that similar processor organization and concepts can be found in an asynchronous superscalar processor SCALP, where the instructions are also statically allocated to FUs, which in turn simplifies some run-time activities, such as instruction issuing and result forwarding (see p. 328).

DTN Dataflow Computer (*Veen and van den Born* [312], 1990): The DTN Dataflow Computer (developed at the Dutch company Dataflow Technology Nederland) is a high-performance workstation well suited for applications containing relatively small computing-intensive parts with enough parallelism to execute efficiently on the dataflow engine. The DTN Dataflow Computer contains a standard general purpose host, a graphical subsystem (four microprogrammable systolic arrays), and a dataflow engine. The dataflow engine consists of eight identical modules interconnected by a token routing network. Each module contains four NEC Image Pipelined Processors, an interface chip, and image memory.

FDA (*Quénot and Zavidovique* [236], 1991): A Functional Dataflow Architecture (FDA) has been developed at ETCA (Arcueil, France) and is dedicated to real-time processing. Two types of data-driven PEs, dedicated respectively to low- and mid-level processing, are integrated in a regular 3D array. Its design relies on close integration of dataflow principles and functional programming. For the execution of low-level functions, a custom dataflow processor

(DFP) with six bi-directional I/O ports was developed. As in DDA (see p. 72) the performance of FDA can be improved by using a more complex graph mapping algorithm, which was inspired by previous work on graph mapping (*Robič et al.* [242]). The core of the DFP processor is interfaced to the outside world through three input stacks and three output stacks. Each stack is an eight 9-bit word, 25 MB/s bandwidth, synchronous FIFO queue and acts as a build-in token balancing buffer. A 3-stage pipelined datapath is inserted between the input and output stacks. The first stage decodes the input data types to generate commands for the following stages. During the second stage, 8-bit operations (input shifts and multiplications) are performed. 16-bit operations (general-purpose arithmetic and logical, absolute values, minimum, maximum, shift) are performed at the third stage. Up to 50 million 8-bit or 16-bit operations per second can be performed by one DFP processor.

2.4 Lessons learned from Dataflow

The latest generation of superscalar microprocessors displays an out-of-order dynamic execution that is referred to as *local dataflow* or *micro dataflow* by microprocessor researchers.

In 1995, in their first paper on the Pentium Pro [53], *Colwell and Steck* described the instruction pipeline as follows: *"The flow of the Intel Architecture instructions is predicted and these instructions are decoded into micro-operations (µops), or series of µops, and these µops are register-renamed, placed into an out-of-order speculative pool of pending operations, executed in dataflow order (when operands are ready), and retired to permanent machine state in source program order."* That is, after a branch prediction (to remove control dependences) and register renaming (to remove antidependences and output dependences), the instructions (or µops) are placed in the *instruction window* of pending instructions, where µops are executed in dataflow fashion, and then in a reorder buffer that restores the program order and execution states of the instructions. The instruction window and reorder buffer may coincide. State-of-the-art microprocessors typically provide 20 (Intel Pentium II), 32 (MIPS R10000), or 56 (HP PA-8000) instruction slots in the instruction window. Each instruction is ready to be executed as soon as all operands are available. A 4-issue superscalar processor issues up to four executable instructions per cycle to the execution units, provided that resource conflicts do not occur. Issue and execution determine the out-of-order section of a microprocessor. After execution, instructions are retired in program order.

Comparing dataflow computers with such superscalar microprocessors (see Chap. 4) reveals several similarities as well as differences which are briefly discussed below.

While a single thread of control in modern microprocessors often does not incorporate enough fine-grained parallelism to feed the multiple functional units of today's microprocessors, the dataflow approach resolves any threads of control into separate instructions that are ready to execute as soon as all required operands become available. Thus, the *fine-grained parallelism potentially utilized* by a dataflow computer is far larger than the parallelism available for microprocessors.

Data and control dependences potentially cause pipeline hazards in microprocessors that are handled by complex forwarding logic. Due to the continuous context switches in fine-grain dataflow computers and in cycle-by-cycle interleaving machines (see Sect. 6.3.3), pipeline hazards are avoided albeit with the disadvantage of poor single thread performance.

Antidependences and output dependences are removed by register renaming that maps the architectural registers to the physical registers of the microprocessor. The microprocessor thereby generates internally an instruction stream that satisfies the single-assignment rule of dataflow. Modern microprocessors remove antidependences and output dependences on-the-fly, and avoid the high memory requirements and the often awkward solutions for data structure storage and manipulation, and for loop control caused by the single-assignment rule in dataflow computers.

The main difference between the dependence graphs of dataflow and the code sequence in an instruction window of a superscalar microprocessor is *branch prediction and speculative execution.* The accuracy of branch prediction is surprisingly high: more than 95 % is reported by *Chang et al.* [46] for single SPEC benchmark programs. However, rerolling execution in the case of a wrongly predicted path is costly in terms of processor cycles, especially in deeply pipelined microprocessors.

The idea of branch prediction and speculative execution has never been evaluated in the dataflow environment. The reason for this may be that dataflow was considered to produce an abundance of parallelism[11] while speculation leads to speculative parallelism which is – because of instruction discarding when the branch is mispredicted – inferior to "real" parallelism.

Due to the single thread of control, a high degree of data and instruction locality is present in the machine code of a microprocessor. The locality allows the use of a *storage hierarchy* that stores the instructions and data which may potentially be executed in upcoming cycles, close to the executing processor. Due to the lack of locality in a dataflow graph, a storage hierarchy is difficult to apply in dataflow computers.

The operand *matching* of executable instructions in the instruction window of microprocessors is restricted to a part of the instruction sequence. Because of the serial program order, the instructions in this window are likely to become executable soon. Therefore, the matching hardware can be restricted to a small number of instruction slots. In dataflow computers the number of tokens waiting for a match can be very high. A large waiting-

[11] This is due to dataflow languages which are inherently fine-grain parallel – each statement is parallel to the other and constrained only by data dependences.

matching store is required. Due to the lack of locality the likelihood of the arrival of a matching token is difficult to estimate, so the caching of tokens to be matched soon is difficult in dataflow.

A large instruction window is crucial for current and future superscalar microprocessors in order to find enough instructions for parallel execution. However, the control logic for very large instruction windows gets so complex that it hinders higher cycle rates. Therefore an alternative instruction window organization is needed. *Palacharla et al.* [222] proposed a multiple FIFO-based organization. Only the instructions at the heads of a number of FIFO buffers can be issued to the execution units in the next cycle. Total parallelism in the instruction window is restricted in favor of a less costly issue that does not slow down the processor cycle rate. The potential fine-grained parallelism is thereby limited – a technique somewhat similar to the threaded dataflow approaches described above. It might also be interesting to look at dataflow matching store implementations and dataflow solutions like threaded dataflow as exemplified by the repeat-on-input technique in the Epsilon-2, and the strongly connected arcs model of EM-4, or the associative switching network in the ADARC, etc. For example, the repeat-on-input strategy issues very small compiler-generated code sequences serially (in an otherwise fine-grained dataflow computer). Transferred to the local dataflow in an instruction window, an issue string might be used where a series of data dependent instructions are generated by a compiler and issued serially after the issue of the leading instruction. However, the high number of speculative instructions in the instruction window remains.

3. CISC Processors

Out-of-order execution is not a new concept – it existed twenty years ago on [CISC] IBM and CDC computers – but it is innovative for single-chip implementations...

Mark Brehob, Travis Doom, Richard Enbody, William H. Moore, Sherry Q. Moore, Ron Sass, Charles Severance
Beyond RISC – The Post-RISC Architecture
(Technical Report TR96-11, Michigan State University, March 1996)

3.1 A Brief Look at CISC Processors

Modern superscalar processors, which will be covered extensively in Chap. 4, use multiple FUs. To keep these FUs as busy as possible situations must be allowed where instructions are executed in a different order from that of the original program. Techniques supporting such an out-of-order execution were developed even in the mid-1960s in some *complex instruction set computers* (CISC) which were large mainframe computers at that time. It would take too much space to describe CISC mainframes in detail. Therefore we only briefly itemize some in this chapter and point out those that made a strong impact on the microarchitecture of today's superscalars.

The main characteristics of CISC are a large number of instructions and a high complexity of some of these instructions. Variable instruction formats and over a dozen different addressing modes are used. Since the control in traditional CISC is microprogrammed, a control memory (ROM) is needed. The average CPI is high (between 2 and 15), due to long microcodes used to control the execution of the complex instructions. With few general-purpose registers, many instructions are of the memory-register type, so memory is often accessed for operands. Conventional CISC architectures usually used a unified cache for holding instructions and data (Princeton approach) and, therefore, shared the same data/instruction path.

CISC mainframes influenced microprocessor design in two different ways: First a line of CISC microprocessors, namely the Intel x86 and Motorola MC 680x0 processors, emerged as descendants with ISAs that resemble the CISC mainframe ISAs, in particular, the ISAs of DEC VAX and PDP computer lines. Otherwise, the CISC microprocessors of the 1980s were much simpler than contemporary mainframes. Second, the out-of-order organisation principles of CISC mainframes greatly influenced the out-of-order superscalar microprocessor design in the mid-1990s.

3.2 Out-of-Order Execution

In Chap. 1 we looked at simple pipelined processors that use an *in-order execution* pipeline organization, i.e., instructions are issued to the FUs and execution is initiated in exactly the same order as instructions appear in the program. *Out-of-order execution* is the next step necessary to keep the multiple FUs as busy as possible.

Allowing instructions to complete out of the original program order introduces WAW hazards due to output dependences. Moreover, going one step further and allowing instructions to be issued out of order introduces a new problem. An antidependence can cause a WAR hazard, if a subsequent instruction starts execution and writes back its result before a previous instruction gets its operands. More precisely, WAR hazards may result from antidependences when instructions may be sent to FUs before preceding instructions have read their required operands. For example, suppose an instruction $Inst_1$ has its first operand in register Reg_1 but is waiting for the second operand to appear in Reg_2 which is eventually produced by a long-running instruction $Inst_0$. The next instruction $Inst_2$ may, due to out-of-order execution, proceed to the EX stage and write its result into its destination register Reg_2 before instruction $Inst_0$ terminates (and completes). A solution that correctly considers WAW and WAR hazards is scoreboarding. Tomasulo's scheme is even able to remove WAW and WAR hazards. Both schemes are described below.

To separate dependent instructions and minimize the number of actual hazards and resultant stalls, *scheduling* must be used, i.e., a process which determines when to start a particular instruction, when to read its operands, and when to write its result. In *static scheduling*, compiler techniques are used to minimize stalls by separating dependent instructions so that they will not lead to hazards. In *dynamic scheduling*, the processor tries to avoid stalling when data dependences are already present. Dynamic scheduling can be either:

- *Control-flow* scheduling, when performed centrally at the time of decode. In control-flow scheduling, data and resource dependences are resolved during the decode cycle and the instructions are not issued until the dependences have been resolved. This kind of scheduling was introduced in the CDC 6600 processor, where it was based on the scoreboarding (see Sect. 3.3.1).
- *Dataflow* scheduling, if performed in a distributed manner by the functional units themselves at execute-time. In a dataflow scheduling system, the instructions leave the decode stage when they are decoded and are held

in buffers at the functional units until their operands and the functional unit are available. In dataflow scheduling, instructions are, in a sense, self-scheduled. Dataflow scheduling was first achieved with Tomasulo's register-tagging scheme in the IBM System/360 Model 91 processor (see Sect. 3.3.2).

The recently increasing interest in dynamic scheduling is motivated by ideas that naturally build on it, such as issuing more instructions per clock cycle and speculative execution (see Chap. 4).

3.3 Dynamic Scheduling

In the mid-1960s the CDC 6600 and the IBM System/360 Model 91 were built in strong competition with each other. In the CISC machine CDC 6600 the control of out-of-order execution was centralized (by a technique called scoreboarding), while in another mature but also influent CISC machine, the IBM System/360 Model 91, control was distributed (by Tomasulo's scheme). In the following we describe both machines and the two dynamic scheduling principles.

3.3.1 Scoreboarding

Scoreboarding is a technique for allowing instructions to execute out-of-order when there are sufficient resources and no data dependences. It was introduced in 1963 by *Thornton* [295, 296] in the CDC 6600 processor.

The goal of scoreboarding is to maintain an execution rate of one instruction per clock cycle (when there are no structural hazards) by executing an instruction as early as possible. Thus, when the next instruction to execute is stalled, other instructions can be executed (if they do not depend on any active or stalled instruction).

A scoreboard is a hardware unit that keeps track of the instructions that are in the process of being executed, the functional units that are doing the executing, and the registers that will hold the results of those units. Based on its own data structure and by communicating with the functional units, a scoreboard *centrally* performs all hazard detection and resolution and thus controls the instruction progression from one step to the next (see Fig. 3.1).

While there are many scoreboard variations, we mention only that of *Müller and Paul* [207, 208], which is based on the original scoreboard as introduced in the CDC 6600 and described by *Thornton* [296].

The ID stage of the standard pipeline (see Sect. 1.4) is split into two stages, the *issue* (IS) stage and the *read operands* (RO) stage, while the EX and WB stages are augmented with additional tasks. More precisely:

1. In the IS stage, if there is no structural hazard (i.e., a FU for the instruction is free) and no WAW hazard (i.e., no other active instruction has

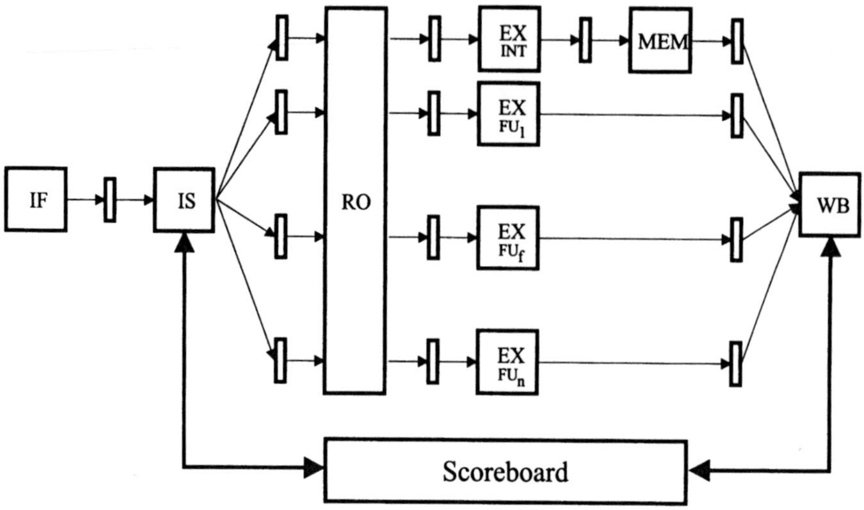

Fig. 3.1. Pipeline with scoreboard

the same destination register), the scoreboard issues the instruction to the FU and updates its internal data structure; otherwise, the instruction issue stalls, and no further instruction is issued until the hazard is cleared.

2. In the RO stage, the scoreboard monitors the availability of the operands, and when they are *all* available,[1] tells the FU to read them from the register file and to proceed to the EX stage. In other words, the scoreboard dynamically resolves RAW hazards; instructions may be dispatched into the EX stage out of order.
3. In the EX stage, the FU begins execution (which may take multiple cycles) and notifies the scoreboard when the result is ready.
4. In the WB stage, once the scoreboard is aware that the FU has completed execution (i.e., the result is ready), the scoreboard checks for WAR hazards and stalls the completing instruction, if necessary (i.e., if there is an instruction that has not read its operand from the destination register of the completing instruction). If a WAR hazard does not exist (any more), the scoreboard tells the FU to store its result in the destination register.

Scoreboarding is a single-issue scheme: only one instruction is issued to the FUs per cycle. A WAW hazard or a structural hazard (an unavailable FU) prevent proceeding the instruction from the IS to the RO stage. No other instruction can be issued (an in-order issue scheme). A RAW hazard prevents an instruction proceeding from the RO to the EX stage and a WAR hazard prevents result write-back. However, other instructions may proceed out of

[1] An operand is available if the register containing the operand is being written by a currently active FU, or if no earlier issued active instruction is going to write it.

order thus preventing interlocking of the pipeline. After the issue to the FU, the FU is blocked until the instruction is dispatched and execution starts. The FU is blocked even longer until the instruction execution terminates and the result is written back to the destination register. No forwarding is applied. Operands can be read in the RO stage after they have been written back to the register set.

Let us describe the scoreboard and the corresponding bookkeeping in more detail.

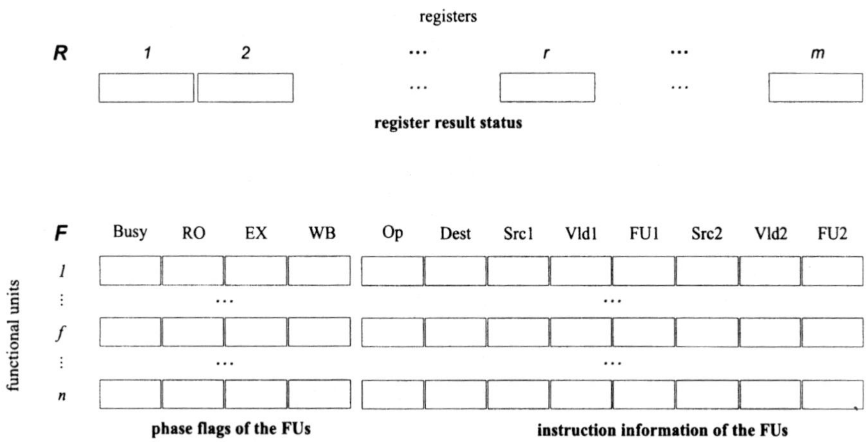

Fig. 3.2. Structure of the scoreboard

Structures in the scoreboard. There are three tables (in two parts, $\boldsymbol{R}$ and $\boldsymbol{F}$) in the scoreboard as described in Fig. 3.2:

- *Register result status*
  This table holds which FU will produce a result in each register (if any). The table is denoted by $\boldsymbol{R}$. The number of entries in $\boldsymbol{R}$ is equal to the number m of registers. If $\texttt{Reg}_r$ is a register, then
  - $\boldsymbol{R}[r] = f$ denotes that $\texttt{Reg}_r$ is currently reserved by FU_f, which is going to produce a result for $\texttt{Reg}_r$,
  - $\boldsymbol{R}[r] = 0$ denotes that no FU has an active instruction whose destination is $\texttt{Reg}_r$.
- *Phase flags of the FUs*
  This table holds the phase of execution for each instruction. To do this, it provides the phase flags *Busy*, *RO*, *EX*, and *WB*, for each functional unit FU_f, with the following meaning:
  - $\boldsymbol{F}[f, \mathit{Busy}] = 1$ when FU_f is reserved for an active instruction, while $\boldsymbol{F}[f, \mathit{Busy}] = 0$ when FU_f is free;

- $\boldsymbol{F}[f, RO] = 1$ indicates that FU_f has read the operands for its current instruction and has switched to the EX phase;
- $\boldsymbol{F}[f, EX] = 1$ indicates that FU_f has finished the computation and has switched to the WB phase;
- $\boldsymbol{F}[f, WB] = 1$ indicates that FU_f has written back the result to the destination register.

- *Instruction information of the FUs*
  This table has one entry per FU, showing which operation the FU is scheduled to do, if any, where its result goes, where its operands come from, and whether those results are available. If an operand is not available, the table tells which FU will produce it. The table provides the following entries that describe the current instruction `Inst : op Dest, Src1, Src2` performed by functional unit FU_f:
  - opcode: $\boldsymbol{F}[f, Op] =$ `op`;
  - destination register: $\boldsymbol{F}[f, Dest] =$ `Dest`;
  - source registers: $\boldsymbol{F}[f, Src1] =$ `Src1` and $\boldsymbol{F}[f, Src2] =$ `Src2`;
  - validity of sources: $\boldsymbol{F}[f, Vld1]$ and $\boldsymbol{F}[f, Vld2]$ are used to check for current data. If $\boldsymbol{F}[f, Vld1] = 0$, the data of $\boldsymbol{F}[f, Src1]$ is not valid yet. (Similarly for $\boldsymbol{F}[f, Vld2]$.)
  - FUs producing sources: $\boldsymbol{F}[f, FU1]$ and $\boldsymbol{F}[f, FU2]$ are used to specify the FUs which are going to produce values in $\boldsymbol{F}[f, Src1]$ and $\boldsymbol{F}[f, Src2]$. If $\boldsymbol{F}[f, FU1] = 0$, then the source was already valid on the instruction issue; otherwise, $\boldsymbol{F}[f, FU1] = g$ denotes that $\boldsymbol{F}[f, Src1]$ will receive the value from FU_g. (Similarly for $\boldsymbol{F}[f, FU2]$.)

  We used $\boldsymbol{F}$ to denote both the phase flags and the instruction information of the FUs. The number of entries in $\boldsymbol{F}$ is equal to the number n of FUs (see Fig. 3.2).

Bookkeeping in the scoreboard. On power-up, the scoreboard is initialized by setting all its entries to zero. The scoreboard then issues the instructions of a program in sequential order to the appropriate FUs, one instruction at a time. For each FU, the scoreboard performs bookkeeping and governs the resources, as follows:

1. In the IS stage, the scoreboard issues the next instruction, for example `Inst : op Dest, Src1, Src2`, as soon as the destination register `Dest` and a FU capable of executing `op` become available. The scoreboard then reserves `Dest` and such a FU, say FU_f. The corresponding register status, FU status, and instruction status are initialized, as follows:

   while `Inst` **not** issued yet **and** previous instruction issued **do**
       if $\boldsymbol{R}[$`Dest`$] = 0$ **and**
           $(\exists\ FU_f : FU_f$ capable of executing `op` **and** $\boldsymbol{F}[f, Busy] = 0)$
       then do_in_the_same_cycle
               Choose such FU_f;
               /* initialize register status */

```
        R[Dest] := f;
        /* initialize FU status */
        F[f, Busy] := 1; F[f, RO] := F[f, EX] := F[f, WB] := 0;
        /* initialize instruction status */
        F[f, Op] := op;
        F[f, Dest] := Dest;
        F[f, Src1] := Src1;
        if R[Src1] = 0 then F[f, Vld1] := 1 else F[f, Vld1] := 0;
        F[f, FU1] := R[Src1];
        F[f, Src2] := Src2;
        if R[Src2] = 0 then F[f, Vld2] := 1 else F[f, Vld2] := 0;
        F[f, FU2] := R[Src2]
    enddo
```

2. In the RO stage, after issuing **Inst** to FU_f, that unit tries to read operands from source registers **Src1** and **Src2**. As long as $\boldsymbol{F}[f, RO] = 0$, it checks whether the registers to be read are both up to date (valid), i.e., whether $\boldsymbol{F}[f, Vld1] = \boldsymbol{F}[f, Vld2] = 1$. If so, the source registers are read and their validity flags are cleared. Since FU_f no longer waits for its operands from other FUs, $\boldsymbol{F}[f, FU1]$ and $\boldsymbol{F}[f, FU2]$, which indicate these FUs, are cleared[2]. The phase flag $\boldsymbol{F}[f, RO]$ is set, thus allowing FU_f to proceed to the EX stage.

```
while F[f, RO] = 0 do
    if F[f, Vld1] = 1 and F[f, Vld2] = 1
    then do_in_the_same_cycle
            Read operands;
            F[f, FU1] := F[f, FU2] := 0;
            F[f, RO] := 1
        enddo
```

3. FU_f remains in the EX stage until its computation is completed, i.e., when FU_f's *result ready* flag is set. Then, $\boldsymbol{F}[f, EX]$ is set indicating that FU_f proceeds to its WB stage.

```
while F[f, RO] = 1 and F[f, EX] = 0 do
    do_in_the_same_cycle
        F[f, EX] := result ready flag of FU_f
    enddo
```

4. In the WB stage, after the FU has run to completion, the scoreboard postpones the write-back (in order to solve WAR hazards) until no FU_g, $g \neq f$, exists such that FU_f's destination register **Dest** is a source register

[2] This is the first improvement to the original scoreboard from the CDC 6600 to avoid deadlocks, as suggested by *Müller and Paul* [207, 208]. The original scoreboard cleared validity flags $\boldsymbol{F}[f, Vld1]$ and $\boldsymbol{F}[f, Vld2]$ instead of $\boldsymbol{F}[f, FU1]$ and $\boldsymbol{F}[f, FU2]$.

of FU_g *and* FU_g still has to read[3] the old (i.e., valid) value of register `Dest`.

$$\begin{array}{l}
\textbf{while } \boldsymbol{F}[f, EX] = 1 \textbf{ and } \boldsymbol{F}[f, WB] = 0 \textbf{ do} \\
\quad \textbf{if not } \Big(\exists\ \text{FU}_g,\ g \neq f : \\
\qquad\qquad ((\boldsymbol{F}[g, Src1] = \boldsymbol{F}[f, Dest] \textbf{ and } \boldsymbol{F}[g, Vld1] = 1) \textbf{ or} \\
\qquad\qquad (\boldsymbol{F}[g, Src2] = \boldsymbol{F}[f, Dest] \textbf{ and } \boldsymbol{F}[g, Vld2] = 1)) \\
\qquad\qquad \textbf{and } \boldsymbol{F}[g, RO] = 0 \Big) \\
\quad \textbf{then do_in_the_same_cycle} \\
\qquad\quad \text{Write results to } \boldsymbol{F}[f, Dest]; \\
\qquad\quad \boldsymbol{F}[f, WB] := 1; \\
\qquad\quad \boldsymbol{R}[\boldsymbol{F}[f, Dest]] := 0 \\
\qquad \textbf{enddo}
\end{array}$$

5. After the WB stage, all FUs with a source depending on a result of FU_f are notified. Their corresponding valid flags are set.

$$\begin{array}{l}
\textbf{while } \boldsymbol{F}[f, WB] = 1 \textbf{ and } \boldsymbol{F}[f, Busy] = 1 \textbf{ do} \\
\quad \textbf{forall } \text{FU}_g,\ g \neq f \\
\qquad \textbf{do_in_the_same_cycle} \\
\qquad\quad \textbf{if } \boldsymbol{F}[g, FU1] = f \textbf{ then } \boldsymbol{F}[g, Vld1] = 1; \\
\qquad\quad \textbf{if } \boldsymbol{F}[g, FU2] = f \textbf{ then } \boldsymbol{F}[g, Vld2] = 1; \\
\qquad\quad \boldsymbol{F}[f, Busy] := 0 \\
\qquad \textbf{enddo}
\end{array}$$

Example. Let us illustrate the bookkeeping in a scoreboard with the following sequence of instructions:

```
mul Reg1, Reg3, Reg5      /* Reg1 ← Reg3  * Reg5
sub Reg2, Reg4, Reg3      /* Reg2 ← Reg4  - Reg3
div Reg6, Reg1, Reg4      /* Reg6 ← Reg1  / Reg4
add Reg4, Reg2, Reg3      /* Reg4 ← Reg2  + Reg3
```

Assume there are three functional units: FU_1 is an *adder* (capable of performing addition as well as subtraction), FU_2 is a *multiplier*, and FU_3 is a *divider*. The execution latencies of the multiplier and divider are several times larger than the latency of the adder.[4] During execution the scoreboard constantly changes its state. Three of these states are depicted in Fig. 3.3.

Figure 3.3a shows the scoreboard when all the instructions are in the pipeline, with `mul` and `sub` in the EX stage. The `add` is stalled because of the structural hazard, i.e., the adder is not available at that moment. Notice,

[3] Testing "valid and unread" instead of testing only "valid" is the second correction to the original scoreboard from the CDC 6600 to avoid deadlocks, also suggested by *Müller and Paul* [207, 208].

[4] For latencies of real microprocessors see, e.g., Table 4.4 on p. 172, Table 4.8 on p. 187, or Table 4.9 on p. 190.

R	1	2	3	4	5	6
	2	1	0	1	0	3

F	Busy	RO	EX	WB	Op	Dest	Src1	Vld1	FU1	Src2	Vld2	FU2
1	1	1	1	0	sub	2	4	1	0	3	1	0
2	1	1	1	0	mul	1	3	1	0	5	1	0
3	1	0	0	0	div	6	1	0	2	4	1	0

(a)

R	1	2	3	4	5	6
	0	0	0	1	0	3

F	Busy	RO	EX	WB	Op	Dest	Src1	Vld1	FU1	Src2	Vld2	FU2
1	1	1	1	0	add	4	2	1	0	3	1	0
2	0	0	0	0	--							
3	1	0	0	0	div	6	1	1	0	4	1	0

(b)

R	1	2	3	4	5	6
	0	0	0	0	0	3

F	Busy	RO	EX	WB	Op	Dest	Src1	Vld1	FU1	Src2	Vld2	FU2
1	0	0	0	0	--							
2	0	0	0	0	--							
3	1	1	1	0	div	6	1	1	0	4	1	0

(c)

Fig. 3.3. Three snapshots of the scoreboard

however, that an additional adder would not prevent stalling the **add** at that moment, because **add** is data dependent on **sub**. Such a data dependency of **div** on **mul** stalls **div**, despite the divider being available at that moment.

Figure 3.3b shows the scoreboard several cycles later, when **mul** has just completed its WB stage and **sub** has completed its WB stage several cycles earlier due to the shorter latency. The **add** is in its EX stage, and **div** is just about to enter its RO stage in the next cycle.

Finally, Fig. 3.3c shows the scoreboard when only **div** is in the pipeline (in the EX stage).

CDC 6600 Architecture

The CDC 6600 was delivered in 1964 by Control Data Corporation (*Thornton* [295, 296]). The developers introduced several enhancements in pipelining and designed the first processor to make extensive use of multiple functional units. These parallel units allowed instructions to complete out of the original program order. The CDC 6600 processor introduced scoreboarding to achieve control-flow dynamic scheduling of instructions. It also introduced a register-register instruction set (load/store architecture) with a 3-address instruction format, and peripheral processors that used a time-shared pipeline.

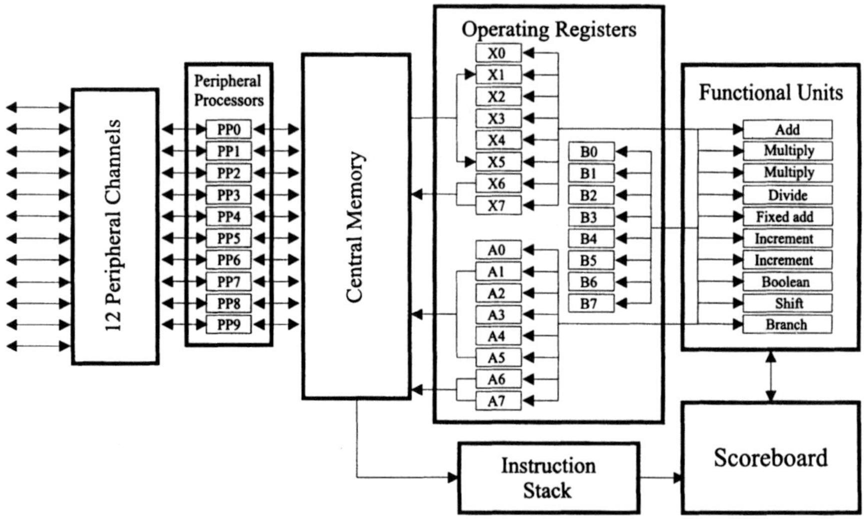

Fig. 3.4. The CDC 6600 processor

In the CDC 6600 processor ten FUs appeared as a multiple execution pipeline (see Fig. 3.4). Of the ten FUs, which were able to operate simultaneously, four carried out 60-bit floating-point arithmetic among eight 60-bit operand registers, while the other six FUs carried out logic, indexing, and program control on the eight 18-bit address registers and eight 18-bit increment/index registers. The processor had instruction buffers for each FU. Instructions were issued to available FUs regardless of whether register input data were available. The instruction's control information would then wait in a buffer for its data to be produced by other instructions. To control the correct routing of data between FUs and registers, the CDC 6600 used a centralized control unit, the scoreboard described above, which keeps track of the registers needed by instructions waiting for the various FUs. When all registers had valid data, the scoreboard enabled the instruction execution. Similarly, when a FU finished, it signaled the scoreboard to release the resources.

3.3.2 Tomasulo's Scheme

Tomasulo's scheme is another dynamic scheduling scheme that allows execution to proceed in the presence of hazards. It was invented by *Tomasulo* [298] and implemented in the IBM System/360 Model 91 in 1967. The scheme addressed the following issues: a small number of floating point registers, long memory latency, cost effectiveness of FU hardware, and the performance penalties of name dependences (i.e., antidependences and output dependences).

Assume a name dependence occurs between two instructions, $Inst_1$ and $Inst_2$, that use the same register (or memory location) *Reg*, but there are no data transmitted between $Inst_1$ and $Inst_2$. If *Reg* is renamed so that $Inst_1$ and $Inst_2$ do not conflict, the two instructions can execute simultaneously or be reordered. The technique that dynamically eliminates name dependences to avoid WAR and WAW hazards is called *register renaming.*

The Tomasulo's scheme combines register renaming with the key elements of scoreboarding. However, the centralized scoreboard is now replaced with *distributed* control within the processor. Each FU has one or more *reservation stations* (see Fig. 3.5). Each reservation station (RS) is given a unique number called a *tag.* A RS may be empty or hold an instruction, which is indicated by the *empty* flag *Empty.* The instruction is either awaiting availability of all of its sources, awaiting availability of the FU, or it is executing at that FU. For each source operand, RS contains either its value (in field *Src*1 or *Src*2) or, in case that the value is not available yet, a tag of the RS where the value will be produced (in field *RS*1 or *RS*2). *Valid* flags (*Vld*1 and *Vld*2) are used to indicate whether the values are available or not. Once the operand values are available (i.e., both *Vld*1 and *Vld*2 are set), and the FU can start executing the next instruction, pipelined execution of that instruction is initiated. When the execution of the instruction actually starts, the *InFU* flag in the corresponding RS is set and remains set until the completion of the instruction. After completion of the instruction from RS tagged s, the result token[5] $\langle s, result\rangle$ is formed and passed on the *common data bus* (CDB) to the register file and, by snooping, directly to all RSs (thus eliminating the need to get the operand value from a register). The RS with tag s (i.e., corresponding to the instruction whose result has been placed on the CDB) is cleared by setting its flag *Empty.* The traffic passing on the CDB is continually monitored (*snooped*) by all reservation stations of all FUs. When the result token $\langle s, result\rangle$ is observed, the *result* is copied into all RSs awaiting it (e.g., having flag *Vld*1 cleared and *RS*1 equal to s) and the

[5] We use the term *token* in order to point out the similarity with tagged-token dataflow (see Chap. 2).

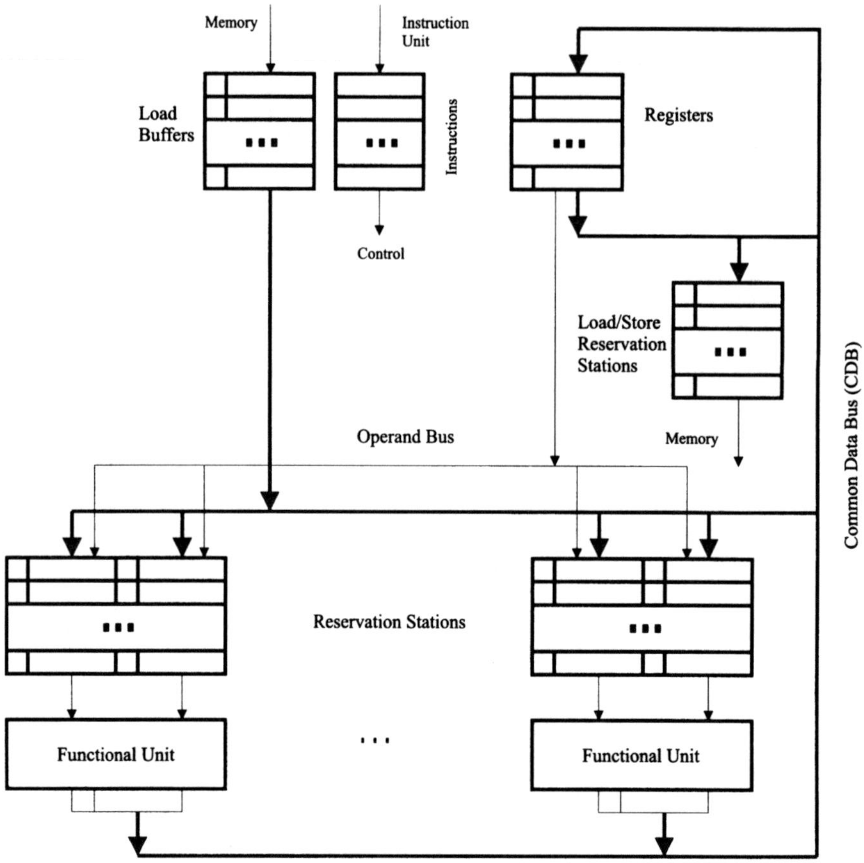

Fig. 3.5. Tomasulo's scheme

associated valid flags are set to indicate the presence of the operand value. Note that the CDB allows all units that are waiting for an operand to be loaded simultaneously, i.e., the result is forwarded to all waiting instructions. Hence, the RS fetches and buffers an operand as soon as it becomes available (dataflow principle). The *load buffers* and *load/store reservation stations* hold data coming from memory, respectively addresses and data going to memory. The *registers* are connected by a pair of buses to the FUs and by a single bus to the load/store reservation stations.

To summarize, a decoded instruction is first sent to the *instruction queue*, proceeds from there to an empty RS based on its opcode, and is subjected to the following three pipeline stages:

1. In the *issue* (IS) stage, the instruction is retrieved from the instruction queue and structural hazards are checked (i.e., if there is no empty RS in the case of a floating-point instruction, or no empty load/store buffer in the case of a load/store instruction). In the case of a structural hazard,

the instruction is stalled until a RS or a buffer is freed. If there is no structural hazard and the instruction is floating-point, the instruction is sent to a free RS while its operands are sent to the same RS if they are in the registers. Similarly, a load/store instruction is sent to a free buffer. Register renaming is performed in this stage since the instruction's register specifiers for pending operands are renamed to the tags (i.e., unique numbers) of those RSs where the instructions producing these operands are located. WAW and WAR hazards are removed in this stage by register renaming.

2. In the *execute* (EX) stage, if at least one of the operands is missing, the CDB is monitored (snooped) while waiting for the operand to be computed by one of the FUs or supplied via a load buffer. When the operand is available, it is placed into the corresponding RS. When both operands are available, the execution of the operation is started. Hence, this step checks for RAW hazards.
3. In the *write-back* (WB) stage, when the result is available, it is put on the CDB and from there written into the registers and RSs waiting for it.

Structures in Tomasulo's scheme. Let us describe Tomasulo's scheme in more detail (Fig. 3.6). First, we introduce two[6] data structures for updating the status of the registers and reservation stations belonging to the FUs:

- *Register status table*
  For each register, it is specified whether the register contains valid data; if not, the RS whose instruction will produce that data is specified. The table is denoted by $\boldsymbol{R}$. The number of entries in $\boldsymbol{R}$ is equal to the number m of registers. If $\texttt{Reg}_r$ is a register, then
  - $\boldsymbol{R}[r, \mathit{Value}]$ is the value contained in the register $\texttt{Reg}_r$.
  - $\boldsymbol{R}[r, \mathit{Vld}]$ is 1 (0) if $\boldsymbol{R}[r, \mathit{Value}]$ is valid (not valid).
  - $\boldsymbol{R}[r, \mathit{RS}] = s$ points to the current source (i.e., the s-th RS) of the register value (if the latter is not present).
- *Reservation station table*
  For each FU_f there is a set $\boldsymbol{S}_f$ of RSs. Let `Inst : op Dest, Src1, Src2` be the instruction issued to the RS which is tagged s and belongs to FU_f. Then:
  - $\boldsymbol{S}_f[s, \mathit{Empty}] = 1$ indicates that the RS is empty;
  - $\boldsymbol{S}_f[s, \mathit{InFU}] = 1$ if and only if FU_f is executing `Inst`;
  - $\boldsymbol{S}_f[s, \mathit{Op}] =$ `op`;
  - $\boldsymbol{S}_f[s, \mathit{Dest}] =$ `Dest`;
  - $\boldsymbol{S}_f[s, \mathit{Src1}] =$ `Src1` and $\boldsymbol{S}_f[s, \mathit{Src2}] =$ `Src2`;

[6] Actually, there are two more data structures, the *load buffer status table* and the *load/store reservation station table*. Since they are similar to the *register status table* and the *reservation station table*, we omit their description.

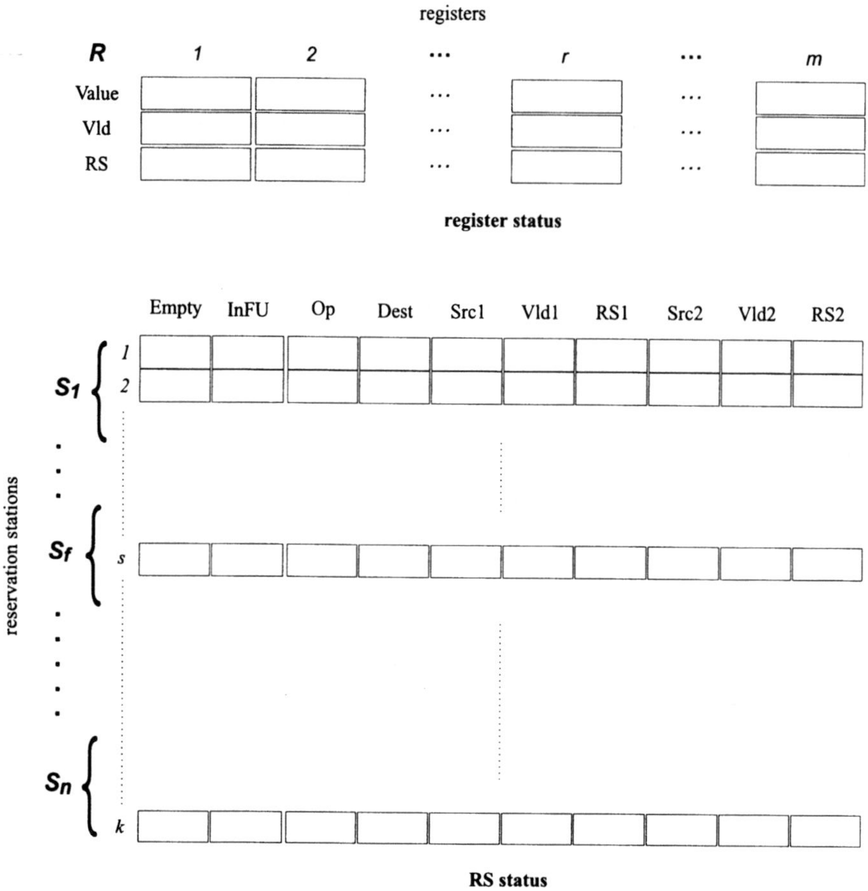

Fig. 3.6. Data structures in Tomasulo's scheme

- $\boldsymbol{S}_f[s, Vld1]$ and $\boldsymbol{S}_f[s, Vld2]$ are used to check for current data. If $\boldsymbol{S}_f[s, Vld1] = 0$, the data of $\boldsymbol{S}_f[s, Src1]$ is not valid yet. (Similarly for $\boldsymbol{S}_f[s, Vld2]$.)
- $\boldsymbol{S}_f[s, RS1]$ and $\boldsymbol{S}_f[s, RS2]$ are used to specify the RSs which are going to produce values in $\boldsymbol{S}_f[s, Src1]$ and $\boldsymbol{S}_f[s, Src2]$. Hence, $\boldsymbol{S}_f[s, RS1] = t$ denotes that $\boldsymbol{S}_f[s, Src1]$ will receive the value from the t-th RS. (Similarly for $\boldsymbol{S}_f[s, FU2]$.)

Bookkeeping in Tomasulo's scheme. Bookkeeping is performed by the following algorithms:

1. In the IS stage, the next instruction, for example `Inst : op Dest, Src1, Src2`, is issued to an empty RS that belongs to a FU_f capable of executing `op`. We denote that RS by s.

while `Inst` **not** issued yet **and** previous instruction issued **do**
  if $\exists f, s$: FU_f capable of executing `op` **and** $\boldsymbol{S}_f[s, \mathit{Empty}] = 1$
  then do_in_the_same_cycle
      Choose such pair f, s;
      /* initialize register status */
      $\boldsymbol{R}[\mathtt{Dest}, \mathit{RS}] := s$;
      $\boldsymbol{R}[\mathtt{Dest}, \mathit{Vld}] := 0$;
      /* initialize reservation station status */
      $\boldsymbol{S}_f[s, \mathit{Empty}] := 0$;
      $\boldsymbol{S}_f[s, \mathit{InFU}] := 0$;
      $\boldsymbol{S}_f[s, \mathit{Op}] := \mathtt{op}$;
      $\boldsymbol{S}_f[s, \mathit{Dest}] := \mathtt{Dest}$;
      if $\boldsymbol{R}[\mathtt{Src1}, \mathit{Vld}] = 1$ **then** $\boldsymbol{S}_f[s, \mathit{Src1}] := \boldsymbol{R}[\mathtt{Src1}, \mathit{Value}]$;
      $\boldsymbol{S}_f[s, \mathit{Vld1}] := \boldsymbol{R}[\mathtt{Src1}, \mathit{Vld}]$;
      $\boldsymbol{S}_f[s, \mathit{RS1}] := \boldsymbol{R}[\mathtt{Src1}, \mathit{RS}]$;
      if $\boldsymbol{R}[\mathtt{Src2}, \mathit{Vld}] = 1$ **then** $\boldsymbol{S}_f[s, \mathit{Src2}] := \boldsymbol{R}[\mathtt{Src2}, \mathit{Value}]$;
      $\boldsymbol{S}_f[s, \mathit{Vld2}] := \boldsymbol{R}[\mathtt{Src2}, \mathit{Vld}]$;
      $\boldsymbol{S}_f[s, \mathit{RS2}] := \boldsymbol{R}[\mathtt{Src2}, \mathit{RS}]$
    enddo

2. In the EX stage, FU_f can start executing instruction `Inst` contained in the s-th RS if `Inst` has not been started yet, i.e., $\boldsymbol{S}_f[s, \mathit{InFU}] = 0$, and `Inst` has collected both operands, i.e., $\boldsymbol{S}_f[s, \mathit{Vld1}] = 1$, $\boldsymbol{S}_f[s, \mathit{Vld2}] = 1$.

while $\boldsymbol{S}_f[s, \mathit{Empty}] = 0$ **and** $\boldsymbol{S}_f[s, \mathit{InFU}] = 0$ **do**
  if $\boldsymbol{S}_f[s, \mathit{Vld1}] = 1$ **and** $\boldsymbol{S}_f[s, \mathit{Vld2}] = 1$
    if FU_f can start executing another instruction, say, `Inst`
      then do_in_the_same_cycle
          $\boldsymbol{S}_f[s, \mathit{InFU}] := 1$;
          FU_f gets $s, \boldsymbol{S}_f[s, \mathit{Op}], \boldsymbol{S}_f[s, \mathit{Src1}], \boldsymbol{S}_f[s, \mathit{Src2}]$
        enddo

3. In the WB stage, after completion of instruction `Inst`, the *result* is written into register `Dest`. The result token is formed and passed on the CDB where it is available to RSs (see snooping on CDB below).

while FU_f completed `Inst` from RS tagged s **do**
  if FU_f can gain control of the CDB
    then do_in_the_same_cycle
        $\mathit{token.tag} := s$; $\mathit{token.data} := \mathit{result}$;
        $\boldsymbol{S}_f[s, \mathit{Empty}] = 1$;
        $\boldsymbol{R}[\mathtt{Dest}, \mathit{Value}] = \mathit{token.data}$;
        $\boldsymbol{R}[\mathtt{Dest}, \mathit{Vld}] = 1$;
        $\boldsymbol{R}[\mathtt{Dest}, \mathit{RS}] = 0$
      enddo

4. Snooping on the CDB allows all units that are waiting for an operand to be loaded simultaneously. In particular, when the result token $\langle tag, data \rangle$ is observed, the value from *token.data* is copied into all RSs awaiting it.

while $token.tag \neq 0$ **and** $\boldsymbol{S}_f[s, Empty] = 0$ **and**
    $(\boldsymbol{S}_f[s, Vld1] = 0$ **or** $\boldsymbol{S}_f[s, Vld2] = 0)$
  if $\boldsymbol{S}_f[s, RS1] = token.tag$ **or** $\boldsymbol{S}_f[s, RS2] = token.tag$
    then do_in_the_same_cycle
      if $\boldsymbol{S}_f[s, RS1] = token.tag$ **then**
        $\boldsymbol{S}_f[s, Src1] = token.data;$
        $\boldsymbol{S}_f[s, Vld1] = 1;$
        $\boldsymbol{S}_f[s, RS1] = 0$
      endif
      if $\boldsymbol{S}_f[s, RS2] = token.tag$ **then**
        $\boldsymbol{S}_f[s, Src2] = token.data;$
        $\boldsymbol{S}_f[s, Vld2] = 1;$
        $\boldsymbol{S}_f[s, RS2] = 0$
      endif
    enddo

Example. The bookkeeping in Tomasulo's scheme will be illustrated using the same sequence of instructions as in the case of scoreboarding:

```
mul Reg1, Reg3, Reg5
sub Reg2, Reg4, Reg3
div Reg6, Reg1, Reg4
add Reg4, Reg2, Reg3
```

Again, there are three functional units: an *adder* (capable of performing addition as well as subtraction), a *multiplier*, and a *divider*. The adder has two RSs (with tags 1 and 2), the multiplier has one RS (with tag 3), and the divider also has one RS (with tag 4). Three snapshots of Tomasulo's scheme data structures are given in Fig. 3.7.

Figure 3.7a shows the situation when all the instructions are in the pipeline. Since there was no lack of RSs, all the instructions were issued to appropriate RSs, in particular, `sub`, `add`, `mul`, and `div` were issued to RSs tagged 1, 2, 3, and 4, respectively. The `mul` and `sub` are in the EX stage. The `add` is stalled because of the structural hazard on the adder (i.e., $\boldsymbol{S}_{adder}[1, InFU] := 1$ at that moment). The `div` is stalled because of the unresolved data dependency on `mul`; that is, $\boldsymbol{S}_{divider}[4, Vld1] := 0$ meaning that `Reg1` does not yet contain a valid operand for `div`. That operand will be provided by the RS tagged 3 (multiplier), i.e., $\boldsymbol{S}_{divider}[4, RS1] := 3$.

Figure 3.7b shows the situation just after snooping on the CDB has been finished following the completion of the WB stage of the `mul`. The WB stage set $\boldsymbol{R}[1, Vld] = 1$ and $\boldsymbol{R}[1, RS] = 0$ (i.e., `Reg1` contains a valid value). Snooping changed $\boldsymbol{S}_{divider}[4, Src1]$ to the contents of the `Reg1`, and set

R	1	2	3	4	5	6
Value	--	--	(Reg3)	(Reg4)	(Reg5)	--
Vld	0	0	1	1	1	0
RS	3	1	0	0	0	4

		Empty	InFU	Op	Dest	Src1	Vld1	RS1	Src2	Vld2	RS2
S_{adder}	1	0	1	sub	2	(Reg4)	1	0	(Reg3)	1	0
	2	0	0	add	4	2	0	1	(Reg3)	1	0
$S_{multiplier}$	3	0	1	mul	1	(Reg3)	1	0	(Reg5)	1	0
$S_{divider}$	4	0	0	div	6	1	0	3	(Reg4)	1	0

(a)

R	1	2	3	4	5	6
Value	(Reg1)	(Reg2)	(Reg3)	--	(Reg5)	--
Vld	1	1	1	0	1	0
RS	0	0	0	2	0	4

		Empty	InFU	Op	Dest	Src1	Vld1	RS1	Src2	Vld2	RS2
S_{adder}	1	1	0	--	--	--	0	0	--	0	0
	2	0	1	add	4	(Reg2)	1	0	(Reg3)	1	0
$S_{multiplier}$	3	1	0	--	--	--	0	0	--	0	0
$S_{divider}$	4	0	0	div	6	(Reg1)	1	0	(Reg4)	1	0

(b)

R	1	2	3	4	5	6
Value	(Reg1)	(Reg2)	(Reg3)	(Reg4)	(Reg5)	--
Vld	1	1	1	1	1	0
RS	0	0	0	0	0	4

		Empty	InFU	Op	Dest	Src1	Vld1	RS1	Src2	Vld2	RS2
S_{adder}	1	1	0	--	--	--	0	0	--	0	0
	2	1	0	--	--	--	0	0	--	0	0
$S_{multiplier}$	3	1	0	--	--	--	0	0	--	0	0
$S_{divider}$	4	0	1	div	6	(Reg1)	1	0	(Reg4)	1	0

(c)

Fig. 3.7. Three snapshots of Tomasulo's scheme

$S_{divider}[4, Vld1] = 1$ and $S_{divider}[4, RS1] = 0$. Observe that **sub** completed its WB stage long before that due to its shorter latency. The **add** is in its EX stage, and **div** is just about to enter its EX stage in the next cycle.

Finally, Fig. 3.7c shows the situation when only **div** is in the EX stage.

The IBM System/360 Model 91 Floating-Point Unit

The IBM System/360 Model 91 belongs to the family of the IBM System/360 architecture (*Amdahl et al.* [10], *Anderson et al.* [11], *Flynn* [88]) and, therefore, shares the ISA with this highly influential CISC machine. It introduced many new concepts, including tagging of data, register renaming, dynamic detection of memory hazards, and generalized forwarding.

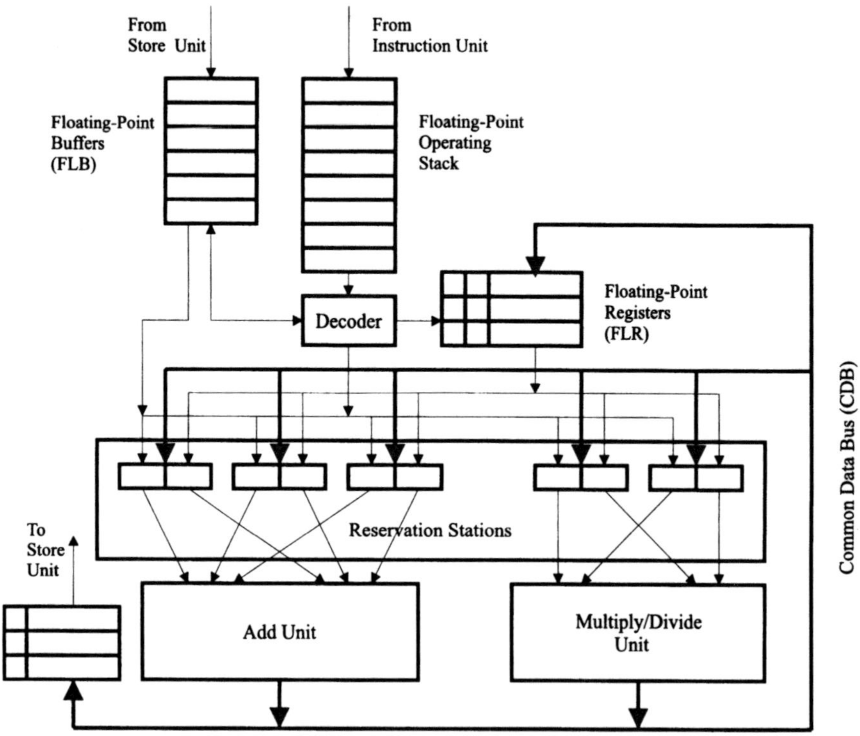

Fig. 3.8. IBM System/360 Model 91 floating-point unit

The IBM System/360 Model 91 was deeply pipelined with an overall pipeline length of 20 stages. No cache was available yet. The floating-point execution unit (see Fig. 3.8) consisted of two separate, fully pipelined floating-point FUs, the *adder* and the *multiplier/divider*. The FUs could be used concurrently. Addition took 2 cycles, multiplication 3 cycles, and division 11 cycles. There were three *reservation stations* (RS) associated with the adder, and two with the multiplier/divider. The RSs associated with a FU made the FU behave like several virtual FUs. Results from the two FUs were sent back via a *common data bus* (CDB) to the memory (via *store buffers*), *register file*, or RSs. The processor used Tomasulo's register-tagging scheme to achieve dataflow dynamic scheduling of instructions to virtual FUs. (See earlier in this subsection for details.) A speculative branch prediction was used that speculated the target would be taken, when the branch target instruction was within the last eight instructions in the *operation stack*. Memory had a 10-cycle access, it was fully buffered and 32-way interleaved. The processor could have up to 32 memory accesses pending in order to reduce latency.

3.3.3 Scoreboarding versus Tomasulo's Scheme

In eliminating stalls, scoreboarding and Tomasulo's scheme are limited by several factors. The first general one is the amount of parallelism available in code. The second factor is the in-order issue of scoreboarding and of Tomasulo's scheme that prevents instruction to proceed. The third factor is the number and types of FUs (respectively RSs in Tomasulo's scheme) since contention for FUs (RSs) leads to structural hazards. Finally, a scoreboard may be limited by the presence of antidependences and output dependences, which may lead to WAR and WAW stalls.

The main advantages of Tomasulo's scheme over scoreboarding is its removal of WAW and WAR hazards (which may lead to instruction stalling in the scoreboarding scheme) and result forwarding (results are available as input operands one cycle earlier than in the scoreboarding scheme). The major drawback of Tomasulo's scheme is its complexity, which requires a reasonable amount of hardware. Namely, the CDB must interact with all the pipeline hardware, so, for the snooping process to be efficient, a complex control logic is needed, as well as *associative* stores.

The full power of dynamic register renaming, as introduced by Tomasulo's scheme, finds expression in the execution of loops. If a loop is unrolled and (statically) scheduled to avoid interlocks, many registers may be required. Distinct from loop unrolling, Tomasulo's scheme supports the overlapped execution of multiple copies of the loop with only a small number of registers used by the program, since the RSs extend the real register set via the renaming process.

For the superscalar pipeline the techniques such as register renaming and dynamic scheduling are crucial. Some state-of-the-art superscalar processors, namely the PowerPC processors, directly enhance the single-issue Tomasulo scheme to a modified four-issue Tomasulo scheme.

3.4 Some CISC Microprocessors

In the following, we give a very brief historical account of some CISC-type microprocessor families from DEC, Intel, Motorola, Zilog, and National Semiconductor. These CISC microprocessors inherited the ISA principles from the CISC mainframes, but not the dynamic scheduling which was not introduced in microprocessors until the mid-1990s.

DEC LSI-11

The DEC PDP-11 was the most popular in the 16-bit PDP (Programmed Data Processors) line of CISC minicomputers, a successor to the previously popular PDP-8 and remained in production for over 25 years, until the end of 1997. The LSI-11 (introduced in 1975) was a popular microprocessor implementation of the PDP-11 using the Western Digital MCP1600 microprogrammable CPU, and the architecture influenced the Motorola MC 68000, National Semicoductor NS 320xx, and Zilog Z-8000 microprocessors, in particular. There was also a 32-bit PDP-11 plan as far back as its 1969 introduction. The PDP-11 was finally replaced by the 32-bit VAX architecture.

Intel x86 family

Intel was and still is a leader in the microprocessor industry, primarily known for its x86 CISC-type family (see Table 3.1). The Intel 8086 (announced by Intel in 1978 as the first in the x86 family) was based on the design of the 8080/8085 (source compatible with the 8080) with a similar register set, but was expanded to 16 bits (*Liu and Gibson* [185]). The instruction lengths varied from one to four bytes. The instruction stream was fed to the execution unit through a small prefetch queue, so fetch and execution were concurrent – a primitive form of pipelining. The 80286 (introduced in 1982) added a protected mode, which extended the directly addressed memory space to 16 MB. However all memory access was still restricted to 64 k segments until the 80386 (in 1985), which included much improved addressing (with paging support in addition to segmented addressing). The 80386 was a 32-bit architecture with 32-bit registers and 32-bit address space. It also had several processor modes for compatibility with the previous design. The 80486 (1989) added full pipelines, single on-chip 8 kB cache, integrated FPU (based on the stack-oriented register set with eight 80-bit registers in the 80387 FPU), and clock doubling versions (like the Z-280). Intel's x86 family superscalar descendants Pentium (late 1993), Pentium Pro (late 1995), Pentium II (April 1997), and Pentium III (February 1999) will be covered in Sect. 4.9.1, as well as the next generation 64-bit processor, code named Merced, based on IA-64 ISA, which was defined jointly by Intel and Hewlett-Packard.

Table 3.1. The Intel x86 family

Introduction Month Year	Type	Transistors (x 1000)	Technology (μm)	Clock (MHz)	Word format	Addressable/Virtual Memory
November 1971	4004	2.3	10	0.108	4-bit	640 bytes
April 1972	8008	3.5	10	0.2	8-bit	16 kB
April 1974	8080	6.0	6	2	8-bit	64 kB
March 1976	8085	6.5	3	5	8-bit	
June 1978	8086	29	3	8-10	16-bit	1 MB
June 1979	8088	29	3	8-10	16-bit	
February 1982	80286	134	1.5	6-12	16-bit	16 MB/1 GB
October 1985	Intel386 DX	275	1.5/1	16-33	32-bit	4 GB/64 TB
June 1988	Intel386 SX	275	1.5/1	16-33	32-bit	16 MB/256 GB
April 1989	Intel486 DX	1200	1/0.8	25-50	32-bit	4 GB/64 TB
October 1990	Intel386 SL	855	1	20-25	32-bit	4 GB/64 TB
April 1991	Intel486 SX	1185	1/0.8	16-33	32-bit	4 GB/64 TB
March 1992	IntelDX2	1200	0.8	50-66	32-bit	4 GB/64 TB
November 1992	Intel486 SL	1400	0.8	20-33	32-bit	
March 1994	IntelDX4	1600	0.6	75-100	32-bit	4 GB/64 TB

Motorola MC 6800 and MC 68000 family

Motorola started with the 8-bit MC6800 family (in 1974) and continued with the 16/32-bit family in 1979. The initial 8 MHz MC68000 was actually a 32-bit architecture internally, but only had a 16-bit data bus and a 24-bit address bus (with version MC68008 having reduced the data bus to 8 bits and the address bus to 20 bits). The 68010 added virtual memory support and a special loop mode – small decrement-and-branch loops could be executed from the instruction fetch buffer. The MC68020 (announced in 1984) expanded the external data and address bus to 32 bits, and had a simple 3-stage pipeline with a 256 byte I-cache. The MC68030 added 256 byte D-cache and brought the MMU onto the chip, which supported two level pages (logical and physical, rather than the segment/page mapping of the Intel 80386). The 68040 (1989/90) extended both I-cache and D-cache to 4 kB, deepened the pipeline to six stages, and added on-chip FPU. The MC 68060 (*Circello et al.* [50], 1995) expanded the design to a two-issue superscalar, like the Intel Pentium and National Semiconductor's Swordfish before it.

Zilog Z-8000 and Z-80000 family

Zilog Co. was another producer of CISC-type microprocessors. The Zilog Z-8000 was introduced not long after the Intel 8086, but had superior features. It was basically a 16-bit processor, but could address up to 23 bits in some versions. The Z-8000 was one of the first to feature two modes, one for the operating system and one for user programs. A later version, the Z-80000, was introduced at about the beginning of 1986, about the same time as the 32-bit Motorola MC68020 and Intel 80386 CPUs, though the Z-80000 was appreciably more advanced. It was fully expanded to 32 bits internally. Finally, the Z-80000 was fully pipelined (six stages), while the fully pipelined Intel 80486 and Motorola MC68040 were not introduced until 1991. There was a radiation resistant military version, and a CMOS version of the Z-80000 (the Z-320). However despite being technically advanced, the Z-8000 and Z-80000 series never met mainstream acceptance, due to initial bugs in the Z-8000 (the complex design did not use microcode – it used only 17 500 transistors) and delays in the Z-80000.

National Semiconductor NS 320xx family

The National Semiconductor CISC-type NS 320xx family consisted of a CPU which was 32-bit internally, and either 32-, 16-, or 8-bit externally. It was similar to the Motorola MC 68000 in basic features, such as byte addressing, 24-bit address bus, memory-to-memory instructions, etc. The NS 320xx also had a coprocessor bus and coprocessor instructions for MMU and a floating-point unit. The series found use mainly in embedded processor products, such as the Swordfish (introduced 1991), a two-issue superscalar microcontroller (*Hintz and Tabak* [136]).

3.5 Conclusions

With all the preceding components of RISC in place (see Chap. 1), several advantages of RISC over CISC become apparent. Optimizing compilers can be developed that improve efficiency by better utilizing the register file. Furthermore, having only simple instructions reduces the complexity and overhead which occurs when several variations of the same instruction are provided. The equal length of all instructions in RISC is advantageous for the implementation of the instruction fetch stage of a pipeline.

In contrast, the CISC trend has always been toward more complicated and feature-rich ISAs. This has been mainly due to the introduction of high-level languages and the subsequent effort to minimize the semantic gap between the HLL constructs and the machine instruction set. The reasoning behind the CISC approach is that complex instructions will execute faster if they are implemented in the microcode as opposed to in the software. In addition, it should be easier to write compilers when more high-level instructions are provided as part of the processor's ISA, and theoretically the machine programs should be shorter since more complex instructions are provided in the ISA. Besides, early CISC machines of the 1960s/1970s had to cope with extremely expensive (core) main memory. As a result, code had to be dense and hence complex, favoring complex addressing modes and encoded ISA with variable instruction lengths.

However, RISC designers found that complicated instructions are more difficult to utilize since the compiler must tune the source code to match these instructions, which are frequently more complicated than is necessary. CISC instructions can always be replaced by a sequence of simpler instructions. Sometimes the sequence of simpler instructions was executed faster than the corresponding single CISC instruction. CISC compilers use simple instructions most of the time anyway and seldom take advantage of the complex instructions. Furthermore, the long opcodes of CISC instructions, which are due to both the greater number of instructions and more complex addressing modes, tend to increase the overall size of the program. In complex designs of CISC, specialized instructions may demand the use of more complicated and time-consuming microcode for what would otherwise be a simple operation. Compared with CISC, the implementation of a RISC ISA uses much less chip space due to the elimination of the microcode control store which is necessary to implement the many instructions and addressing modes of a CISC ISA. RISC is also much easier to implement in VLSI and performs at least as well as comparable CISC chips.

For RISC to be successful, it must be able to outperform significantly comparable CISC technology while maintaining its simplicity and low price. The latest advancement in RISC has come with the ability to execute programs in 64-bit mode. 64-bit processors use an enlarged address space, 64-bit general purpose registers and internal data paths. This dramatically extends the capabilities of the CPU as far as data handling, memory management, and I/O operations. 64-bit processors deal with larger and more accurate numbers, maintain buses 64 bits wide and greater, and can access huge amounts of media space which is not likely to be exceeded for several years. 64-bit RISC processors are today's choice for high-performance workstations and server computers, while 32-bit processors are found in the PC class of machines.

While RISC has many definite advantages over CISC, the larger instruction set and the more compact machine code of CISC still have their merits. As generally occurs when two such opposing designs compete, a hybrid combination of the two emerges. It is yet to be seen what the end product will be, but even now each of the two technologies is taking components of the other and uses them to increase its performance. Principles and techniques that have been developed by the CISC approaches have become very important in today's multiple-issue (RISC) processors (see Chap. 4). Such a principle is out-of-order execution which allows instructions to complete out of the original program order.

Modern microprocessors use ideas from RISC and CISC approaches. If unhindered by a legacy CISC ISA, RISC ISA principles are applied such as a load/store architecture, fixed instruction length, and simple addressing modes only. Multiple-issue and dynamic (out-of-order) scheduling are crucial techniques to keep the various functional units of contemporary microprocessors busy. We have demonstrated that out-of-order execution is not a new concept – it existed in the mid-1960s in CISC machines CDC 6600 (which used scoreboarding) and IBM System/360 Model 91 (which used Tomasulo's scheme) – but it was innovative for single-chip microprocessors in the mid-1990s. We remark that an out-of-order scheduling implementation is quite similar to dataflow architecture (*Brehob et al.* [37]). It is referred to as *micro dataflow* by microprocessor researchers. However, the two major problems of dataflow implementations, that is token matching and handling complex data structures, are trivialized in the restricted internal environment of a microprocessor. Namely, determining which instruction can be executed (matching) is easy since all operands are within registers of a microprocessor. Since the core of a microprocessor deals with only simple data types, the problem of handling complex data structures does not exist.

4. Multiple-Issue Processors

What is the limitation of a multiple-issue approach? If we can issue five operations per clock cycle, why not 50? Limits on available instruction level parallelism are the simplest and most fundamental ...
... What is clear is that some level of multiple issue is here to stay and will be included in all processors in the foreseeable future.

John L. Hennessy and David A. Patterson
Computer Architecture A Quantitative Approach
(Morgan Kaufmann Publishers, 1996)

4.1 Overview of Multiple-Issue Processors

Superscalar processors started to conquer the microprocessor market at the beginning of the 1990s with dual-issue processors. The principal motivation was to overcome the single issue of scalar RISC processors by providing the facility to fetch, decode, issue, execute, and write back results of more than one instruction per cycle. In fact, the first commercially successful superscalar microprocessor was the Intel i960 RISC processor which hit the market in 1990. Further first generation dual-issue superscalar RISC processors were the Motorola 88110, Alpha 21064, and the HP PA-7100.[1] Other superscalar RISC processors of the mid-1990s era were the IBM POWER2 RISC System/6000 processor, its offspring PowerPC 601, 603, 604, 750 (G3), and 620, the DEC Alpha 21164, the Sun SuperSPARC and UltraSPARC, the HP PA-8000, and the MIPS R10000. Today's superscalar RISC processors MIPS R12000, HP PA-8500, Sun UltraSPARC-II, IIi and III, Alpha 21264, IBM POWER2-Super-Chip (P2SC) are 4-issue or 6-issue processors.

The commercially dominating Intel line of superscalar microprocessors continued the legacy Intel x86 ISA with the dual-issue Pentium processor

[1] The Intel i860 of 1989 was not superscalar, it was rather a special kind of VLIW. A dual-instruction mode (sometimes called superscalar mode at that time) allowed the execution of two instructions simultaneously. However, dual-instruction mode instructions are marked using a bit in the instruction word by the compiler. In combination with dual-operation instructions, up to three operations were executed simultaneously per cycle.

of 1993, the Pentium Pro, the Pentium II and its newer offspring Celeron, Klamath, and Pentium III (Katmai). Because of their ISA features these processors are viewed as CISC microprocessors. A number of companies designed Intel-compatible processors like AMD with its K5, K6, K6-2, and K6-3 processors, and Cyrix with its 6x86, M II, and MXi. These CISC microprocessors feature a slightly more complex pipeline than the superscalar RISC processors with additional stages that generate so-called RISC86ops or µops from x86 instructions. All these x86-based microprocessors feature a 32-bit architecture targeted for use in personal computers, while the newer ones of the superscalar RISCs are 64-bit architecture machines and mainly used in servers.[2] Intel announced its first 64-bit processor with the Merced (P7) processor.

Multiple-issue comprises superscalar and the even older, but previously not very successful, *very long instruction word* (VLIW) technique that now has a strong renaissance in the area of signal processors, multimedia processors, and with the Hewlett-Packard and Intel's *explicitly parallel instruction computing* (EPIC) instruction format that has been proposed for the Merced in the general-purpose processor field.

Components of a state-of-the-art superscalar processor. Let us first look at the principal components of a superscalar processor as shown in Fig. 4.1.

Such a superscalar RISC microprocessor features a load/store architecture with a fixed instruction format of 32-bit instruction length. The processor consists of an instruction fetch unit, an instruction decode and register rename unit, an issue unit, several independently executing functional units (FUs), a retire unit, 32 general-purpose registers, 32 floating-point registers, additional rename registers, separate I-cache and D-cache that are connected via a bus interface unit with the external memory bus or an external secondary cache unit, and additional internal buffers like the instruction buffer and a reorder buffer. The FUs usually comprise:

- A load/store unit that loads a data value from the D-cache into one of the general-purpose or floating-point registers, or, vice versa, stores a value from a register to the D-cache. Generation of load and store addresses is supported by a memory management unit (MMU) that comprises a translation look-aside buffer (TLB) to translate logical addresses into physical addresses. In the case of a D-cache miss, a new cache line is automatically loaded via the bus interface unit, while the load or store operation that triggered the miss is stalled. State-of-the-art non-blocking caches allow the execution of succeeding load and store operations that are independent of the missing cache line. Moreover, typically load/store units allow loads to overtake stores if the data addresses are different.
- One or more floating-point units perform floating-point operations loading their operands from floating-point registers. The floating-point unit

[2] PowerPC 601, 603, 604, and 750 are also 32-bit processors.

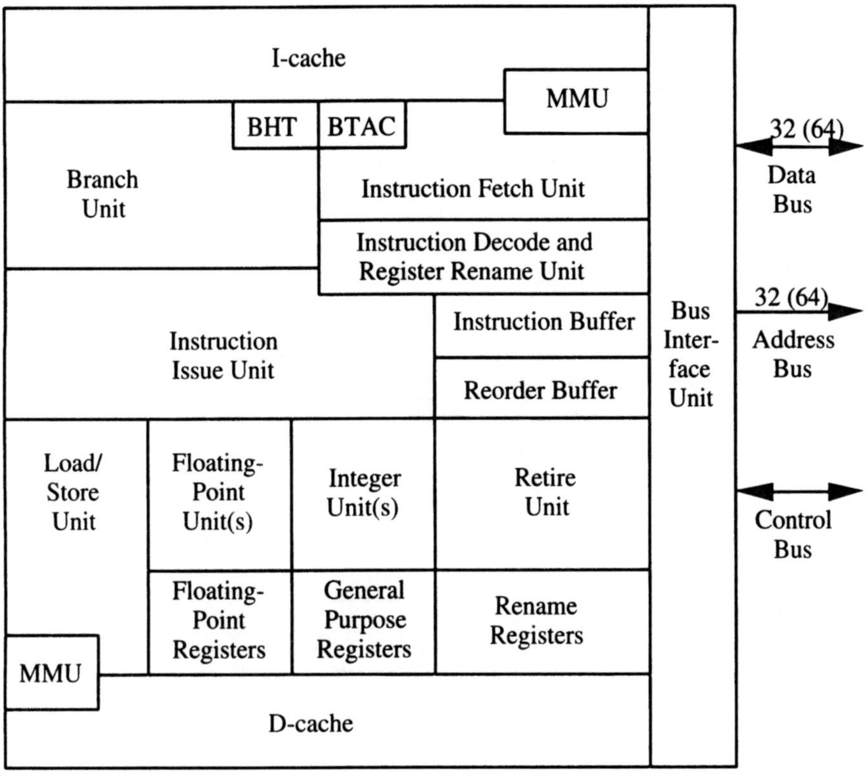

Fig. 4.1. Components of a superscalar processor

is pipelined with a 3-stage pipeline with latency 3 and throughput of 1. Sometimes more complex operations like a combined multiply-add remain longer in the pipeline and lead to less throughput.

- One or more integer units that execute the arithmetic and logical instructions on general-purpose register values. Depending on the complexity of the operation, integer units can be single-stage units with latency of 1 or, for example, 3-stage pipelined units with latency 3 and throughput 1. Sometimes division or square root units exist that fall out of the scheme by not being pipelined, due to long latencies of 17 and more cycles.
- A multimedia unit in state-of-the-art processors performs several arithmetic, masking, selection, reordering, and conversion instructions on 8-bit, 16-bit or 32-bit values in parallel, accessing either the integer or the floating-point registers.
- A branch unit controls execution of branch instructions. After fetching a branch instruction, the branch target may be unknown for several cycles. In such a situation, when it is not yet known whether the branch will be taken or not, a speculative fetch, decode, and execution of instructions is performed using a static or dynamic branch prediction technique. Today's

superscalar processors usually employ dynamic branch prediction based on the history of previous executions of the program paths. A branch target address cache (BTAC) contains jump and branch target addresses and a branch history table (BHT) monitors previous branch outcomes. The task of the branch unit is to determine the branch outcome, monitor branch history, and reroll speculatively executed instructions in the case of a mispredicted branch.

The type and number of FUs vary depending on the specific processor.

Superscalar processor pipeline. A superscalar pipeline features the same stages as a simple RISC pipeline, but several instructions are fetched, decoded, and executed, and several results are written back, simultaneously. Moreover, additional issue and retire stages are necessary, and additional buffers decouple pipeline stages (see Fig. 4.2).

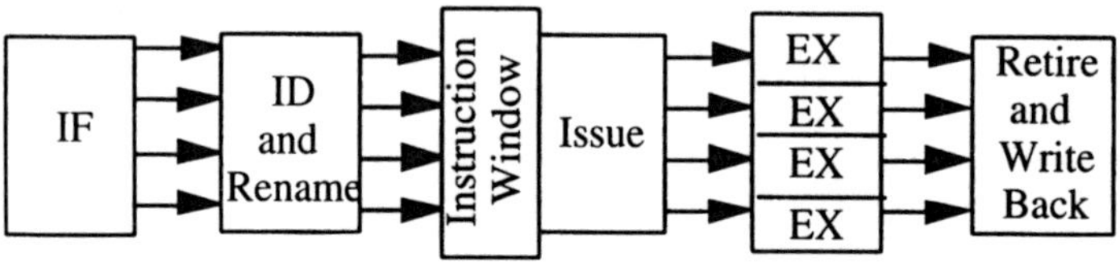

Fig. 4.2. Superscalar pipeline

The pipelining starts with the instruction fetch (IF) stage that fetches several instructions from the I-cache into a fetch buffer. Typically at least as many instructions as the maximum issue rate are fetched at once. To avoid pipeline interlocking due to jump or branch instructions, the BTAC contains the jump and branch target addresses that are used to fetch instructions from the target address. The fetch buffer decouples the fetch stage from the decode stage.

In the instruction decode (ID) stage, a number of instructions are decoded (typically as many as the maximum-issue bandwidth). The operand and result registers are renamed, i.e., available physical registers are assigned to the architectural registers specified in the instructions. Then the instructions are placed in an instruction buffer, often called the *instruction window*. Instructions in the instruction window are free from control dependences due to branch prediction, and free from name dependences due to register renaming. So, only data dependences and structural conflicts remain to be solved.

The issue logic examines the waiting instructions in the instruction window and simultaneously assigns ("issues") a number of instructions to the FUs up to a maximum-issue bandwidth. The program order of the issued instructions is stored in the reorder buffer. Instruction issue from the instruction window can be in order (only in program order) or it can be out of order. It can be either subject to simultaneous data dependences and resource constraints, or divided into two (or more) stages, checking structural conflict in the first and data dependences in the next stage (or vice versa). In the case of structural conflicts first, the instructions are issued to reservation stations (buffers) in front of the FUs where the issued instructions await missing operands. In contrast to the Tomasulo algorithm (see Sect. 3.3.2), several instructions can be issued simultaneously when space is available in the respective reservation stations. Depending on the specific processor, reservation stations can be central to a number of FUs (see, for example, the Intel Pentium II), or each FU has one or more of its own reservation stations (see IBM/Motorola/Apple PowerPC 604). In the latter case, a structural conflict arises if more than one instruction is issued to the reservation stations of the same FU simultaneously. In this case only one instruction can be issued within a cycle.

The instructions await their operands in the reservation stations, as in the Tomasulo algorithm. An instruction is then said to be *dispatched* from a reservation station to the FU when all operands are available, and execution starts. The dispatch sends operands to the execution unit.[3] If all its operands are available during issue and the FU is not busy, an instruction is immediately dispatched, starting execution in the cycle following issue. Thus, the dispatch is usually not a pipeline stage. An issued instruction may stay in the reservation station for zero to several cycles. Dispatch and execution are performed out of program order.

When the FU finishes the execution of an instruction and the result is ready for forwarding and buffering, the instruction is said to *complete*. Instruction completion is out of program order. During completion the reservation station is freed and the state of the execution is noted in the reorder buffer. The state of the reorder buffer entry can denote an interrupt occurrence. The instruction can be completed and still be speculatively assigned, which is also monitored in the reorder buffer.

After completion, operations are committed in order. An instruction can be *committed*:

- if instruction execution is complete,
- if all previous instructions due to the program order are already committed or can be committed in the same cycle,
- if no interrupt occurred before and during instruction execution, and
- if the instruction is no longer on a speculative path.

By or after commitment, the result of an instruction is made permanent in the architectural register set, usually by writing the result back from the rename register to the architectural register. This is often done in a stage of its own, after the commitment of the instruction, with the effect that the rename register is freed one cycle after commitment.

[3] In the literature, the meanings of the terms *dispatch* and *issue* are often interchanged or even indistinguishable.

If an interrupt occurred, all instructions that were in program order before the interrupt-signaling instruction are committed, and all later instructions are removed. Thus, a so-called *precise exception* is guaranteed. Precise exception means that all instructions before the faulting instruction are committed and those after it can be restarted from scratch. Depending on the architecture and on the type of exception, the faulting instruction should be committed or removed without any lasting effect.

We use the term *retired*, in conformity with *Shriver and Smith* [258], when the reorder buffer slot of an instruction is freed either because the instruction commits (the result is made permanent) or because the instruction is removed (without making permanent changes).

Superscalar. The term *superscalar* was first time coined by *Agerwala and Cocke* [4], here cited after *Diefendorff and Allen* [65]: *Superscalar machines are distinguished by their ability to (dynamically) issue*[4] *multiple instructions each clock cycle from a conventional linear instruction stream.*

The meaning of superscalar can be explained as follows:

- Instructions are issued from a sequential stream of "normal" instructions.
- The instructions that are issued are scheduled dynamically by the hardware.
- More than one instruction can be issued each cycle (motivating the term "superscalar" instead of "scalar").
- The number of issued instructions is determined dynamically by hardware, that is, the actual number of instructions issued in a single cycle can be zero up to a maximum instruction issue bandwidth.
- The dynamic instruction issue complicates the hardware scheduler of a superscalar processor. The scheduler complexity increases when multiple instructions are issued out of order from a large instruction window.
- It is a presumption that multiple FUs are available. The number of available FUs is at least the maximum-issue bandwidth, but often higher to diminish potential resource conflicts.
- One important point is that the superscalar technique is a *microarchitecture* technique, not an architecture technique. Recall from Sect. 1.2 that the architecture of a processor is defined as the ISA, i.e., everything that is seen outside of a processor. In contrast, the microarchitecture comprises implementation techniques. Code that is generated for a scalar microprocessor can also be executed on a superscalar microprocessor of the same architecture, and vice versa. This is the case for the scalar microSPARC-II and the superscalar SuperSPARC and UltraSPARC processors.

The term superscalar is often used in a less precise fashion to describe a processor with multiple parallel pipelines or a processor with multiple FUs.[5]

[4] The term *issue* is used here instead of the term *dispatch* in the original definition.

[5] In 1991, *Johnson* defined superscalar as follows [150]: A superscalar processor reduces the average number of cycles per instruction beyond what is possible in

Both definitions do not allow superscalar to be distinguished from VLIW (see Sect. 4.10).

The ability to execute instructions out of order partitions a superscalar pipeline into three distinct sections:

- an *in-order section* with the instruction fetch, decode, and rename stages – the issue is also part of the in-order section (in the case of an in-order issue),
- an *out-of-order section* starting with the issue in the case of an out-of-order issue processor, the execution stage, and usually the completion stage,
- and again an *in-order section* that comprises the retirement and write-back stages.

Another aspect of superscalar is that instruction pipelining and superscalar techniques both exploit fine-grain (instruction-level) parallelism. While pipelining (see also superpipelining in Sect. 1.7.4) utilizes "temporal" parallelism, the superscalar technique also utilizes "spatial" parallelism. Performance can be increased by temporal parallelism, using longer pipelines (deeper pipelining) and faster transistors (a faster clock). Provided that enough fine-grain parallelism is available, performance can also be increased by spatial parallelism using more FUs and a higher issue bandwidth by applying more transistors in the superscalar case.

4.2 I-Cache Access and Instruction Fetch

A so-called *Harvard architecture* (separate instruction and data memory and access paths) is used internally in a high-performance microprocessor with separate on-chip primary I-cache and D-cache. The I-cache is less complicated to control than the D-cache, because the I-cache is read-only[6], and it is not subjected to cache coherence in contrast to the D-cache. Typically the primary I-cache consists of 8–32 kB cache size, organized as direct-mapped or 2-way set-associative in 32-byte cache lines holding eight 32-bit instructions each. If an instruction format like the x86 ISA allows variable length instructions, the fetch block contains a varying number of instructions and the beginning of the instructions is yet to be determined – this complicates instruction decode and needs more sophisticated instruction fetch techniques.

Sometimes the instructions in the I-cache are predecoded (see also Sect. 4.4) on their way from the memory interface to the I-cache to simplify the decode stage (e.g., due to predecoding in the PowerPC 620 the

a pipelined, scalar RISC processor by allowing concurrent execution of instructions *in the same* pipeline stage, as well as concurrent execution of instructions in different pipeline stages. The term superscalar emphasizes multiple, concurrent operations on scalar quantities, as distinguished from multiple, concurrent operations on vectors or arrays as is common in scientific computing.

[6] Self-modifying code is usually not, or at least not efficiently, supported in today's processors.

decode, rename and issue can be done in a single pipeline stage, instead of two separate stages in the PowerPC 604).

The main problem of instruction fetching is the control transfer performed by jump, branch, call, return, and interrupt instructions. The sequential addressing of instructions by the PC is disrupted. Moreover, the disruption may occur in the middle or shortly after the beginning of a fetched instruction block, rendering all fetched instructions after the disruption useless. *Wallace and Bagherzadeh* [319] show that an 8-issue superscalar processor with simple fetching hardware could only expect to fetch less than four usable instructions per cycle with programs of the SPECint95 benchmark suite.

A straightforward technique for simple instruction fetch from the I-cache is to fetch as many instructions per portion as the cache line size. However, if the starting PC address is not the address of the cache line, fewer instructions than the fetch width are returned. As with all fetching techniques, if there is a control transfer instruction, then instructions after it are invalidated.

If the cache line size is extended beyond the width of the fetch block, the number of instructions that will be lost when fetching after a control transfer instruction with an unaligned target address is reduced.

However, the problem with target instruction addresses that are not aligned to the cache line addresses can be solved completely in hardware using a *self-aligned instruction cache*. Such an I-cache reads and concatenates two consecutive lines within one cycle to be able always to return the full fetch bandwidth. A self-aligned I-cache can be implemented either by use of a dual-port I-cache, by performing two separate cache accesses in a single cycle, or by a two-banked I-cache. Using a two-banked I-cache is preferred for both space and timing reasons (*Wallace and Bagherzadeh* [319]).

All these techniques can be used in conjunction with instruction prefetching. Prefetching improves the instruction fetch performance, but fetching is still limited because instructions after a control transfer must be invalidated. Here instruction fetch prediction helps to determine the next instructions to be fetched from the memory subsystem. Instruction fetch prediction is applied in conjunction with branch prediction which foretells the outcome of conditional branch instructions.

A multiple cache lines fetch from different locations may be needed in future, very wide-issue processors where more than one branch will often be contained in a single contiguous fetch block. It may also be useful to support eager execution of both sides of a branch or to support multithreaded processors.

4.3 Dynamic Branch Prediction and Control Speculation

Excellent branch handling techniques are essential for current and future microprocessors. Many instructions are in different stages in the pipeline of a wide-issue superscalar processor. Instruction issue also works best with a large instruction window, leading to even more instructions that are "in flight" in the pipeline. However, approximately every seventh instruction in an instruction stream is a branch instruction which potentially interrupts the instruction flow through the pipeline.

The task of high performance branch handling consists of the following requirements:

- an early determination of the branch outcome (the so-called branch resolution),
- buffering of the branch target address in a BTAC after its first calculation and an immediate reload of the PC after a BTAC match,
- an excellent branch predictor (i.e., branch prediction technique) and speculative execution mechanism,
- often another branch is predicted while a previous branch is still unresolved, so the processor must be able to pursue two or more speculation levels,
- and an efficient rerolling mechanism when a branch is mispredicted (minimizing the branch misprediction penalty).

An early branch resolution is supported by forwarding as soon as possible to the branch instruction the results of compare instructions that may be stored in a general-purpose register or in a special condition-code register. Branch testing could be moved forward in the pipeline as far as the ID stage, as was demonstrated in Sect. 1.6.3. Previous calculations of branch target addresses are cached in a BTAC (see Sect. 1.6.3 and the next section) and accessed during the IF stage.

The performance of branch prediction depends on the prediction accuracy and the cost of misprediction. Prediction accuracy can be improved by inventing better branch predictors. In Sect. 1.6.3 we have already seen some static branch prediction techniques. An alternative to static prediction is dynamic branch prediction which usually has superior performance.

When a branch is not predicted correctly, there is rarely a penalty of less than two cycles, even in simple RISC pipelines. However, the misprediction penalty depends on many organizational features: the pipeline length (favoring shorter over longer pipelines), the overall organization of the pipeline, whether misspeculated instructions can be removed from internal buffers, or have to be executed and can only be removed in the retire stage. Further dynamic aspects that influence the misprediction penalty are the number of speculative instructions in the instruction window or the reorder buffer. Typically only a limited number of instructions can be removed each cycle. Therefore, rerolling when a branch is mispredicted is typically expensive, for example, 11 or more cycles in the Pentium II or the Alpha 21264 processors. The high misprediction penalty in current and prospective future microprocessors shows the importance of excellent branch prediction mechanisms for the overall performance of a processor.

Other techniques to handle branches are predication using so-called predicated or conditional instructions that allow the removal of the branch from the instruction flow, and eager execution of both branch sides (see Sect. 4.3.4).

Eager execution is especially effective when the branch direction changes in an irregular fashion which means the branch is not predictable. In that case the expensive rerolling mechanism slows down execution. However, eager execution is not possible with today's superscalar processors because the ability to pursue two instruction streams in parallel is necessary.

4.3.1 Branch-Target Buffer or Branch-Target Address Cache

The branch target address is needed at the same time as the prediction. In particular, it should be known already in the IF stage whether the as-yet-undecoded instruction is a (conditional or unconditional) branch to allow an instruction fetch at the target address in the next cycle. The *branch-target buffer* (BTB) or *branch-target (address) cache* (BTAC) is a branch-prediction cache that stores the predicted address for the next instruction after a branch (*Lee and Smith* [175]). The BTB is accessed during the IF stage. It consists of a table with branch addresses, the corresponding target addresses, and prediction information (see Fig. 4.3 for a simple BTB). The PC for the next instruction to fetch is compared with the entries in the BTB. If a matching entry is found in the BTB, fetching can start immediately at the target address.

The BTB stores branch and jump target addresses. Branch target addresses are predicted addresses, while jump target addresses always transfer control. Jumps (unconditional branches) are usually much less frequent than conditional branches.

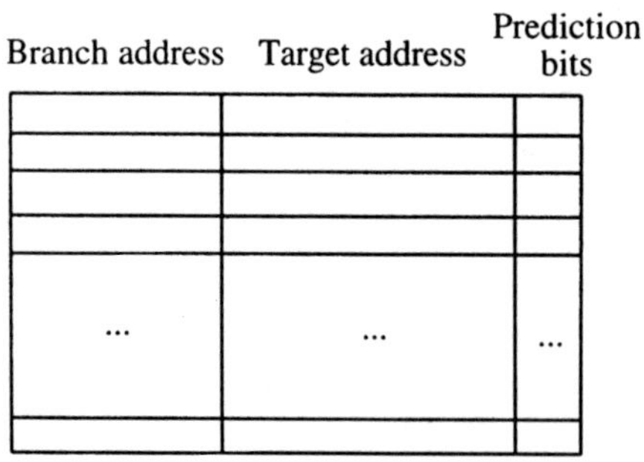

Fig. 4.3. Branch-target buffer

Fetching instructions from a new target address is fast if the fetch address hits in the I-cache. A variation of the BTB that was popular for older processors without on-chip I-caches is to store one or more target instructions additionally to the target address. Such a BTB is often called a *branch-target cache* (BTC) instead of a *branch-target (address) cache* (BTAC).

Moreover, for procedure calls and returns a small stack of return addresses is often used in addition to, and independent of, a BTB. Such a *return address stack* (RAS) appears, for example, in the Alpha 21164 organized as a 12-entry circular buffer that makes the last 12 return addresses available.

4.3.2 Static Branch Prediction Techniques

Static branch prediction is a simple prediction technique which either always uses a fixed prediction direction or allows the compiler to determine the prediction direction. The prediction direction of a branch instruction is never changed.

Simple hardware-fixed direction mechanisms can be:

- *Predict always not taken*: This is the simplest scheme because the assumption is a straight instruction flow. Unfortunately, due to frequent loops in an instruction flow, this technique is not very effective. This prediction technique should not be confused with the delayed branch technique (see Sect. 1.6.3). The instruction in the delay slot is always executed, while the predict-not-taken-technique executes the instructions after the branch speculatively and squashes the instruction execution in the case of misprediction.
- *Predict always taken*: Here branches at the end of a loop iteration are correctly predicted as long as the loop loops. The branch target address has to be stored within the instruction fetch unit to allow a zero delay.
- *Backward branch predict taken, forward branch predict not taken*: Here the idea is that branches with branch target addresses pointing backwards stem from loops and should be predicted taken, while other kind of branches are preferably not taken.

Sometimes a bit in the branch opcode allows the compiler to decide the prediction direction either directly (bit set means "predict taken", bit not set means "predict not taken") or by reversing the hardware-determined direction.

The compiler may use several techniques for a good compiler-based static prediction. It may either:

- examine the program structure for prediction (branches at the end loop iteration code should be predicted as taken, if-then branches predicted as not taken),
- relegate prediction to the programmer by compiler directives, or
- use a profile-based prediction by predicting the branch directions based on prior runs of the program with recording of the branch behavior.

The profile-based prediction is nearly always better than the simpler direction-based predictions.

4.3.3 Dynamic Branch Prediction Techniques

In a dynamic branch prediction scheme, prediction is decided on the computation history of the program. After a start-up phase of the program execution, where a static branch prediction might be effective, historical information is gathered and dynamic branch prediction becomes more effective. In general, dynamic gives better results than static branch prediction, but at the cost of increased hardware complexity.

One-bit predictor. The simplest dynamic branch prediction scheme is a simple *branch prediction buffer* or *branch history table* (BHT). The BHT is a small buffer memory containing branch addresses indexed by the lower bits of the address of a branch instruction. Each entry of the BHT contains one bit that indicates whether the branch was recently taken or not. If the bit is set, the branch is predicted taken. If the bit is not set, the branch is predicted not taken. In the case of a misprediction, the bit state is reversed and so is the prediction direction.

One-bit predictors can also be implemented in the BTB by only storing the target addresses of predicted taken branches.

The prediction states of a one-bit predictor are shown in Fig. 4.4 (T stands for taken and NT stands for not taken).

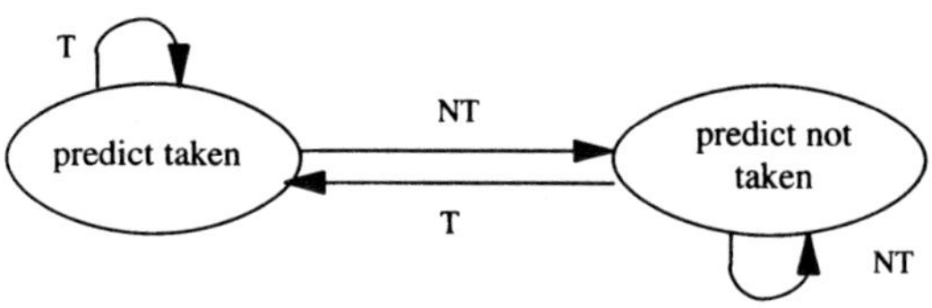

Fig. 4.4. One-bit predictor states

Such a one-bit predictor correctly predicts a branch at the end of a loop iteration, as long as the loop does not exit. However, in nested loops, a one-bit prediction scheme will cause two mispredictions for the inner loop: one at the end of the loop, when the iteration exits the loop instead of looping again, and one when executing the first loop iteration, when it predicts exit instead of looping. Such a double misprediction in nested loops is avoided by a two-bit predictor scheme.

Two-bit predictors. In a two-bit prediction scheme two bits instead of one are assigned to each entry in the BHT. The two bits stand for the prediction states "predict strongly taken", "predict weakly taken", "predict strongly not taken", "predict weakly not taken". In the case of a misprediction in the "strongly" state cases, the prediction direction is not changed, rather the prediction goes into the respective "weakly" state. A prediction must miss twice before it is changed when a two-bit prediction scheme is applied.

Two kinds of two-bit prediction schemes are used: the saturation up-down counter scheme demonstrated in Fig. 4.5 and the scheme given in Fig. 4.6.

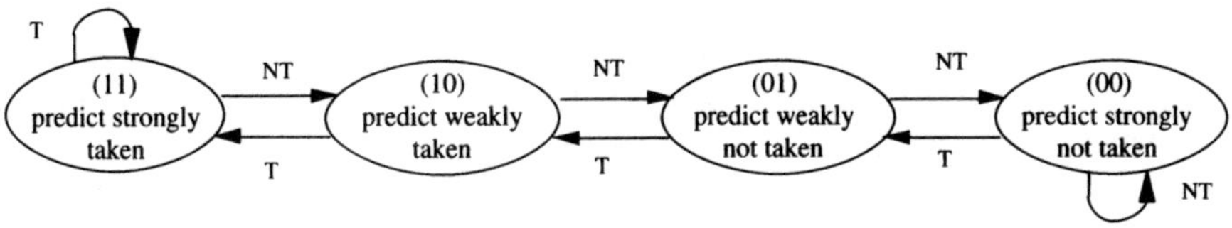

Fig. 4.5. Two-bit predictor saturation counter states

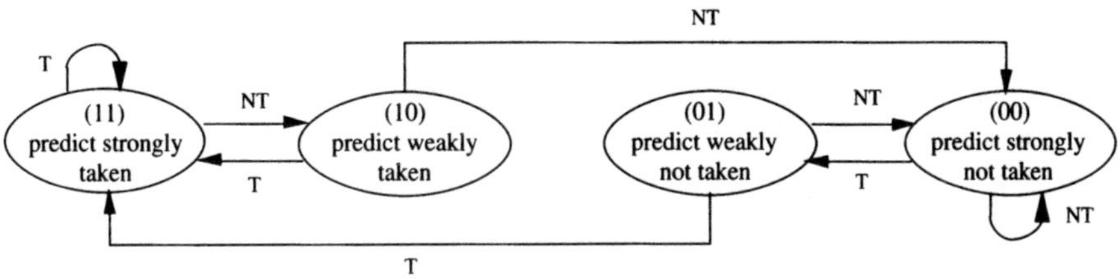

Fig. 4.6. Two-bit predictor states

In the two-bit saturation up-down counter scheme, the counter is incremented for each taken branch occurrence, and decremented each time the branch is not taken. The counter is saturating, i.e., it is not decremented past 0, nor is it incremented past 3. The most significant bit determines the prediction.

The other scheme given in Fig. 4.6 differs from the saturation up-down counter scheme by changing directly from the "weakly" to the "strongly" states in the case of a second misprediction. This scheme is applied in the UltraSPARC-I processor. Branches without prediction are initialized by the UltraSPARC-I processor to "predict weakly not taken" (*Tremblay and O'Connor* [301]).

Hennessy and Patterson [134] showed that the mispredictions of SPEC89 programs vary from 1 % (`nasa7`, `tomcatv`) to 18 % (`eqntott`), with `spice` at 9 % and `gcc` at 12 %, assuming a 4 096-entry BHT.

The two-bit prediction scheme is extendable to a n-bit scheme. However, studies have shown that a two-bit prediction scheme does almost as well as a n-bit scheme with $n > 2$.

Two-bit predictors can be implemented in the BTB, assigning two state bits to each entry in the BTB. Another solution, which is proposed for the PowerPC 604 and 620, is to use a BTB for target addresses and a BHT as a separate prediction buffer. While the BTB is accessed in the IF stage, the BHT prediction is performed in the PowerPC 604 and 620 one cycle later in the ID stage and may override the previous BTB prediction.

A mispredict in the BHT occurs for two reasons: either a wrong guess for that branch, or the branch history of a wrong branch is used because of the way the table is indexed. In an indexed table lookup, part of the instruction address is used as an "index" to identify a table entry. Instruction addresses with the same bit pattern used as an index share the same table entry, leading to frequent mispredicts if the table is small.

Two-bit predictors work well for scientific floating-point intensive programs which contain many frequently executed loop-control branches. Shortcomings of the two-bit prediction schemes arise from dependent (correlated) branches, which are frequent in integer-dominated programs.

The following example of two branches, one dependent on the other, demonstrates that one-bit and two-bit predictors can potentially mispredict every time. Let us look at the following program (see [134]):

```
if (d == 0)    /* branch b1 */
    d = 1;
if (d == 1)    /* branch b2 */
    ...
```

In assembly language notation the program can be given as follows (variable d is assigned to register $R1$):

```
     bnez R1,L1      ; branch b1 (d ≠ 0)
     addi R1,R0,#1   ; d == 0, so d = 1
L1:  subi R3,R1,#1
     bnez R3,L2      ; branch b2 (d ≠ 0)
     ...
L2:  ...
```

Consider a sequence where d alternates between 0 and 2 which generates a sequence of NT-T-NT-T-NT-T for branches b1 and b2. The execution behavior is given in the following table:

initial d	$d \stackrel{?}{==} 0$	b1	d before b2	$d \stackrel{?}{==} 1$	b2
0	yes	NT	1	yes	NT
2	no	T	2	no	T

If we apply a one-bit predictor which is initialized to "predict taken" for branches b1 and b2, then every branch is mispredicted. The same behavior is shown for the two-bit predictor of Fig. 4.5 starting from the state "predict weakly taken". The two-bit predictor of Fig. 4.6 mispredicts every second branch execution of b1 and b2. A (1,1)-correlating predictor (see below) can take advantage of the correlation between the two branches; it mispredicts only in the first iteration when $d = 2$.

Correlating branch predictors usually reach higher prediction rates for integer-intensive programs than the two-bit predictor scheme and require only a small increase in hardware cost.

Correlation-based predictors. The two-bit predictor scheme only uses the recent behavior of a single branch to predict the future of that branch. Correlations between different branch instructions are not taken into account. Let us also look at the recent behavior of other branches rather than just the branch we are trying to predict.

The so-called *correlation-based predictors* or *correlating predictors* developed by *Pan et al.* [223] are branch predictors that additionally use the behavior of other branches to make a prediction. While two-bit predictors use self-history only, the correlating predictor uses neighbor history as well. Many integer workloads feature complex control-flows whereby the outcome of a branch is affected by the outcomes of recently executed branches. In other words, the branches are correlated [223].

A correlation-based predictor denoted as an *(m,n)-correlation-based predictor*, or in short an *(m,n)-predictor*, uses the behavior of the last m branches to choose from 2^m branch predictors, each of which is a n-bit predictor for a single branch. The global history of the most recent m branches can be recorded in a m-bit shift register – called a *branch history register* (BHR) – where each bit records whether the branch was taken or not taken. Each time a branch in execution resolves, its sign bit is shifted into the BHR. The contents of the BHR are used to address (index) the entries in a so-called *pattern history table* (PHT).[7] Typically two-bit predictors are used in PHTs.

A (1,1)-predictor uses the behavior of the last branch to choose between a pair of one-bit predictors, and a correlation-based predictor denoted as a (2,2)-predictor uses a BHR of two bits to choose among four 2-bit prediction tables. A two-bit predictor (without global history) can simply be denoted as a (0,2)-predictor.

Figure 4.7 shows the implementation of correlation-based predictor, a type (2,2)-predictor with four 1 k-entry PHTs. The BHR bit pattern selects the specific PHT. The entries of the 1 k-entry PHTs are generally accessed by using the lower order 10 bits of the branch address. Depending on the implementation, the PHTs may alternatively be accessed using 10 bits of the address of the instruction immediately prior to the branch under consideration (*Pan et al.* [223]). The four 1 k-entry PHTs can also be viewed as a single 4 k-entry PHT. Then 12 bits are required for the PHT lookup. Therefore, two bits from the BHR are concatenated with 10 bits from the branch address.

[7] *Pan et al.* used the terms "branch prediction table" (BPT) instead of "pattern history table" (PHT), and "m-bit shift register" instead of "branch history register" (BHR).

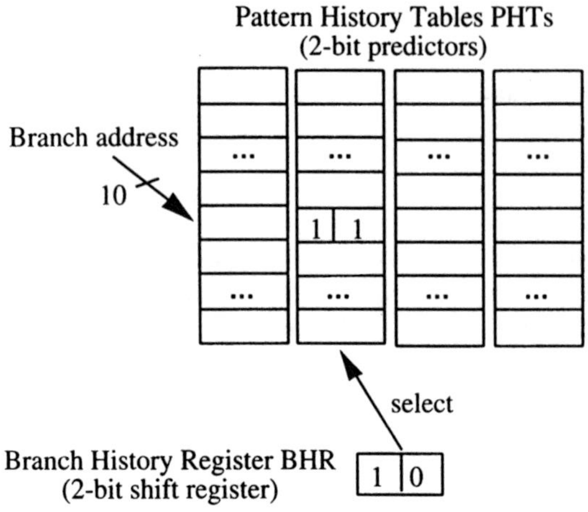

Fig. 4.7. Implementation of a (2,2)-predictor

Two-level adaptive predictors. The two-level adaptive predictor was developed by *Yeh and Patt* [331] at the same time as the closely related correlation-based prediction scheme. There are several variations of the two-level adaptive prediction scheme (*Yeh and Patt* [332]).

The basic two-level predictor uses a single "global" branch history register (BHR) of k bits to index in a pattern history table (PHT) of 2-bit counters. The global BHR is updated with outcomes from all branches. Thus, not only the history of a branch, but also the history of other branches, influence the prediction of the branch. All schemes that use a single global BHR are called *global history schemes* and correspond to *Pan et al.*'s correlation-based predictor schemes.

In the simplest case there is a single global BHR (denoted G) and a single global PHT (denoted g), this simple predictor is called GAg (A stands for "adaptive"). All PHT implementations of *Yeh and Patt* use 2-bit predictors. An implementation of a GAg-predictor with a 4-bit BHR length (therefore, also denoted as GAg(4)) is shown in Fig. 4.8.[8] The BHR is implemented as a simple shift register shifting right to left with the sign (1 for branch taken, 0 for branch not taken) of the last resolved branch at the rightmost bit position.

In the GAg predictor scheme the PHT lookup depends entirely on the bit pattern in the BHR and is completely independent of the branch address. The advantages of the "degenerate" GAg scheme are its simple implementation and the fact that the predicted outcome of a branch can be known long before the execution of that branch [223].

A simple GAg(k)-predictor often performs better on integer programs than a 2-bit-predictor (with a saturation up-down counter scheme).

[8] The GAg scheme is called the "degenerate case" of the correlation scheme by *Pan et al.* [223].

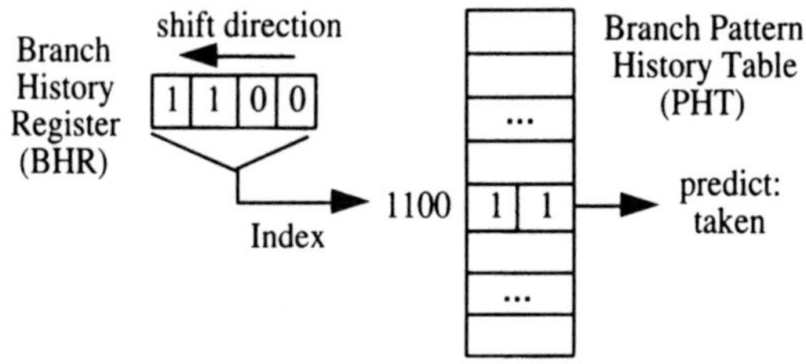

Fig. 4.8. Implementation of a GAg(4)-predictor

However, GAg-predictors still suffer from branch patterns that emerge several times within a computation. Two code sequences may have the same bit pattern in the BHR and thus index the same pattern in the PHT. Since the branch behavior of the two code sequences may differ, the shared pattern may lead to the wrong predictions.

Such wrong predictions can be restrained by additionally using:

- the (full) branch address to distinguish multiple PHTs (called per-address PHTs);
- a subset of branches (e.g., defined by part of the branch address) to distinguish multiple PHTs (called per-set PHTs);
- the (full) branch address to distinguish multiple BHRs (called per-address BHRs);
- a subset of branches to distinguish multiple BHRs (called per-set BHRs); or
- a combination scheme.

In the first two cases, a single global BHR is combined with multiple per-address selected PHTs, denoted as GAp, or with multiple per-set addressed PHTs, denoted as GAs. A GAp predictor with a 4-bit BHR, denoted as GAp(4), is shown in Fig. 4.9, and a GAs predictor with a 4-bit BHR, denoted as GAs(4,2^n), is shown in Fig. 4.10. In the GAs(4,2^n) predictor n bits of the branch address are used to define 2^n different branch sets corresponding to 2^n PHTs with 2^4 entries each. Branches of the same branch set share the same PHT in a GAs predictor.

The three two-level adaptive predictors GAg, GAp, and GAs use a single global BHR and together form the *global history scheme* predictors. These predictors are closely related to the correlation-based predictor.

In fact, by rotating Fig. 4.7 90 degrees to the right and assuming a 4-bit BHR, it can be seen that a correlation-based (4,2)-predictor is equivalent to a GAs(4) predictor, assuming $n = 10$ bits in the branch address (compare with Fig. 4.10).

A second scheme class is defined as the *per-address history schemes* where the first-level branch history refers to the last k occurrences of the same branch instruction (using self-history only!). Therefore, a BHR is associated with each branch instruction to distinguish the branch history information of

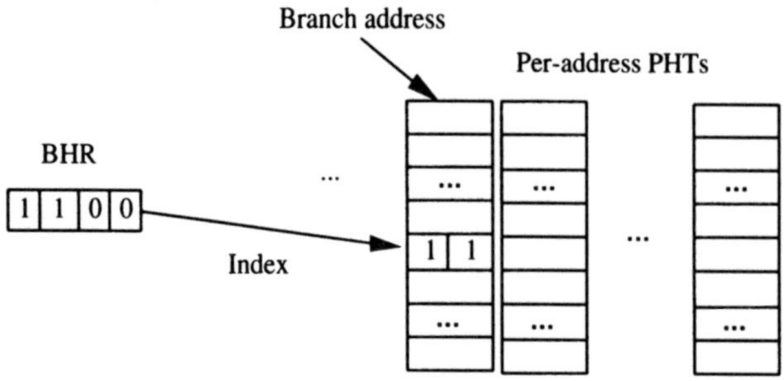

Fig. 4.9. Implementation of a GAp(4) predictor

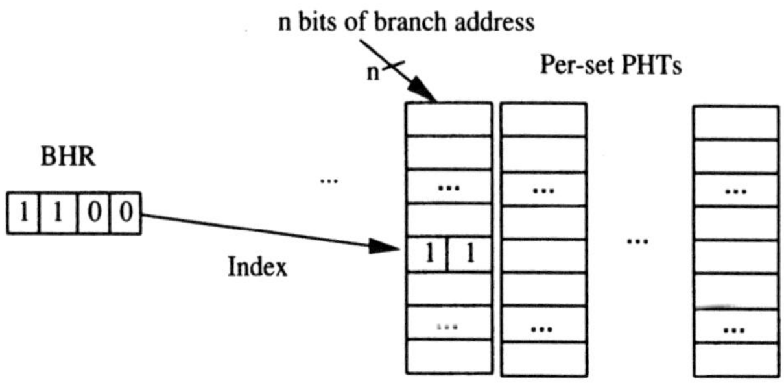

Fig. 4.10. Implementation of a GAs(4,2^n) predictor

each branch. The BHRs record self-history in contrast to the neighbor-history recording BHR used in global history schemes. The per-address branch history registers are combined in a table which is called the per-address branch history table (PBHT) by *Yeh and Patt.*

In the simplest per address history scheme, the BHRs index into a single global PHT. Such a two-level adaptive predictor is denoted as PAg (multiple per-address indexed BHRs, and a single global PHT). An implementation of a PAg(4) predictor is shown in Fig. 4.11. Two different branches with the same BHT bit pattern select the same PHT entry leading to unnecessary misprediction.

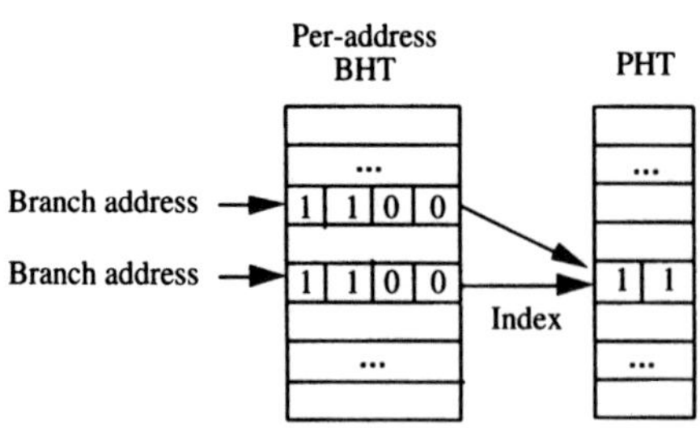

Fig. 4.11. Implementation of a PAg(4) predictor

The combination of multiple per-address BHRs with multiple per-address PHTs, denoted as a PAp predictor, and of multiple per-address BHRs with multiple per-set PHTs, denoted as a PAs predictor, is also possible. In the PAp scheme each branch has its own BHR and its own PHT. So the number of BHRs in the per-address BHT and the number of PHTs is equal. However, the numbers are not fixed. They depend on the number of branches in the program.[9] Conceptually, the BHR content is used as an index to select an entry in its PHT. The PHT is selected by the branch instruction address (PAp) or by the branch set (PAs). An implementation of a PAp(4) predictor is shown in Fig. 4.12. The figure shows the case of two branches with the same BHT bit pattern that indexes the same line in the per-address PHTs. However, the branch addresses select different PHTs and thus different PHT entries.

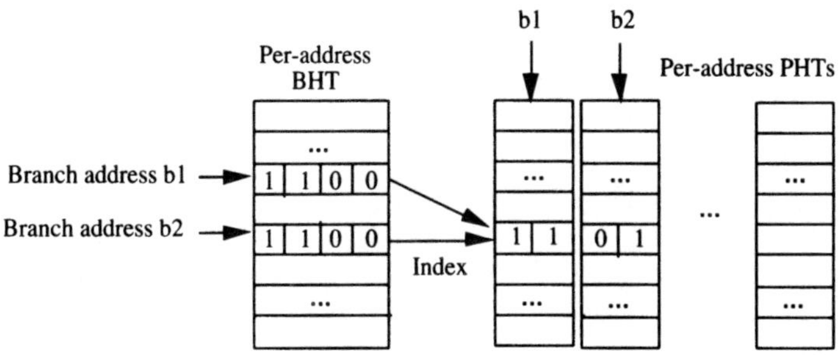

Fig. 4.12. Implementation of a PAp(4) predictor

In the per-address history schemes only the execution history of the branch itself has an effect on its prediction. The branch prediction is non-correlating – independent of the execution history of other branches.

In the *per-set history schemes* the first-level branch history means the last k occurrences of the branch instructions from the same subset. Each BHR is associated with a set of branches. The set attributes of a branch can be determined by the branch opcode, the branch class which is assigned by the compiler, or by part of the branch address. Since a per-set addressed BHR is potentially updated with history from all branches in the same set, the prediction of a branch is influenced by other branches in the same set (*Yeh and Patt* [332]). Again the three variations are determined by the variations in the organization of the second-level, namely SAg, SAs, and SAp. Implementations of a SAg(4) and a SAs(4) predictor are shown in Fig. 4.13 and Fig. 4.14. Figure 4.13 shows that the SAg-predictor may suffer from branch patterns that emerge several times within a computation (the same bit pattern in the BHRs select the same PHT entry in the global PHT). Moreover, in all per-set

[9] The PAp predictor is mainly of theoretical interest, because the variable numbers of BHRs and PHTs cause implementation problems.

history schemes, branches which fall into the same set (e.g., having the same n bits in the branch address) select the same entry in the BHT (and/or the same PHT). This is demonstrated in Fig. 4.14.

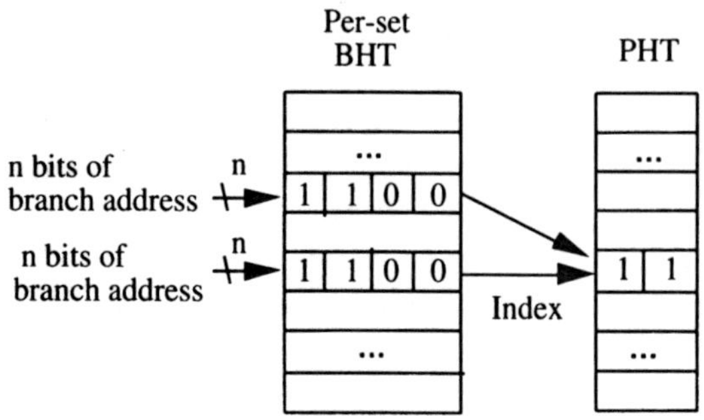

Fig. 4.13. Implementation of a SAg(4) predictor

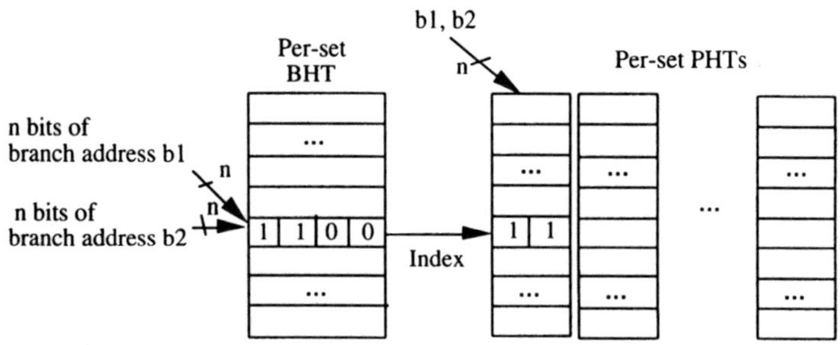

Fig. 4.14. Implementation of a SAs(4) predictor

The full table of *Yeh and Patt's* two-level adaptive branch predictors is given as follows [332]:

	global PHT	per-set PHTs	per-address PHTs
global BHR	GAg	GAs	GAp
per-address BHT	PAg	PAs	PAp
per-set BHT	SAg	SAs	SAp

The denotation of the two-level adaptive branch predictors are derived from the following table which gives a simplified estimation of the hardware costs [332]:

Scheme name	BHR Length	No. of PHTs	Hardware Cost
GAg(k)	k	1	$k + 2^k \times 2$
GAs(k,p)	k	p	$k + p \times 2^k \times 2$
GAp(k)	k	b	$k + b \times 2^k \times 2$
PAg(k)	k	1	$b \times k + 2^k \times 2$
PAs(k,p)	k	p	$b \times k + p \times 2^k \times 2$
PAp(k)	k	b	$b \times k + b \times 2^k \times 2$
SAg(k)	k	1	$s \times k + 2^k \times 2$
SAs(k,$s \times p$)	k	p	$s \times k + p \times 2^k \times 2$
SAp(k)	k	b	$s \times k + b \times 2^k \times 2$

In the table b is the number of PHTs or entries in the BHT for the per-address schemes. p and s denote the number of PHTs or entries in the BHT for the per-set schemes, assuming that different per-set schemes are possible for BHR selection and for PHT selection.

The simulations of *Yeh and Patt* [332] using the SPEC89 benchmarks show that the performance of the global history schemes is sensitive to the branch history length. Interference of different branches that are mapped to the same pattern history table is decreased by lengthening the global BHR leading to better prediction accuracy. Similarly adding PHTs reduces the possibility of pattern history interference by mapping interfering branches into different tables.

In general, the global history schemes are better than the per-address schemes for the integer SPEC89 programs, while the per-address schemes are better for the floating-point intensive programs. The phenomenon is due to the ability of the global history schemes to utilize branch correlation, which is often the case in the frequent if-then-else statements in integer programs, while the per-address schemes are better in predicting loop-control branches which are frequent in the floating-point SPEC89 benchmark programs. The per-set history schemes are in between other schemes.

Comparing the cost effectiveness of the different schemes using the formulas in the table given above and a fixed hardware budget of 8 k bits, the most cost-effective global history scheme is GAs(7,32), the best per-address scheme is PAs(6,16), and for per-set schemes SAs(6,4×16) scores best. From these three configurations PAs(6,16) achieves the highest average prediction accuracy.

When given a higher hardware budget of 128 k bits, the most cost-effective global history scheme is GAs(13,32), the best per-address scheme is PAs(8,256), and the best per-set scheme is SAs(9,4×32). Of these configurations GAs(13,32) achieves the highest measured prediction accuracy of 97.2 %.[10]

[10] Prediction accuracy measured for SPECint95 or OLTP (online transaction processing) programs is much lower than for SPEC89 benchmarks (see Table 4.1).

Yeh and Patt conclude that global history schemes perform better than other schemes on integer-dominated programs but require higher implementation costs to be effective overall. However, in the global history schemes, the pattern history of different branches interfere with each other if they map to the same PHT. Therefore, long BHRs and/or many PHTs should be used.

Per-address history schemes perform better than other schemes on floating-point programs. Per-set history schemes have a performance that is similar to global history schemes on integer programs and similar to per-address schemes on floating-point intensive programs.

gselect and gshare predictors. *McFarling* [196] analyzed the two-bit predictors and correlation-based predictor schemes and introduced a number of new predictors. One set of new correlation-based predictors uses a hash function into the PHT instead of indexing the PHT to reduce conflicts.

Recall that in the correlation-based predictor scheme the (2,2)-predictor shown in Fig. 4.7 requires 12 bits for a PHT table lookup (assuming a single unified PHT instead of the four PHTs); two bits from the BHR are concatenated with 10 bits from the branch address. *McFarling* calls this bit concatenation in a correlation-based or GAs predictor the *gselect* predictor which concatenates some lower order bits of the branch address and of the bit pattern in the BHR.

In contrast to simple indexing, *McFarlings*'s *gshare* predictor uses the bitwise exclusive OR of part of the branch address and the BHR as a hash function. To demonstrate the ability of both predictor types, *McFarling* uses the following table:

Branch Address	BHR	*gselect* 4/4	*gshare* 8/8
00000000	00000001	00000001	00000001
00000000	00000001	00000000	00000000
11111111	00000000	11110000	11111111
11111111	10000000	11110000	01111111

Strategy *gselect* 4/4 concatenates the lower order 4 bits of the branch address with the lower order 4 bits of the BHR. Strategy *gshare* 8/8 uses the bitwise XOR of all 8 bits of both the branch address and the BHR. Comparing *gshare* 8/8 and *gselect* 4/4 shows that only *gshare* is able to separate all four cases. The *gselect* predictor cannot take advantage of the distinguishing history in the upper four bits of the BHR.

Hybrid predictors. The second strategy proposed by *McFarling* is to combine multiple separate branch predictors, each tuned to a different class of branches. Different branch prediction schemes have different advantages. Hopefully, such a *combining predictor* achieves an even better prediction accuracy than either of the predictors used for combination. To predict a given branch, typically two or more predictors and a predictor selection mechanism are necessary in a combining predictor. In principle, all kinds of branch predictors are candidates for combination of predictors.

McFarling combined the two-bit predictor[11] with the *gshare* two-level adaptive predictor, and concluded that, in this combination, global information can be used if it is worthwhile; otherwise, the usual branch direction as predicted by the two-bit predictor can be used. Another combination proposed by the same author is the combination of a PAp predictor[12] with the *gshare* scheme. Simulations with SPEC89 benchmarks showed that both hybrid predictors outperform the *gshare* which itself is better than *gselect* and all other predictors for a given counter array size.

Another kind of *hybrid predictor* proposed by *Young and Smith* [333] combines a compiler-based static branch prediction with a dynamic predictor of the two-level adaptive type. Profiling is used to collect the static prediction information [307]. Numerous other selector and hybrid predictor types are evaluated and reported in the research literature. *Patt et al.* proposed a multi-hybrid branch predictor for an advanced superscalar processor of the one giga-transistor chip era (see Sect. 5.2).

Grunwald et al. [110] compared the SAg, *gshare* and *McFarling*'s combining predictor (combining a two-bit predictor with the gshare predictor) using the SPECint95 benchmarks. The results are reported in the Table 4.1. The table shows that for SPECint95 benchmark programs about every sixth instruction of the trace (the executed and committed instructions) is a branch instruction and in the mean misprediction rate the combining predictor performs best with 8.1 % mispredictions. Further simulation of *Grunwald et al.* showed that the processor typically issued 20–100 % more instructions than actually commit, due to speculative execution [110].

Other simulations by *Keeton et al.* [160] using an OLTP (online transaction workload) on a Pentium Pro multiprocessor reported a misprediction ratio of 14 % with a branch instruction frequency of about 21 %. The speculative execution factor, given by the number of instructions decoded divided by the number of instructions committed, is 1.4 for the database programs.

Two different conclusions may be drawn from these simulation results: branch predictors should be improved further and/or branch prediction is only effective if the branch is predictable. If a branch outcome is dependent on irregular data inputs, as is often the case in OLTP applications or game-playing programs, the branch often shows an irregular behavior. This may be the reason for the high misprediction rate of the SPECint95 benchmark program `go`.

---

[11] called a bimodal predictor by *McFarling*

[12] called a local predictor by *McFarling*, per-address scheme in *Yeh and Patt*'s nomenclature.

Table 4.1. SAg, *gshare* and *McFarling*'s combining predictor

Application	committed instructions (in millions)	conditional branches (in millions)	taken branches (%)	misprediction rate (%) SAg	gshare	combining
compress	80.4	14.4	54.6	10.1	10.1	9.9
gcc	250.9	50.4	49.0	12.8	23.9	12.2
perl	228.2	43.8	52.6	9.2	25.9	11.4
go	548.1	80.3	54.5	25.6	34.4	24.1
m88ksim	416.5	89.8	71.7	4.7	8.6	4.7
xlisp	183.3	41.8	39.5	10.3	10.2	6.8
vortex	180.9	29.1	50.1	2.0	8.3	1.7
jpeg	252.0	20.0	70.0	10.3	12.5	10.4
mean	**267.6**	**46.2**	**54.3**	**8.6**	**14.5**	**8.1**

4.3.4 Predicated Instructions and Multipath Execution

Confidence estimation. If a branch is not, or is not easily, predictable, its irregular behavior will frequently yield costly misspeculations. The predictability of branches can be assessed by additionally measuring the *confidence* in the prediction. A *low confidence branch* is a branch which frequently changes its branch direction in an irregular way making its outcome hard to predict or even unpredictable.

Confidence estimation is a technique for assessing the quality of a particular prediction. If applied to branch prediction, a confidence estimator attempts to assess the prediction made by a branch predictor. Because each branch is eventually determined to have been predicted correctly or incorrectly, the confidence estimator assigns a "high confidence" (HC) or a "low confidence" (LC) to each prediction. In combination with the two prediction outcomes "correctly predicted" (C) and "incorrectly predicted" (I), four confidence classes can be measured:

- correctly predicted with high confidence C(HC);
- correctly predicted with low confidence C(LC);
- incorrectly predicted with high confidence I(HC); and
- incorrectly predicted with low confidence I(LC).

When a branch is actually resolved, the branch can be classified as belonging to one of these classes (*Grunwald et al.* [110]).

To implement a confidence estimator, information from the branch prediction tables is used. *Smith* [268] proposed already in 1981 to use saturation counter information to construct a confidence estimator. The concept was to speculate more aggressively when the confidence level is higher [269]. *Jacobsen et al.* [148] used a miss distance counter table (MDC) in addition to the branch predictor. Each time a branch is predicted, the value in the MDC is compared to a threshold. If the value is above the threshold, then the branch is considered to have high confidence, and low confidence otherwise. *Tyson et al.* [305] observed that a small number of branch history patterns typically leads to correct predictions in a PAs predictor scheme. Their confidence estimator assigned high confidence to a fixed set of patterns and low confidence to all others [110].

Confidence estimation can be used for speculation control provided that ways other than branch speculation can be used to utilize the processor resources. Such alternative ways can be, for example, thread switching in multithreaded processors (see Chap. 6) or multipath execution where instructions from both branch directions are fetched and executed, and the wrong path instructions are afterwards discarded. In a simultaneous multithreaded processor (see Sect. 6.4), it may be more cost effective to switch threads than speculatively evaluate a branch of low confidence. In a multipath execution model both branch paths of a low confidence branch may be evaluated, whereas a conventional branch speculation may be employed to high confidence branches. Both techniques need the ability of a processor to pursue two different instruction streams simultaneously. Because of the limitation of a single instruction pointer in today's superscalar processors, such techniques are confined to multithreaded processors and related processor techniques such as multiscalar (Sect. 5.4) and trace processors (Sect. 5.5).

Predicated instructions. One technique that allows us to "evaluate" two branch paths in a multiple-issue processor is *predication* (see *Mahlke et al.* [188], *August et al.* [19, 20], *Hwu* [142]). Using this technique, the ISA of a processor is enhanced by so-called *predicated* or *conditional instructions* and one or more *predicate registers*. The Boolean result of a condition test is recorded in a (one-bit) predicate register. Predicated instructions use a predicate register as an additional input operand.

Predication is demonstrated by the following source code sequence:

if $(x == 0)\{$ /* branch b1 */
  $a = b + c;$
  $d = e - f; \}$
$g = h * i;$ /* instruction independent of branch b1 */

Translation of the example source code sequence, using a branch instruction for the alternative, would lead to a speculative execution with instruction $g = h*i$ and all later instructions on the speculative path of branch b1. In the case of a misspeculation temporary results of this and all later instructions would be unnecessarily discarded.

However, the source code is translated in the following code sequence using predicated instructions (each line represents a single machine operation):

$(Pred = (x == 0))$ /* branch b1: *Pred* is set to true if x equals 0 */
if *Pred* **then** $a = b + c$; /* The operations are only performed */
if *Pred* **then** $d = e - f$; /* if *Pred* is set to true */
$g = h * i$;

As can be seen from the example, predication is able to eliminate a branch and, therefore, the associated branch prediction, increasing the distance between mispredictions. Also the run length of a code block is increased which allows better instruction scheduling by an optimizing compiler. However, the compiler must assure that the exception behavior is not changed by moving the instruction across a set-predicate instruction.

Predication affects the instruction set, adds a port to the register file, and complicates instruction execution. Predication is most effective when control dependences can be completely eliminated, such as in an if-then with a small then body, and when the condition can be evaluated early. The use of predicated instructions is limited when the control flow involves more than a simple alternative sequence. Moreover, predicated instructions that are discarded still consume processor resources; the fetch bandwidth is especially affected.

If the full instruction set is predicated (a so-called full predication model), predication bits in the opcode are additionally needed for each instruction to denote a predicate register. Thus, often only a few instructions of the ISA, in most cases the load instructions, are predicated instructions.

Most signal processors, high-performance microcontrollers and some contemporary superscalar processors employ predication. As examples, the ARM processor ISA is fully predicated; Alpha, MIPS, PowerPC, and SPARC processors use conditional move instructions, and the Intel Merced will be fully predicated (see Sect. 4.10.2).

Predicated instructions are fetched, decoded, and placed in the instruction window like nonpredicated instructions. It depends on the processor architecture how far a predicated instruction proceeds speculatively in the pipeline before its predication is resolved:

- A predicated instruction executes only if its predicate is true, otherwise the instruction is discarded. In this case predicated instructions are not executed before the predicate is resolved.
- Alternatively, as reported for Intel's IA-64 ISA, the predicated instruction may be executed, but commits only if the predicate is true, otherwise the result is discarded (*Dulong* [71]).

The latter case is similar to the eager or multipath execution model described below.

Eager execution. With the *eager* or *multipath* execution model, execution proceeds down both paths of a branch, and no prediction is made. When a branch resolves, all operations on the non-taken path are discarded. Consequently, eager execution with unlimited resources, which can be characterized as "oracle execution", would give the same theoretical maximum performance as a perfect branch prediction. With limited resources, the eager execution strategy must be employed carefully. Resource consumption rises exponentially with each level of branches that are executed eagerly. Therefore, instead of employing full eager execution, a mechanism is required that decides when to employ prediction and when eager execution.

One decision mechanism is the use of a confidence estimator. If a branch prediction can be made with high confidence, branch prediction and single path speculative execution is employed; when low confidence is the case, eager execution spares the misprediction penalty.

Until now, the eager execution strategy has rarely been implemented, except for limited applications, such as instruction fetch in the SuperSPARC processor and in the IBM 360/91 (*Uht et al.* [307]) and subsequent IBM mainframes, for example, the IBM 3090 processor.

The "nanothreaded" DanSoft processor implements a multipath-execution model using confidence information from a static branch prediction mechanism (see Sect. 6.3.5).

A number of research projects have surveyed eager execution. The Polypath architecture (*Klauser et al.*) [163] enhances a superscalar processor by a limited multipath execution feature to employ eager execution. *Heil and Smith* [130] propose selective dual path execution; and *Tyson et al.* [305] propose a limited dual path execution. *Wallace et al.* [318] survey threaded multipath execution, employing eager execution in a simultaneous multithreaded processor model.

Unger et al. [308, 309] propose a compiler technique called *simultaneous speculation scheduling* in combination with a "minimal" multithreaded execution model to enable speculative execution of alternative program paths. The technique is only applicable for architectures that fulfill certain requirements of a base multithreaded processor model:

- First, the processor must be able to pursue two or more threads of control concurrently, i.e., it must provide two or more independent program counters.
- All concurrently executed threads of control share the same address space, preferably the same register set.
- The instruction set must provide a number of thread-handling instructions: Here the minimal requirements for multithreading are an instruction for creating a new thread (**`fork`**) and an instruction that conditionally stops its own execution or the execution of some other threads (**`sync`**).
- Creating a new thread by the **`fork`** instruction and joining threads by the **`sync`** instruction must be extremely fast, preferably single-cycle operations.

Uht and Sindagi [306, 307] propose the *disjoint eager execution* technique. The idea is to assign resources to branch paths whose results are most likely to be used, i.e., branches with the highest cumulative execution probability. *Uht and Sindagi's* notion of branch execution probability is closely related to the confidence in a branch prediction, for which they use the branch prediction accuracy, i.e., the percentage of taken or untaken executions of a branch.

While a branch path is speculatively executed, further branches may be encountered before the first branch resolves, often resulting in a branch speculation level of 4 or more. The cumulative execution probability accumulates the prediction accuracies of a branch and of the pending (predicted but yet to be resolved) branches of previous speculation levels. If all branches in such a sequence of pending branches are simply assumed to be independent of each other, the single prediction accuracies can be multiplied to determine the cumulative execution probability of the last branch in the sequence.

Thus in the disjoint eager execution model, all branches are predicted, the cumulative prediction accuracy is computed and compared to the accuracies of all branch paths that were yet to be chosen for speculative execution. The branch path with the highest cumulative prediction accuracy is executed, leading to either another single path speculative execution or an eager execution.

The three different possibilities of single path speculative execution as produced by the usual speculation methods described above, full eager execution, and disjoint eager execution are demonstrated in Fig. 4.15 [307]. Each line with an arrow represents a branch path marked by its cumulative probability. For illustration, branch prediction accuracy is 70 % for each individual branch. All branches are pending. Branch paths with circled numbers are in execution, branch paths that are not chosen by the prediction are the paths without circled numbers. Circled numbers indicate the order of the resource assignment, i.e., the order in which the paths are speculatively assigned. Figure 4.15(c) shows that the disjoint eager execution strategy allocates resources to more likely branch paths than the single path and the eager execution models.

4.3.5 Prediction of Indirect Branches

All branch prediction techniques reported above are directed towards prediction of direct branches, whose targets are encoded in the instruction itself. Indirect branches, which transfer control to an address stored in a register, are even harder to predict accurately. Though indirect branches are not as frequent as direct branches in C- or FORTRAN-benchmark programs, indirect branches occur with higher frequency in machine code compiled from object-oriented programs like C++ and Java. Virtual function tables, used in C++ and Java compilers to implement late binding of subroutine invocations, execute an indirect branch for every polymorphic call. A simple BTB is a poor predictor for branches with changing targets. One simple possibility is to update the PHT to include the branch target addresses.

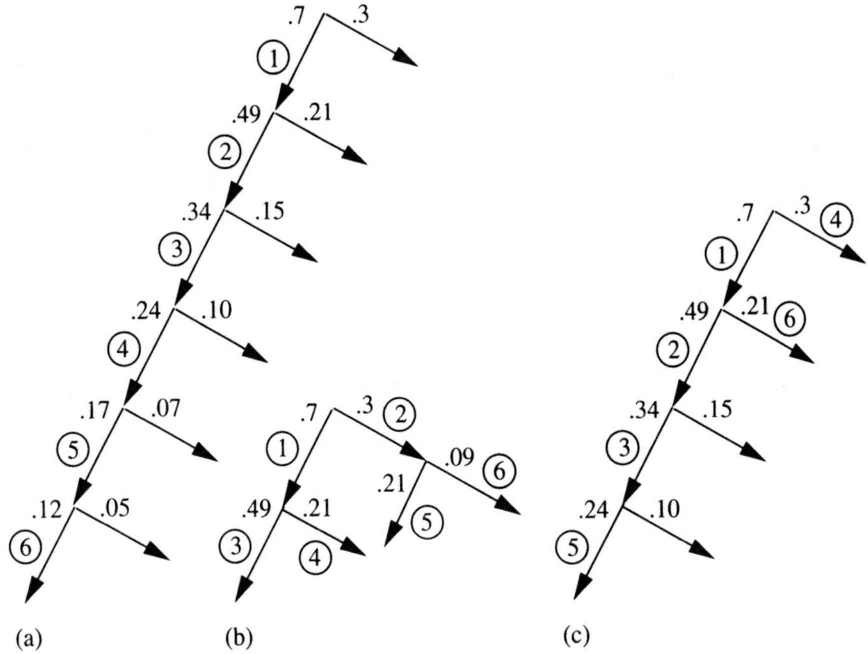

Fig. 4.15. (**a**) Single path speculative execution (**b**) full eager execution (**c**) disjoint eager execution

Driesen and Hoelzle [70] reported an indirect branch frequency of once every 50 instructions for several large object-oriented C++ programs. They investigated two-level and hybrid indirect branch predictors and reported a misprediction rate of 9.8 % with a 1 k-entry table, 7.3 % with an 8 k-entry table, 8.98 % for a 1 k-entry hybrid predictor, and 5.95 % in the 8 k-entry hybrid predictor case.

4.3.6 High-Bandwidth Branch Prediction

Future microprocessors will require more than one prediction per cycle starting speculation over multiple branches in a single cycle. Here the GAg scheme is able to predict multiple branches without knowing the branch instruction address. However, the instruction fetch is also affected. When multiple branches are predicted per cycle, instructions must be fetched from multiple target addresses per cycle, complicating I-cache access. A trace cache (see Sect. 5.5) in combination with next trace prediction is able to solve both problems by fetching from a dynamically assembled trace line, rather than from I-cache.

A combination of branch handling techniques will most likely be applied, such as a multi-hybrid branch predictor (*Evers et al.* [80], *Patt et al.* [229];

see Sect. 5.2) combined with support for context switching, indirect jumps, and interference handling.

Table 4.2 shows some branch handling techniques and their implementations in state-of-the-art microprocessors.

Table 4.2. Branch handling techniques and implementations

Technique	Implementation examples
No branch prediction	Intel 8086
Static prediction:	
always not taken	Intel i486
always taken	Sun SuperSPARC
backward taken, forward not taken	HP PA-7x00
semistatic with profiling	early PowerPCs
Dynamic prediction:	
1-bit	DEC Alpha 21064, AMD-K5
2-bit	PowerPC 604, MIPS R10000, Cyrix 6x86 and M2, NexGen 586
two-level adaptive	Intel Pentium Pro, Pentium II, AMD-K6
Hybrid prediction	DEC Alpha 21264
Predication	Intel/HP Merced and most signal processors as, e.g., ARM processors, TI TMS320C6201 and many other
Eager execution (limited)	IBM mainframes: IBM 360/91, IBM 3090
Disjoint eager execution	none yet

4.4 Decode

For good performance, the processor must fetch and decode instructions at a higher bandwidth than it can execute them. If the instruction window is kept full, the deeper instruction lookahead allows more instructions to be issued to the functional units. Moreover, the processor fetches and decodes more (today about twice as many) instructions than it commits, because it discards instructions on mispredicted branch paths.

Typically the decode bandwidth is the same as the instruction fetch bandwidth. Multiple instruction fetch and decode is supported by a fixed instruction length.

If the instruction length varies, which is often the case for legacy CISC instruction sets such as the Intel x86 ISA, a multistage decode is necessary. The first stage determines the instruction limits within the instruction stream provided by the fetch unit, and delivers a number of instructions to the second decode stage. The second stage decodes the instructions generating one or several μops from each instruction. Complex CISC instructions are split into μops which resemble ordinary RISC instructions.

The advantage of CISC instructions over RISC instructions is the denser code (especially if a Huffman encoding is used), the disadvantage is the more complex decode. The same argument is valid for a stack register instruction set.

If the opcode organization is adequate, the IF stage can analyze part of the opcode and use it for prediction. If a partial decode is done when the instructions are transferred from memory or secondary cache to the I-cache (see PowerPC 620 or MIPS R10000), the decode stage is simpler.

For instance, the MIPS R10000 predecodes each 32-bit instruction into a 36-bit format stored in the I-cache. The four extra bits indicate which functional unit should execute the instruction. The predecoding also rearranges operand- and destination-select fields to be in the same position for every instruction, and modifies opcodes to simplify decoding of integer or floating-point destination registers (*Yeager* [330]). Thus the decoder can decode this expanded format more rapidly than the original instruction format.

4.5 Rename

The aim of register renaming is to remove antidependences and output dependences dynamically by the processor hardware. *Register renaming* is the process of dynamically associating specific ***physical registers*** (also called *rename registers*) with the ***architectural registers*** (also called ***logical registers***) referred to in the instruction set of the architecture. The physical registers are internal registers that cannot be accessed directly by the programmer or compiler.

Register renaming is implemented by allocating a new physical register for every destination register specified in an instruction. If the same architectural register is used by a preceding instruction either as an operand or destination register, that register is mapped to another physical register, thus dynamically removing antidependences and output dependences from the instruction flow. Succeeding instructions that use the same architectural register as an operand register access the newly allocated physical register as an input register. After the mapping, register data dependences are simply detected by comparing physical register numbers, no longer considering instruction order.

Each physical register is written only once after each assignment from the free list of available registers. If a subsequent instruction needs its value, that instruction must wait until it is written (data dependence). After the register is written, it is ready, and its value never changes. When a subsequent instruction updates the corresponding architectural register, the result is written into a newly assigned physical register.

Typically there are more physical registers than architectural registers. For instance, the Pentium Pro ISA defines 8 architectural integer registers, but contains 40 physical registers. Separate sets of physical registers are provided to rename integer registers and floating-point registers. The MIPS R10000 ISA defines 33 architectural integer registers (including `Hi` and `Lo` registers used for integer divides) and 32 architectural floating-point registers, but provides 64 physical integer registers and 64 physical floating-point registers.

There are two principal techniques to implement renaming:

- Separate sets of architectural registers and rename (physical) registers are provided. The physical registers only contain temporary values (of completed but not yet retired instructions), while the architectural registers store the committed values. After commitment of an instruction, copying its result from the rename register to the architectural register is required. The PowerPC 604 and 620 provide both types of registers in hardware, and a separate copy stage after the commit stage. The physical register is freed for re-use when the instruction commits and its result is written back to the corresponding architectural register.
- Only a single set of registers is provided and architectural registers are dynamically mapped to physical registers. The physical registers contain committed values and temporary results. After commitment of an instruction, the physical register is made permanent and no copying is necessary. This is the mode implemented in the Pentium II and in the MIPS R10000. The old physical register can be freed for re-use when a subsequent instruction writes to the corresponding architectural register.

The physical registers can be implemented in the reservation stations as is the case in Tomasulo's scheme of renaming or they can be separate from the reservation stations.

Another alternative to dynamic renaming is the use of a large register file, as defined for the Intel Merced. Antidepencences and output dependences can be removed by a static register mapping by the compiler. One problem of this approach is that more registers need more bits in the instruction format to specify the register numbers. This extra space may not be available in a 32-bit instruction format. Moreover, the more complex register access may limit cycle time or lead to a two-stage register access and write-back, thus increasing pipeline length by an additional stage. On the other hand, the renaming hardware is saved, leading to less hardware complexity.

Register mapping is often not a pipeline stage on its own, but is combined with the decode stage. After renaming the instruction is written into the instruction window, awaiting issue to functional units.

4.6 Issue and Dispatch

The notion of the *instruction window* comprises all the waiting stations between the decode (rename) and execute stages. The instruction window isolates the decode/rename from the execution stages of the pipeline. The decode stage continues to decode instructions regardless of whether they can be executed immediately or not. The decode stage places the decoded (and renamed) instructions in the instruction window as long as there is room in the window.

The instructions in the instruction window are free from control dependences, which are removed by branch prediction, and free from antidependences or output dependences, which are removed by renaming. Thus only data dependences and resource conflicts remain to be taken into consideration.

Instruction issue is the process of initiating instruction execution in the processor's functional units.[13] The *instruction-issue policy* is the protocol used to issue instructions. The processor's *lookahead* capability is the ability to examine instructions beyond the current point of execution in the hope of finding independent instructions to execute (*Johnson* [150]).

Hennessy and Patterson [134] distinguish dynamic from static issue, and dynamic from static scheduling. In this terminology, *superscalar* is characterized by a *dynamic issue*, whereby it is decided by hardware which instructions are issued and the issue of a varying number of instructions per clock cycle is possible. Dynamic issue can be *statically scheduled* or *dynamically scheduled*, meaning instructions must be issued in program order as defined by the compiler, or the issue can also be performed out of order, i.e., (dynamically) scheduled by the hardware.

VLIW is characterized by a *static issue* whereby a fixed number of instructions is issued each cycle, which are statically scheduled by the compiler. The instructions can be organized as one large instruction or as a fixed instruction packet.

To summarize, *Hennessy and Patterson* distinguish dynamic (superscalar) from static (VLIW) issue, and dynamic (out-of-order) from static (in-order) scheduling. In-order issue was the rule for superscalar processors until approximately 1995. Out-of-order issue reaches better IPC (instructions per cycle) and is adopted by all state-of-the-art superscalar microprocessors.

Today, superscalar microprocessors are able to issue up to four or six instruction per cycle out of order from a 16-entry to 56-entry instruction window. A large instruction window and excellent branch prediction is necessary to reach an IPC value that is close to the maximum-issue bandwidth.[14]

[13] We use the term issue for an issue to a FU or a reservation station and the term dispatch, if a second issue stage exists, to denote when an instruction begins execution in the functional unit.

[14] There exists a strong relation to the dataflow scheme (see Chap. 2). Dynamic scheduling can be viewed as a kind of "local dataflow" or "windowed dataflow".

One way to implement the instruction window is to centralize the window buffering of every instruction for every FU in a common window. The problem with a large central instruction window is that an issue from a single large instruction window limits future cycle rate increase. In each processor cycle, the availability of operands has to be updated, instructions that are ready to issue have to be selected, and availability of the appropriate resources must be checked. The necessary resources for an instruction to be issued comprise a free functional unit (or a free entry in its reservation station) and an entry in the reorder buffer. The update and select complexity rises extremely quick as the instruction window is made larger.

Moreover, each instruction issued in a single cycle must be accompanied by all its required operands. If the instruction window is distributed to individual buffers to each FU, these buffers are called reservation stations. As in the Tomasulo scheme a single reservation station is able to host a single instruction.[15]

There exist several alternatives to a single-stage issue from a central instruction window:

- The first alternative is a multistage issue. Operand availability and resource availability checking is split into two separate stages. A resource dependent issue can be performed first to reservation stations which are arranged in front of each FU or in front of a group of related FUs. In the second stage, the instruction is dispatched to the FU when operands are available and instruction execution starts.
  In principle, the two stages could also be organized in reverse order. Data dependence checking is performed first and instructions are issued to (possibly decoupled) reservation stations. Execution of an instruction starts when the appropriate FU is free.
- A second alternative is the decoupling of instruction windows. A small number of instruction windows (or reservation stations) is provided. Each instruction window is shared by a group of (usually related) FUs. In the HP PA-8000, separate floating-point and integer/general-purpose windows are provided; in the MIPS R10000 floating-point, integer, and address windows are distinguished.
  Data dependence checking is simplified, because data dependences are mostly limited to each of the instruction windows. A slower exception mechanism may be provided for the few dependences between floating-point and integer instructions, for example.

In contrast to the superscalar approach, no renaming is necessary in the dataflow scheme because of the single-assignment rule, and no branch prediction or speculative execution is taken into consideration. In principle, the "instruction window" in the dataflow scheme is the capacity of the matching store, and enabling of instructions is tested by the matching unit.

[15] In contrast to the definition above, a reservation station is sometimes defined in literature as a multiple-entry reservation station – an instruction buffer containing several entries. We adhere to the original definition given by Tomasulo.

- The third alternative is a combination of multistage issue and decoupling of instruction windows. The decoupling can even be extended to a complete distribution as done by reservation stations which are dedicated to individual FUs (e.g., PowerPC). Instructions can only be dispatched to the associated FU. However, the availability of operands may depend on result values delivered from an arbitrary FU.

The issue from each of the instruction windows can be an in-order or an out-of-order issue. In a two-stage issue scheme, with resource dependent issue preceding the data-dependent dispatch, the first stage is still performed in order, while the second stage is performed out of order. Moreover, if operands are available and the FU is idle, the instruction may be dispatched immediately, during issue to the reservation station, thus avoiding the dispatch cycle. The following issue schemes are commonly used:

- single-level issue out of a central window as in the Pentium II processor (see Fig. 4.16),

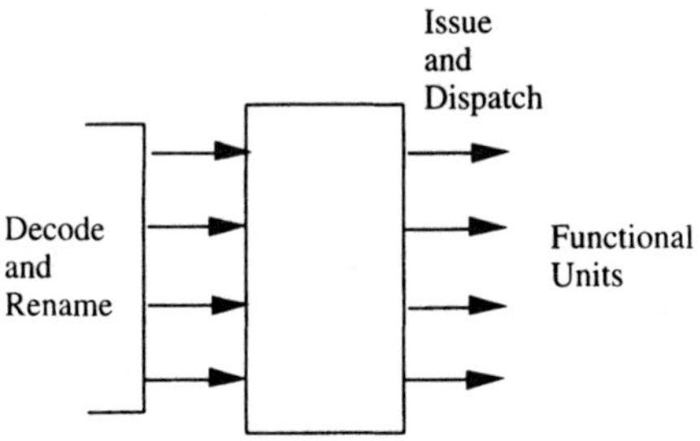

Fig. 4.16. Single-level, central issue scheme

- single-level issue with an instruction window decoupling using two separate windows (most commonly separate floating point and integer windows as in HP 8000 processor; see Fig. 4.17),

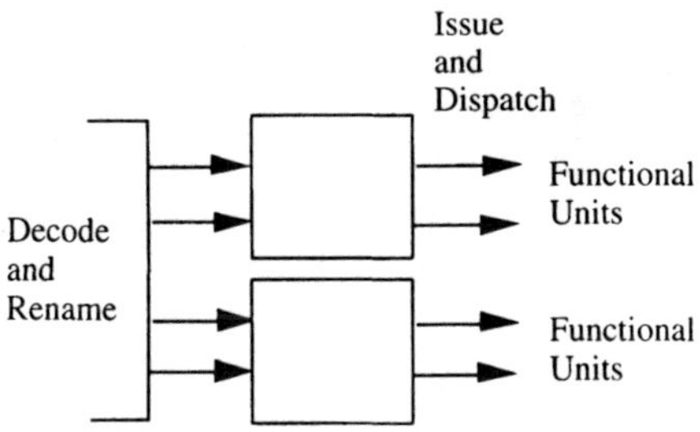

Fig. 4.17. Single-level, two-window issue scheme

- two-level issue with multiple windows with a centralized window in the first stage and separate windows in the second stage. Figure 4.18 shows a two-level issue as performed in the PowerPC 604 and 620 processors. At first resource conflicts (structural hazards) are checked and up to four instructions are issued in order to the respective reservation stations in front of the appropriate FUs. A resource conflict arises when multiple instructions should be issued to reservation stations of the same FU, when the reservation station(s) for the appropriate FUs are full, or when the reorder buffer is full. In a second stage an instruction can be dispatched from a reservation station to its FU as soon as all input operands are available and the FU is not busy. The issue is performed from a central instruction window to reservation stations which are separate for each FU.

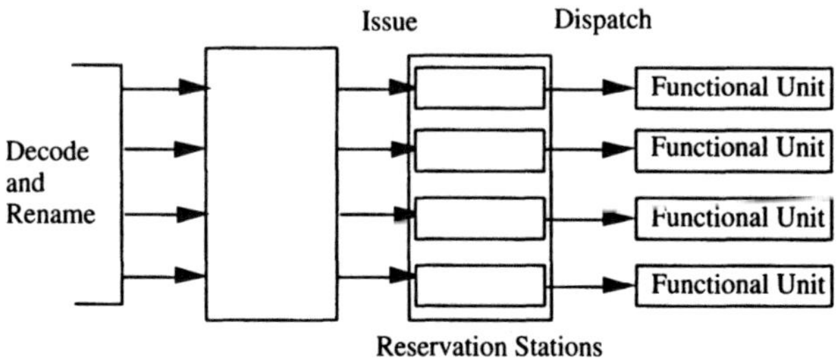

Fig. 4.18. Example of a two-level issue scheme

If several instructions are ready for issue, exceeding the issue bandwidth, an instruction issue strategy is applied. In superscalar processors the strategy is either an oldest-instruction first or a round-robin scheduling concerning the entries in the instruction window. Instruction issue strategies get more interesting in simultaneous multithreaded processors (see Sect. 6.4) that are able to issue instructions of several threads simultaneously and often provide more enabled instructions than the issue bandwidth allows.

In future superscalar processors, the issue may be even more complicated by a speculation beyond data dependences. Operand value prediction, load value and load address prediction, constant value and stride value prediction, and related prediction techniques lead to a speculative execution of data dependent instructions (see Sect. 5.3). This may prove useful when a multi-stage issue is combined with a large issue bandwidth that cannot be filled otherwise.

4.7 Execution Stages

A FU executes an instruction in one or several cycles and in pipelined or nonpipelined fashion until the execution completes which makes the result available for forwarding and buffering.

There exist various types of FUs that may be classified as single-cycle (latency of one) or multi-cycle (latency more than one) units. Single-cycle units are the simplest kinds of FUs which produce a result one cycle after an instruction started execution. Usually they are also able to accept a new instruction each cycle (throughput of 1). Multicycle units perform more complex operations that cannot be implemented within a single cycle under the respective timing constraints of the processor. Multicycle units can be pipelined to accept a new operation each cycle or every other cycle or they are nonpipelined. Another class of unit exists that performs the operations with variable cycle times.

Types of FUs are:

- single-cycle (single latency) units: (simple) integer and (integer-based) multimedia units;
- multi-cycle units that are pipelined (throughput of 1): complex integer, floating-point, and (floating-point-based) multimedia unit (also called multimedia vector units);
- multi-cycle units that are pipelined but do not accept a new operation each cycle (throughput less than 1): often the 64-bit floating-point operations in a floating-point unit;
- multi-cycle units that are not pipelined: division unit, square root units, complex multimedia units;
- variable cycle time units: load/store unit (depending on cache misses) and special implementations of floating-point units, for example.

Simple integer unit. A simple integer unit typically contains an ALU that handles all the 32-bit (or 64-bit) fixed-point addition instructions and the logical instructions.

Complex integer unit. A complex integer unit handles the more complex integer operations as, for example, the 32-bit and 64-bit signed and unsigned integer multiplications. The fully pipelined unit can start a new multiply instruction every clock cycle with an execution latency of three cycles. The multiplier typically uses Booth partial product generators and a Wallace tree to sum the partial products. For integer divisions a dedicated division unit may be present or divisions are performed by the complex integer unit in a nonpipelined fashion. Dividers typically use a radix-4 or radix-8 SRT algorithm (Sweeney-Robertson-Tosher) with a latency depending on the operand type and precision (see *Hennessy and Patterson* [134] for a description of these algorithms). Latencies are typically in the range of 13 to 17 cycles for a single-precision division. The divide unit is most often also used for square root computations, if such an instruction is present in the ISA (as for example in the MIPS R10000).

Floating-point execution unit. Floating-point execution units are fully pipelined and able to perform floating-point operations in ANSI/IEEE 754-1985 single-precision or double-precision formats. Typically a three-cycle latency is needed, although a special shortcut may reduce the latency for the floating-point compare instruction to one cycle. Due to the exceptions in the IEEE floating-point standard, the rounding and normalization can especially be rather complex. Thus, the full IEEE standard is not usually implemented in hardware.

Load/store unit. Load/store units are rather complex units and cannot be covered here in full detail. Therefore only a few basic facts are presented. Primary D-cache access is performed in two cycles which results in a two cycle load latency in the case of a D-cache hit. After address calculation a load instruction simultaneously accesses the TLB for virtual to physical address translation, the cache tag array, and the cache data array. Different queues are present within the load/store unit for loads that only need an address and for stores that need a store address and a value.

To preserve correct branch speculation rerolling and a precise exception mechanism, store operations are allowed to affect the D-cache or memory only when the store instruction commits. Therefore, store instructions may block succeeding load instructions. This can be avoided when loads are allowed to pass store instructions provided that the addresses are different. Thereby processor performance is increased, because instructions that are data dependent on the loaded value can be issued faster. However, the sequential consistency[16] assumed by most programs can no longer be guaranteed.

A similar potential violation of sequential consistency arises with non-blocking or lockup-free caches that are state-of-the-art. In this caching scheme a cache miss does not block further load/store operations provided that the same cache line is not affected. Often up to four memory operations can be outstanding.

[16] *Sequential consistency* was defined by *Lamport*, 1979 [171] for multiprocessor systems as follows: ***The results of any execution is the same as if the operations of all the processors were executed in some sequential order, and the operations of each individual processor were to appear in this sequence in the order specified by its program.*** The principal benefit of sequential consistency as an interface to shared-memory multiprocessor hardware is that sequential consistency is what people expect (*Hill* [135]). Sequential consistency is defined to guarantee that load and store accesses of parallel programs executed on a multiprocessor system arrive in program order at the memory, and the parallel programs are executed as in multiprogramming fashion on a single-processor system. However, sequential consistency prevents reordering of load/store accesses.

Typically a load cannot pass a store operation when the store address has not yet be computed. In this case the load may be executed speculatively, assuming that the load address is different from all addresses of all passed store operations. If that is not the case, the loaded value must be discarded.

The single load/store unit is often a bottleneck for the performance of today's microprocessors, but multiple load/store units are difficult to implement.

Media processing. Media processing (digital multimedia information processing) is the decoding, encoding, interpretation, enhancement, and rendering of digital multimedia information. One important example for media processing is the MPEG-2 video algorithm – the video compression standard defined in ISO/IEC 13818-2 [146] – that has been chosen for digital TV (cable, satellite, terrestrial broadcast), DVD and HDTV. It provides high quality video with data rates of 2–20 Mb/s. The MPEG-2 video decompression (or decoding) can be divided into the following six steps:

1. Header decode provides video sequence parameters such as picture rate, bit rate, image size, structure, and decoding parameters.
2. Huffman decode decodes variable-length codes into fixed length numbers, which represent quantized inverse DCT (IDCT) coefficients, scaling factors, and motion vectors. This step includes run-length decoding of zeros for the DCT coefficients.
3. Inverse quantization multiplies coefficients by quantizer factors to restore them to the original range.
4. IDCT changes each 8×8 block of IDCT coefficients to convert the data from the frequency domain back to the original spatial domain. This gives the actual pixel values for I-blocks, but only the differences for each pixel for P-blocks and B-blocks.
5. Motion compensation adds the differences in the IDCT step to the pixels in the reference block as determined by the motion vector for P blocks, and to the average of the forward and backward reference blocks for B-blocks.
6. Display converts color from YCbCr coordinates to RGB color coordinates, including upsampling Cb and Cr values, and writing to the frame buffer for displaying the decoded video.

Today's video and 3D graphics require high bandwidth and processing performance. This can be achieved by:

- separate special-purpose video chips (e.g., for MPEG-2, 3D-graphics, etc.);
- multi-algorithm video chip sets;
- programmable video processors which are typically very sophisticated, digital signal processors (e.g., TMS320C82, Siemens Tricore, or Hyperstone);
- specialized media processors and media coprocessors (e.g., the Philips Trimedia TM-1 (*Rathnam and Slavenburg* [238]), the MPACT (*Kalapathy*

[152], *Foley* [89]) and MPACT2 (*Yao* [329]) of Chromatic Research, the MicroUnity Media processor (*Hansen* [122]); and
- multimedia units which are multimedia extensions for general-purpose processors.

Multimedia unit. Based on the single instruction multiple data (SIMD) model, media processors and multimedia extensions for general-purpose processors process multiple sets of small operands and obtain multiple results with a single instruction. The same operation as indicated by the opcode is applied to several data items within a register simultaneously, thereby utilizing very fine-grained parallelism which is often referred to as SIMD parallelism or subword parallelism. Such subword parallel instructions deal with arithmetic and logical operations on packed data types, such as 8- by 8-bit bytes, 4- by 16-bit words, or 2- by 32-bit doublewords, all packed inside one 64-bit quadword. Operations include packing and unpacking, arithmetic, comparisons, logic, shifting, and (on SPARC and Alpha machines) motion estimation for motion-video encoding. Figure 4.19 gives an example of a 4- by 16-bit SIMD multiplication.

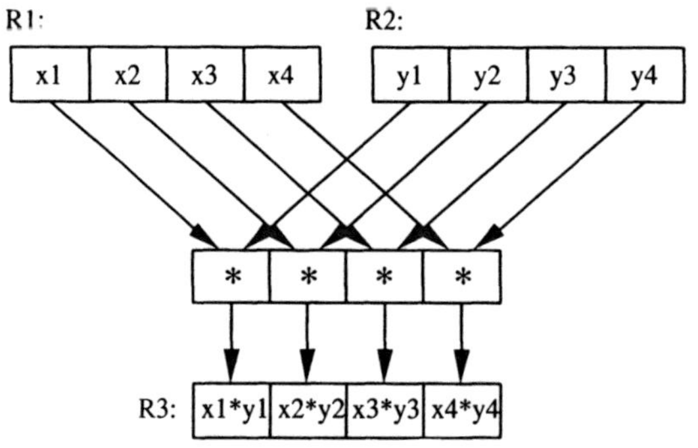

Fig. 4.19. Typical multimedia instruction execution

Multimedia units employ SIMD instructions, saturation arithmetic, and additional fixed-point arithmetic, masking, selection, reordering, and conversion instructions.

The MPEG-2 decompression algorithm does benefit highly from special multimedia instructions. Steps 3 to 6 of the MPEG-2 decompression algorithm can be executed in parallel for all blocks (and macroblocks) of a single image, applying the SIMD instructions. Such multimedia instruction sets have allowed software-only real-time video decompression without extra hardware.

Several multimedia extensions are applied in current microprocessors:

- *visual instruction set* (VIS) for UltraSPARC chips [165, 301];
- *multimedia acceleration extensions* (MAX-1, MAX-2) for HP PA-8000 and PA-8500 (*Lee* [176, 177]);

- *matrix manipulation extensions* (MMX, MMX2) for the Intel x86 (*Peleg et al.* [232, 233]);
- *the AltiVec extensions* for Motorola processors;
- *motion video instructions* (MVI) for Alpha processors; and
- *MIPS digital media extensions* (MDMX) for MIPS processors.

Sun Microsystems was one of the first companies to introduce CPU multimedia features, with its VIS for its UltraSPARC CPU. Now all major processor architectures have defined multimedia instruction extensions, which vary in scope and feature sets. These instructions accelerate calculations and make most 2D and 3D graphics and/or video, audio, voice-processing, and data communications tasks up to several times faster.

On Intel's P55C and Pentium II chips, the ALUs and eight 64-bit media registers are shared with the FPU, so MMX and floating-point instructions cannot be processed simultaneously. This is a problem during rendering operations, where the FPU is doing geometry calculation and MMX instructions are simultaneously trying to do texture mapping. Frequent switching between floating-point and MMX modes can impair performance.

VIS, MVI, and MDMX build on 64-bit RISC architectures with three-operand instructions and large sets of 64-bit registers (which are already present) to facilitate multimedia tasks. With the Alpha, on the other hand, the thirty-two 64-bit integer registers handle multimedia operations simultaneously with other integer instructions. Sun's VIS is similar to DEC's MVI in that it enables operation on an entire 4-by-4 matrix directly to its 32 registers, compared to only 8 registers for Intel's MMX.

3D graphical enhancement. MPEG-2 decompression is a good example for a class of video stream algorithms but not for 3D applications. A floating-point unit is not necessary for MPEG-2, but would be necessary for 3D applications.

The ultimate goal is the integrated real-time processing of multiple audio, video, and 2D and 3D graphics streams on a system CPU, although time is still needed to attain that level of performance.

In the context of multimedia or 3D graphical enhancements, two (or four) paired single-precision floating-point operations are executed in parallel on two (or four) single-precision floating-point values stored in a 64-bit (or 128-bit) register.

To speed up 3D applications by the main processor, fast low-precision floating-point operations are required. Moreover, reciprocal instructions are of specific importance, for example, square root reciprocal with low precision.

Such vector operations are defined by the so-called 3Dnow! extension developed by AMD and by Intel's MMX enhancement *internet streaming SIMD extension* (ISSE) (previously also code-named *Katmai new instructions*, KNI or MMX-2).

The 3DNow! defines 21 new instructions which are mainly paired single-precision floating-point operations, for example, specifying reciprocal and root reciprocal functions with single-precision format. Currently, only AMD's K6-2 series carry the 3DNow! instruction set. In the future AMD's K6-3 and K7, Cyrix's Cayenne (next-generation 6x86MX) and MXi (next generation of MediaGX), and IDT's WinChip 2 3D will all have the 3DNow! instruction set.

While 3DNow! instructions are meant to optimize 3D games and applications, Intel's MMX was primarily focused on image, audio, and video processing. For this reason, Intel introduced the ISSE set of instructions which focus on floating-point-intensive 3D graphics acceleration, and thus have much heavier memory demands. In 1999, the ISSE instruction set was first time implemented in the new Intel Pentium III processor. The 72 new instructions operate on a set of eight additional 128-bit SIMD floating-point registers. Four 32-bit low-precision floating-point operations are performed in parallel by a new SIMD floating-point unit.

Future Directions. In future, FUs which are possibly very complex are envisaged, such as a floating-point vector unit, a high-accuracy inner product floating-point unit, specialized multimedia units like a MPEG unit or a 3D-graphics unit.

FUs may be grouped relating to their usage. An example is the decoupling of integer units and floating-point units leading to separate instruction windows, forwarding paths, and logical and physical register sets. Forwarding is faster within a single group than between groups.

4.8 Finalizing Pipelined Execution

4.8.1 Completion, Commitment, Retirement and Write-Back

An instruction is *completed* when the FU has finished the execution of the instruction and the result is made available for forwarding and buffering. Instruction completion is out of program order.

We use the terms retired, committed, and removed in conformity with *Shriver and Smith* [258] in the following way. After completion, operations are committed in order. *Retiring* an operation does not imply the results of the operation are either permanent or nonpermanent. *Committing* an operation means that the results of the operation have been made permanent and the operation retired from the scheduler. Retiring means removal from the scheduler with or without the commitment of operation results, whichever is appropriate. Timing-wise, commitment and retirement often happen simultaneously. *Shriver and Smith* use the term *removed* to mean the operation is retired from the scheduler without making permanent changes.[17]

A result is made permanent either by making the mapping of architectural to physical register permanent (if no separate physical registers exist) or by copying the result value from the rename register to the architectural register (in the case of separate physical and architectural registers). The latter is often done in a write-back stage of its own after the commitment of the instruction. The main effect when using a separate write-back stage is that the rename register is freed one cycle after commitment.

4.8.2 Precise Interrupts

An interrupt or exception is called *precise* if the saved processor state corresponds with the sequential model of program execution, where one instruction execution ends before the next begins. To be more specific, the saved state should fulfil the following conditions [270]:

- All instructions preceding the instruction indicated by the saved program counter have been executed and have modified the processor state correctly.
- All instructions following the instruction indicated by the saved program counter are unexecuted and have not modified the processor state.
- If the interrupt is caused by an exception condition raised by an instruction in the program, the saved program counter points to the interrupted instruction. The interrupted instruction may or may not have been executed, depending on the definition of the architecture and the cause of the interrupt. Whichever is the case, the interrupted instruction has either ended execution or not started.

If the saved processor state is inconsistent with the sequential architectural model and does not satisfy the above conditions, then the interrupt is *imprecise*. Interrupts belong to two classes:

- Program interrupts or traps result from exception conditions detected during fetching and execution of specific instructions. These exceptions may be caused by illegal opcodes, numerical errors such as overflow, or they may be part of normal execution, e.g., page faults.
- External interrupts are caused by sources outside the currently executing instruction stream (e.g., I/O interrupts and timer interrupts). For such interrupts, restarting from a precise processor state should be made possible.

Typically in superscalar processors, instructions stay in sequence until the time they are issued. Moreover, the processor state is not modified by an

---

[17] Unfortunately, the terms completion, retirement, and commitment are often used interchangeably or with different meaning in the literature. *Hennessy and Patterson* [134] use the terms completed and committed as follows: when an instruction is guaranteed to complete, it is called committed.

instruction before it issues. When an exception condition can be detected prior to issue, instruction issuing is simply halted and the processor waits until all previous issued instructions are retired. Then the processor is in a precise state with the program counter corresponding to the instruction being held in the issue register. Registers and main memory are in a state consistent with this program counter value. Examples of such exceptions are all external interrupts that can be checked at the issue stage, and program interrupts such as illegal opcodes or privileged instruction faults [270].

Processors often have two modes of operation. One mode guarantees precise exception and another mode, which is often 10 times faster, does not. Such processors are the POWER2, Alpha 21064, and MIPS R8000. The faster mode allows more overlap in long-latency floating-point operation, while precise exceptions are usually supported for integer operations [134].

4.8.3 Reorder Buffers

In this chapter we usually assume a reorder buffer implementation to organize an in-order retirement and allow for precise exceptions. The *reorder buffer* keeps the original program order of the instructions after instruction issue and allows result serialization during the retire stage. State bits store whether an instruction is on a speculative path, or, when the branch is resolved, whether the instruction is on a correct path or must be discarded. When an instruction completes, the state is marked in its entry. Also, when a program interrupt occurs, the exception is marked in the reorder buffer entry of the triggering instruction. The reorder buffer is implemented as a circular FIFO buffer. Reorder buffer entries are allocated in the (first) issue stage and deallocated serially when the instruction retires.

During the retire stage a number of instructions at the head of the FIFO queue are scanned and an instruction is committed if all previous instructions are committed or can be committed in the same cycle. In the case of instructions that are on a misspeculated path, the instructions are removed from the reorder buffer and the physical registers freed without making the results permanent or copying back results. The same happens for all subsequent instructions after an interrupted instruction. The fetch unit is notified to restart fetching instructions from the correct path. Typically the retire bandwidth is the same as the issue bandwidth.

There are several differing implementations of the reorder buffer. The reorder buffer may also be defined to hold the result values of completed instructions instead of rename registers (see *Johnson* [150]). The reorder buffer described above does not hold result values but only instruction execution states. It is close to *Johnson*'s description of a reorder buffer in combination with a so-called *future file*. The future file is the working file used for computation by the FUs, i.e., it is similar to the set of rename registers that are separate to the architectural registers. In contrast, *Smith and Pleskun* [270] describe a reorder buffer in combination with a future file, whereby both receive and store results at the same time. Moreover the instruction window can be combined with the reorder buffer into a single buffer unit.

4.8.4 Checkpoint Repair Mechanism and History Buffer

There are also a number of other ways to implement recovery and restart mechanisms. Besides the different reorder buffer variations, which are most common, such ways include checkpoint repair and a history buffer.

In the *checkpoint repair mechanism*, the processor provides a set of logical spaces, where each logical space consists of a full set of software-visible registers and memory. One is used for current execution, the others contain back-up copies of the in-order state that correspond to previous points in the execution. At various times during execution, a check point is made by copying the architectural state of the current logical state to the back-up space. Restarting is accomplished by loading the contents of the appropriate back-up stage into the current logical state [150].

The *history buffer* was proposed by *Smith and Pleskun* ([270], 1985) together with the reorder buffer and future file as another alternative for recovery organization. There are no rename registers in a history buffer organization. Rather the (architectural) register file contains the current state, and the history buffer contains old register values which have been replaced by new values. The history buffer is managed as a LIFO stack, and the old values are used to restore a previous state, if necessary. A history buffer organization was used for the Motorola 88110 microprocessor.

4.8.5 Relaxing In-order Retirement

As described above, retiring is always strictly in program order, thereby guaranteeing result serialization as demanded by the serial instruction flow of the von Neumann architecture. The only relaxation that may exist is in the order of load and store instructions which may arrive at the processor in an order different to the program order. Thus even a fully parallel and highly speculative processor must sometimes look like a simple von Neumann processor when it was state-of-the-art in the 1950s.

Relaxing in-order retirement is not implemented in today's superscalar microprocessors. Nevertheless it is possible. Assume an instruction sequence A ends with a branch that predicts an instruction sequence B, and B is followed by a sequence C which is not dependent on B. Thus C is executed independently from the branch direction. Therefore, instructions in C can start to retire before B. If predication is provided by the processor, B can also be implemented by predicated instructions to remove the branch. Then the instructions in C can be retired before the predicated instructions.

There are at least two complications. An interrupt signaled by one of the instructions in B is hard to implement in a precise manner. Moreover, it is difficult to relax retirement constraints without sacrificing binary code compatibility with legacy code.

4.9 State-of-the-Art Superscalar Processors

4.9.1 Intel Pentium family

Intel's family of CISC microprocessors are the most commercially important microprocessors to date. The architectural line of the Intel ISA started with the CISC microprocessor Intel 8086, successively refined by the scalar processors 8088, 80286, Intel386, Intel486 (see Table 3.1) and then continued with superscalar processors from the families P5 and P6 (see Table 4.3).

The Pentium processor was a member of the P5 family (*Alpert and Avnon* [6]) and was the first Intel 2-issue superscalar processor. In 1995, it was followed by the Pentium Pro, which was the first member of the P6 family (*Colwell and Steck* [53], *Gwennap* [114], *Papworth* [227]). In 1997, Intel introduced MMX into the P5 and P6 families which resulted in the Pentium MMX processor (also called P55C) and the Pentium II processor ([143, 144]; *Bhandarkar and Ding* [27]). In 1999, ISSE instruction set was implemented in the Pentium III processor.

Table 4.3. The Intel P5 and P6 family

Year	Type	Transistors (x1000)	Technology (μm)	Clock (MHz)	Issue	Word format	L1cache	L2cache
1993	Pentium	3100	0.8	66	2	32-bit	2 X 8 kB	
1994	Pentium	3200	0.6	75-100	2	32-bit	2 X 8 kB	
1995	Pentium	3200	0.6/0.35	120-133	2	32-bit	2 X 8 kB	
1996	Pentium	3300	0.35	150-166	2	32-bit	2 X 8 kB	
1997	Pentium MMX	4500	0.35	200-233	2	32-bit	2 X 16 kB	
1998	Mobile Pentium MMX	4500	0.25	200-233	2	32-bit	2 X 16 kB	
1995	PentiumPro	5500	0.35	150-200	3	32-bit	2 X 8 kB	256/512 kB
1997	PentiumPro	5500	0.35	200	3	32-bit	2 X 8 kB	1 MB
1998	Intel Celeron	7500	0.25	266-300	3	32-bit	2 X 16 kB	–
1998	Intel Celeron	19000	0.25	300-333	3	32-bit	2 X 16 kB	128 kB
1997	Pentium II	7500	0.25	233-450	3	32-bit	2 X 16 kB	512 kB
1998	Mobile Pentium II	7500	0.25	300	3	32-bit	2 X 16 kB	512 kB
1998	Pentium II Xeon	7500	0.25	400-450	3	32-bit	2 X 16 kB	512 kB/1 MB
1999	Pentium II Xeon	7500	0.25	450	3	32-bit	2 X 16 kB	512 kB/2 MB
1999	Pentium III	9500	0.25	450-500	3	32-bit	2 X 16 kB	512 kB
1999	Pentium III Xeon	9500	0.25	500-550	3	32-bit	2 x 16 kB	512 kB

In the following, we focus on the P6 family of processors. The first processor in this family was the Pentium Pro. In 1997, it was followed by the Pentium II, in 1998, by the Pentium II Xeon (targeted for servers and workstations) and Celeron (targeted for desktops), and in 1999, by Pentium III and Pentium III Xeon.

Pentium II

Let us describe the Pentium II in more detail. Although it is probably the processor which is most often written about, its full details are still not known. The Pentium II (as other members of P6) implements an out-of-order su-

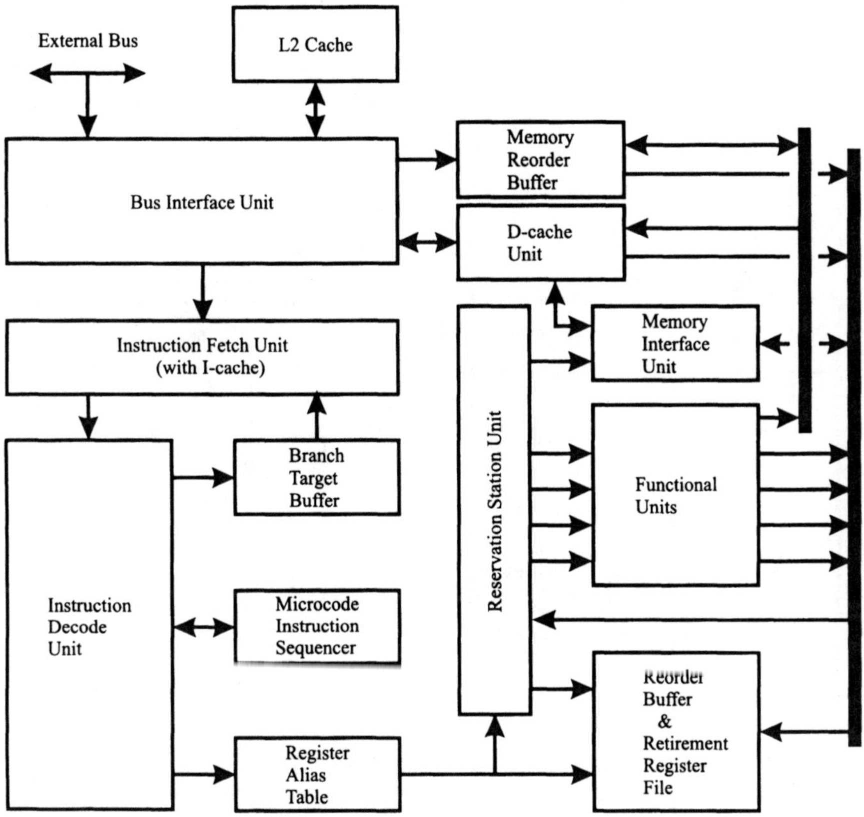

Fig. 4.20. The Pentium II microprocessor

perscalar issue – so-called dynamic execution[18] – which employs register renaming, non-blocking caches, and multiprocessor bus support. The ISA is Intel IA-32, which is basically the x86 instruction set with some extensions. Intel IA-32 instructions begin and end execution in program order. They

[18] This phrase was coined to represent a number of other words that appear in the context of dynamic micro dataflow execution, e.g., out-of-order, speculative execution, superscalar, and superpipelined.

are translated into a sequence of simpler RISC-like micro-operations (μops) which are register-renamed and placed into the central instruction window, the so-called reservation station unit which is described as an out-of-order speculative pool of pending operations. Once the data arguments and the necessary resources are available, the μops are issued for execution in their out-of-order execution engine. After execution has completed, the μops of an instruction are held in the reorder buffer until they can be retired, which may occur only after all previous instructions' μops have been retired, and all of the constituent μops have completed. Up to three μops can be retired per clock cycle, yielding a theoretical minimum of 0.33 cycles per μop.

The processor is organized in three sections: an in-order section, an out-of-order execute section, and an in-order retire section (Fig. 4.20). The two in-order sections guarantee the sequential program semantics for the Pentium II as for an Intel 486 processor.

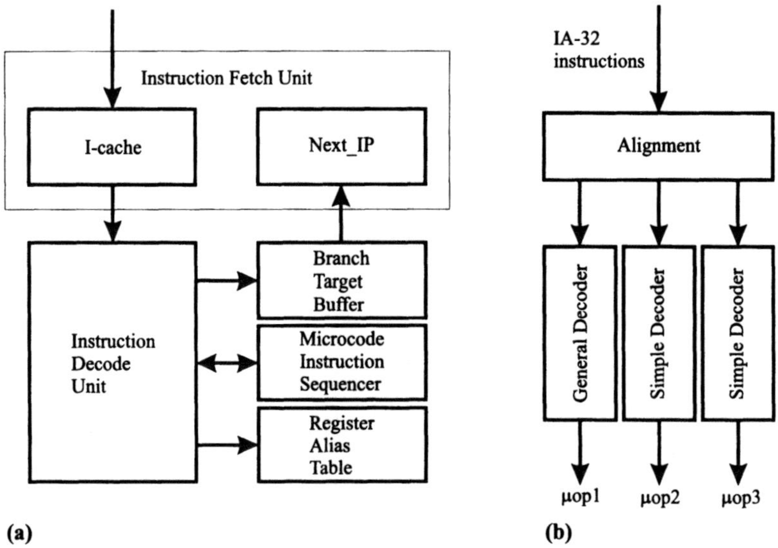

Fig. 4.21. Inside the Pentium II fetch/decode/issue section

The in-order section. This section is depicted in Fig. 4.21a. The *instruction fetch unit* (IFU) contains a non-blocking I-cache and Next_IP unit. The Next_IP unit provides the I-cache index (based on inputs from the BTB), trap/interrupt status, and branch misprediction indications from the integer FUs. The processor implements a branch prediction scheme derived from the two-level adaptive scheme described by *Yeh and Patt* [331] (see Sect. 4.3.3).[19] The BTB, which contains 512 entries, maintains branch history information and the predicted branch target address. Mispredicted branches incur

[19] Exactly which of *Yeh and Patt*'s schemes has been applied is not publicly known.

a penalty of at least 11 cycles, with an average misprediction penalty of 15 cycles [114].

The cache line corresponding to the index from the Next_IP and the next line are fetched from the I-cache. The 16 bytes are aligned in order to mark the beginning and end of each of the fetched IA-32 CISC instructions (which are of variable length and up to 7 bytes long). The *instruction decoder unit* (IDU) is composed of three separate decoders, one for each aligned IA-32 instruction (Fig. 4.21b). A decoder breaks the IA-32 instruction down to µops which are the atomic units of work in the P6 processor, each comprised of an opcode, two source operands, and one destination operand. These µops are of fixed length. Most IA-32 instructions are converted directly into single µops (by any of the three decoders), some instructions are decoded into one to four µops (by the general decoder), while more complex instructions are used as indices into the *microcode instruction sequencer* (MIS) which will generate the appropriate stream of µops. In general, simple register-register and load instructions are only one µop; store and read-modify instructions are translated into two µops; simple register-memory instructions have two to three µops; and simple read-modify-write instructions have four µops. Decoding at most three IA-32 instructions generates at most six µops per clock cycle (four by the general decoder and one by each simple decoder). At most three µops can be forwarded from the IDU into the pipeline. The P6 may be viewed as a three-issue superscalar processor.

The µops are queued, and sent to the *register alias table* (RAT) where register renaming is performed, i.e., the logical IA-32-based register references are converted into references to physical registers. Then, with added status information, µops continue to the *reorder buffer* (ROB) and to the *reservation station unit* (RSU).

The out-of-order execute section. The µops have to go to the ROB to ensure in-order retirement after out-of-order completion. Thus, when the µops flow into the ROB they effectively take a place in line so that it is remembered how to retire them later and keep the sequential program semantics. µops also go to the RSU which forms a central instruction window with 20 *reservation stations* (RS), each capable of hosting one µop. If the status indicates that a µop has all of its operands, and if the FU needed by that µop is also available, the RSU removes that µop and issues it to the FU where it is immediately executed. µops are issued to the FUs according to dataflow constraints and resource availability, without regard to the original ordering of the program. The RSU has five ports and can issue at a peak rate of 5 µops per clock cyle, though a sustained rate of 3 µops per clock cyle is more typical (Fig. 4.22).

Several FUs can be clustered on a port: integer FUs on port 0 and 1, floating-point FUs on port 0, MMX FUs on port 0 and 1, a jump FU (port 1). Ports 2, 3, and 4 are dedicated to memory access with a load FU attached

Table 4.4. Latencies and throughput for different Pentium II FUs

RSU Port	FU	Latency	Throughput
0	Integer arithmetic/logical	1	1
	Shift	1	1
	Integer mul	4	1
	Floating-point add	3	1
	Floating-point mul	5	0.5
	Floating-point div	long	nonpipelined
	MMX arithmetic/logical	1	1
	MMX mul	3	1
1	Integer arithmetic/logical	1	1
	MMX arithmetic/logical	1	1
	MMX shift	1	1
2	Load	3	1
3	Store address	3	1
4	Store data	1	1

to port 2, and two store FUs attached to port 3 and 4. Table 4.4 gives latencies and throughtputs for different FUs.

After completion the result goes to two different places, RSU and ROB. There may be other μops in RSU waiting for the result before they themselves become ready. Therefore, each port has its own write-back path back to the RSU. There is a full crossbar between all those ports so that any returning result could be bypassed to any other FU for the next clock cycle.

The in-order retire section. The other place where the result of a μop goes is the ROB. The retire section (see Fig. 4.23) controls the in-order retirement of μops. A μop can be retired if its execution is completed, if it is its turn in program order, and if no interrupt, trap, or misprediction occurred. Retirement means taking data that was speculatively created and writing it into the *retirement register file* (RRF). Three μops per clock cycle can be retired.

Branch μops are tagged (in the in-order pipeline) with their fall-through address and the destination that was predicted by the BTB. After the branch resolution the branch outcome is compared against what was predicted. If they coincide, the branch μop eventually commits and the speculatively executed μops between it and the next branch in ROB can be retired. If they do not coincide, however, the jump FU changes the status of all of the μops behind the branch to remove them from the ROB. In that case the proper branch destination is provided to the BTB which restarts the whole pipeline from the new target address [145].

The Pentium II pipeline. The flow of μops through the processor is controlled by the pipeline shown in the Fig. 4.24. The pipeline is segmented into three pieces: an in-order pipeline, out-of-order execute pipelines, and an in-

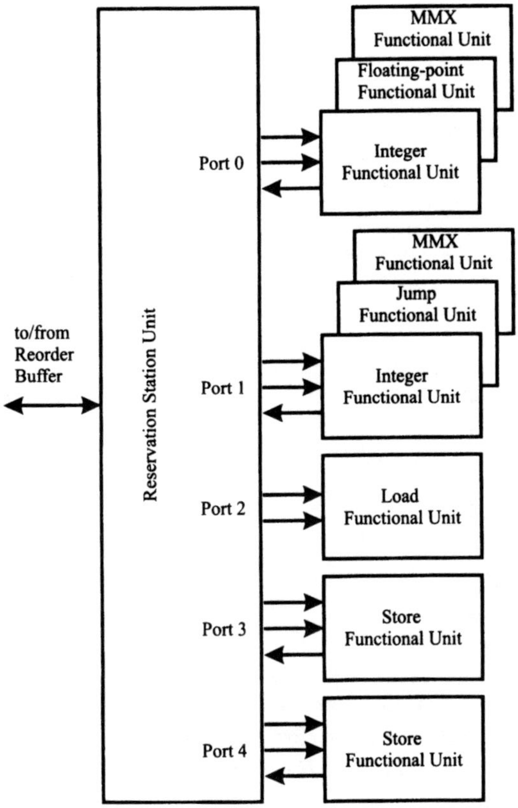

Fig. 4.22. Inside the Pentium II issue/execute/complete section

order retire pipeline.[20] As the μops flow from one segment into the next, the reservation station scheduling and retirement scheduling is performed.

The details of these pipeline sections are:

- The in-order pipeline section involves nine clock cycles (Fig. 4.24a). The first two identify the next instruction pointer (Next_IP). It is the BTB deciding where is the best place to look in the I-cache for the next cache line. The I-cache access and instruction alignment and predecode take the next three clock cycles. The instruction is decoded in the next two cycles using two simple decoders and one general decoder. Register renaming takes the next clock cycle. The final stage of the in-order pipeline, ROB read, can usually be overlapped with at least one of the clock cycles in the next pipeline segment.
- The out-of-order execute pipelines are used to execute integer, jump, floating-point, MMX, and memory access functions (Fig. 4.24b). These pipelines share the beginning stage, RSU write, during which the RSU

[20] This is why Intel uses the term superpipeline.

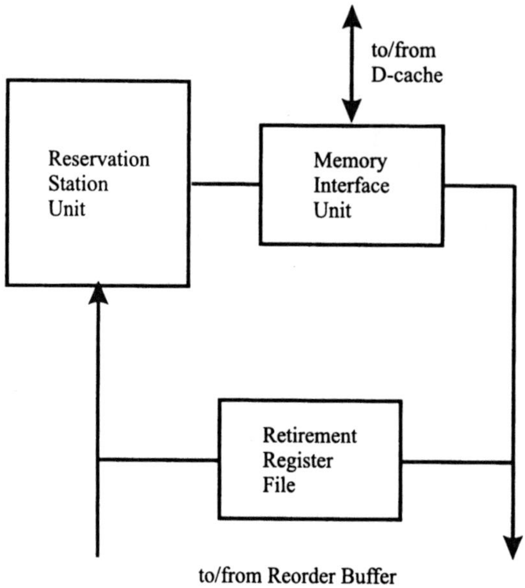

Fig. 4.23. Inside the Pentium II retire section

identifies μops that have all the operands and are ready to be issued to FUs. For an integer μop one cycle is needed for the execution and the return of the results. For a floating-point μop the execution stretches out over several additional cycles. For a load/store μop even more additional clock cycles are needed to perform address calculation, access to D-cache (which is pseudo-dual ported via interleaving with one port dedicated to loads and the other to stores), and L2 cache access, if necessary. Once a FU has created its result, it sends it back to the RS to enable future μops and also into the ROB to enable retirement.

- The in-order retire pipeline takes two clock cycles, one for ROB write-back and one for the retiring (Fig. 4.24c). Since an instruction may be mapped into several μops, the retirement process has to make sure that if any of these μops is retired, all of them are retired automatically. Otherwise the processor might enter an inconsistent state if an interrupt appeared during partial retirement of these μops.

Pentium II offsprings

The next Pentium II core is Pentium III in February 1999, initially at 450 MHz and 500 MHz. Pentium III (initially code-named Katmai) has the *internet streaming SIMD extension* (ISSE)[21] instruction set, which includes floating-point SIMD instructions and eight new 128-bit floating-point SIMD registers

[21] The old name of ISSE was *Katmai new instructions* (KNI) or MMX2.

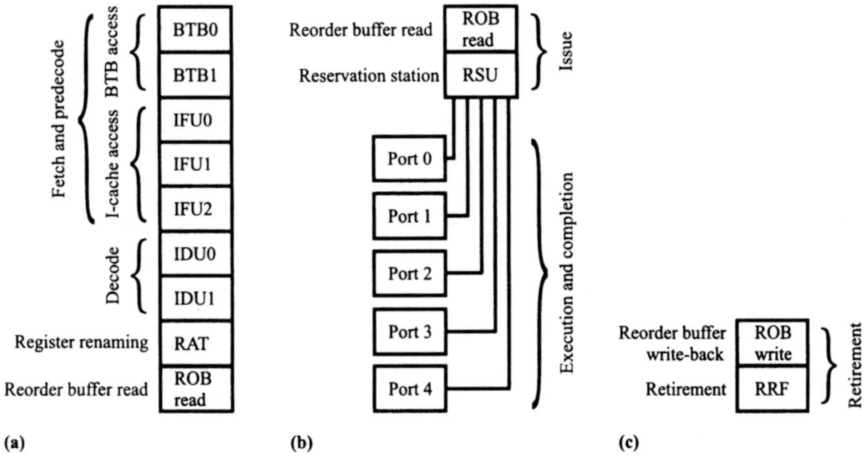

Fig. 4.24. The Pentium II pipeline (a) in-order pipeline (b) out-of-order execution pipelines (c) in-order retire pipeline

to accelerate 3D graphics. The Pentium III is very similar to the Pentium II except for the ISSE ISA enhancement, the new floating-point SIMD registers, and a SIMD FUs. Further extensions are a higher-performant write buffer, cache prefetch instruction, and a non-cached store instruction. Pentium III Xeon (initially code-named Tanner) is Pentium II Xeon with ISSE and starting off with 500 or 550 MHz in March 1999.

Coppermine will be a shrink of Pentium III down to 0.18 µm. *Cascades* will be a cheaper version of Pentium III Xeon with a clock speed of more than 600 MHz, with on-die 256 kB L2 cache. For mid-2000 Intel expects to launch *Merced*, which is to be the first member of the Intel's P7 family of 64-bit processors. The P7 processors will be based on the EPIC design style that was developed by Hewlett-Packard and Intel (see Sect. 4.10.2).

Alternatives to the Pentium

Several (super)scalar processors compete with Intel's Pentium to meet the needs of the basic personal computer market. Among these are mP6 designed by Rise Technology Company, AMD-K6 designed by Advanced Micro Device, M II designed by Cyrix, and WinChip C6 designed by Integrated Device Technology. A comparison of their basic features is given in Table 4.5.

We now describe in more detail the approaches taken by AMD and Cyrix.

4.9.2 AMD-K5, K6 and K7 families

Based on the 29000 series of scalar processors, Advanced Micro Device designed in 1990 a superscalar processor 29050. The processor has a redesigned

Table 4.5. Some of the competitors for the personal computer market

	Pentium	Pentium II	Celeron	AMD-K6	M II	mP6	WinChip C6
Company	Intel	Intel	Intel	AMD	Cyrix	Rise	IDT
Superscalar	Yes	Yes	Yes	Yes	Yes	Yes	No
x86 IPC	2	3	3	2	2	3	1
MMX IPC	2	2	2	1	1	3	1
Pipelined FPU	Yes	Yes	Yes	No	No	Yes	No

floating-point unit so that four instructions can be issued to execute out of order and speculatively.

In late 1995 AMD dropped development of these processors in favor of the more profitable clones of the Intel 80x86 processors. Much of the development of the 29000 superscalar core was shared with the new AMD-K5 processor. This processor, compatible with Intel's Pentium, was able to translate IA-32 instructions to RISC-style instructions. Its performance stems from AMD's independently developed 4-issue superscalar 5-stage pipelined architecture with 6 parallel FUs. This 32-bit processor contains a 16 kB I-cache and an 8 kB D-cache. Several versions of the AMD-K5 processor have been made between 1995 and 1997 operating on frequencies from 50 MHz up to 166 MHz.

In 1997 AMD introduced the AMD-K6 processor (*Shriver and Smith* [258]), which contains parallel decoders, a centralized RISC86 operation scheduler, and seven FUs that support superscalar operation of Intel's IA-32 instructions. AMD's RISC86 microarchitecture implements the IA-32 instruction set by internally decoding IA-32 instructions into the simpler, fixed-length RISC86 operations (RISC86ops).[22]

The L1 cache is two-way set-associative with a separate 32 kB I-cache and a 32 kB D-cache. The 32-byte cache lines are prefetched from main memory using an efficient pipelined burst transaction. As the I-cache is filled, each IA-32 instruction is analyzed using predecoding logic. More precisely, the predecode logic supplies the predecode bits associated with each IA-32 instruction byte. Among other information, the predecode bits indicate the number of bytes to the start of the next IA-32 instruction. These bits are stored in the I-cache beside each IA-32 instruction byte.

Up to 16 bytes (with predecode bits) per clock cycle can be fetched from the I-cache or BTC and sent to a 16-byte instruction buffer which, in turn, feeds them directly into the decoders. The decoders translate up to two IA-32 instructions per clock cycle into RISC86ops. There are four decoders:

- two parallel short decoders, which translate the most commonly used IA-32 instructions into zero, one, or two RISC86ops each, and are designed to

[22] A similar approach is taken by the Intel P6 processors which translate IA-32 instructions into μops.

decode up to two IA-32 instructions per clock cycle (of the two instructions, at most one can be a MMX instruction);

- one long decoder, which handles commonly used IA-32 instructions that can be represented in four or fewer RISC86ops; and
- one vectoring decoder for handling all other translations in concert with RISC86ops sequences fetched from an on-chip ROM.

The instruction scheduling is performed by the *instruction control unit* (ICU), which buffers and manages up to 24 RISC86ops at a time in order to use efficiently the 6-stage pipeline and the 7 parallel FUs. ICU controls the out-of-order execution, data forwarding, register renaming, simultaneous issuing and retirement of multiple RISC86ops, and speculative execution. In one clock cycle, the scheduler accepts up to 4 RISC86ops from the decoders, issues up to 6 operations to FUs (floating-point, multimedia, branch, load, store, and two integer) and retires up to 4 operations. There are 48 physical registers in the register file, 24 of which are general registers and the other 24 are renaming registers. The processor's two-level dynamic branch prediction logic consists of an 8 192-entry BHT, a BTC and a return address stack. Since the BHT does not store predicted target addresses, special *address ALUs* calculate target addresses on-the-fly during instruction decode. The BTC augments predicted branch performance by avoiding a one-cycle cache fetch penalty. This specialized target cache supplies the first 16 bytes of target instructions to the decoders when the branches are predicted.

The AMD-K6 is fabricated in 0.35 and 0.25 μm technology, contains 8.8 million transistors and runs on frequencies from 180 to 233 MHz. Table 4.6 summarizes some of the features of the AMD-K6 and Intel Pentium II.

In 1998, the AMD-K6-2 processor was launched. This is the first AMD processor to feature the new 3DNow! technology which is aimed to enhance floating-point intensive 3D graphics and multimedia performance.[23] The processor operates at clock speeds of 300–400 MHz (450 MHz in 1999), and is fabricated in 0.25 μm technology with 9.3 million transistors. The AMD-K6-3 is a forthcoming improved version of the AMD-K6-2 processor with on-chip 256 kB L2 cache operating at the processor's frequency (at least 450 MHz), and optionally supporting L3 caches. Due to the on-chip L2 cache the AMD-K6-3 will consist of 21.3 million transistors. Table 4.6 summarizes some of the features of the AMD-K6-3 and Intel Pentium III (see *Stiller* [279], 1999).

In 1999, the next generation of AMD-K7 processors with 3DNow! is expected. These processors will run at more than 500 MHz, and will feature a 9-issue superscalar microarchitecture, superscalar pipelined floating-point execution unit, 128 kB of on-chip L1 cache, and support for scalable multiprocessing.

---

[23] A similar direction is taken by the Intel's ISSE instruction set that is used by the Pentium III.

Table 4.6. AMD-K6 vs Pentium II and AMD-K6-3 vs Pentium III

Features	AMD-K6	Pentium II
RISC core	Yes	Yes
Speculative execution	Yes	Yes
Out-of-order execution	Yes	Yes
Data forwarding	Yes	Yes
Register renaming	Yes	Yes
IA-32 decoders	2 short or 1 long or 1 vector	2 simple and 1 general
Execution pipelines	6	5
Branch prediction	Yes	Yes
BHT	8192-entry	512-entry
BTC	16-entry	--
Executes MMX	Yes	Yes
L1 I-cache/D-cache	32/32 kB	16/16 kB
Bus width	64-bit	64-bit
Max. memory bandwidth	528 MB/s	528 MB/s

Features	AMD-K6-3	Pentium III
L1 cache (kB)	2 x 32	2 x 16
L2 cache (kB)	256	512
L2 cycle rate	1:1	1:2
Integer pipeline depth	6	12
Integer units	2	2
Floating-point units	1 (nonpipelined)	1 (pipelined)
MMX units	2	2
Floating-point SIMD units	2	1 + mul/div
Floating-point SIMD registers lenght	64 bit	128 bit
Floating-point SIMD registers	8 x 2 single precision	8 x 4 single precision

4.9.3 Cyrix M II and M 3 Processors

Cyrix was already well known for its x86-based scalar processor 6x86 and its many scalar offspring (e.g., the low-voltage version 6x86L, MMX-supporting versions 6x86MX, 3D application-oriented Cayenne, two-chip low-cost MediaGX and its 2D- and 3D-supporting upgrade MXi). In 1997, however, the Cyrix M II processor appeared, based on the proven low-cost 6x86 processor core, but as a superscalar processor that operates at higher frequences (233 MHz with 0.25 μm technology) and contains two separate pipelines. It features a 64 kB unified L1 cache (4-way associative, dual-port address), a two-level TLB and a 512-entry BTB. The Cyrix M II processor supports the MMX ISA extension. Cyrix's next generation processor M 3 (code-named Jalapeno), is expected to debut in late 1999. Since the 6x86 processor, M 3 will be the first completely new architecture, with an 11-stage pipeline, a completely new floating-point unit, a 3D graphics engine, 256 kB on-chip L2 cache (8-way associative, 8-way interleaved, fully pipelined), on-chip memory controller allowing 3.2 GB/s transfer rate. The M 3 will be produced in 0.18 μm technology and will run in the 600–800 MHz clock speed range. Similarly to AMD-K7 processor, the M 3 will support execution of both MMX and 3DNow! instructions.

4.9.4 DEC Alpha 21x64 family

In the early 1990s DEC introduced the Alpha architecture which is a 64-bit RISC architecture designed with particular emphasis on clock speed and multiple instruction issue (see *Sites* [265]). In the following years, three generations of implementations of Alpha architecture appeared, and the fourth is to come in mid-2000 (see Table 4.7).

The first generation started in 1992 with the 21064 processor, which was a 64-bit 2-issue RISC microprocessor running at 200 MHz. It was fabricated in 0.75 μm CMOS technology and contained 1.68 million transistors (*Dobberpuhl et al.* [69], 1992). The 0.5 μm technology made it possible to use 2.5 million transistors to double the on-chip cache to 32 kB (compared with the 21064) and thus to produce the 21064A processor, which was running at frequencies up to 300 MHz. Another two representatives of the first Alpha generation were the 21066A and 21068 processors. The 21066A was a highly integrated implementation for high-performance, PCI-based systems. On-chip functions included an industry-standard PCI I/O controller and a 21066A-exclusive graphics accelerator. The processor was offered with clock frequency up to 233 MHz. The 21068 processor was a lower-frequency version of the 21066, running at 66 MHz.

The second generation of Alpha implementations introduced 4-issue superscalar processors 21164 and 21164PC (*Edmondson et al.* [74], 1995). The 21164 is a 64-bit in-order issue processor running at frequencies up to 612 MHz, and is fabricated in 0.35 μm technology with 9.3 million transistors. The 21164PC is based on the 21264 and uses DEC's MVI to enhance visual computing and multimedia performance. The operating frequency is up to 533 MHz.

Table 4.7. The Alpha 21x64 family

Year	Type	Technology (μm)	Clock (MHz)	Issue	Out-of-order issue	Word format	Internal caches (kB)	SPECint95	SPECfp95
1992	21064	0.75	200	2	No	64-bit	2 X 8	2.31	3.22
								@ 166 MHz	
1994	21064A	0.5	200-300	2	No	64-bit	2 X 16	4.18	5.78
								@ 266 MHz	
1996	21164	0.35	300-600	4	No	64-bit	2 X 8 + 96	18.0	27.0
								@ 600 MHz	
1997	21164PC	0.35	400-533	4	No	64-bit	8 + 16	12.6	16.1
								@ 533 MHz	
1998	21264	0.35	575	6	Yes	64-bit	2 X 64	30.3	47.7
								@ 575 MHz	
1999	21264	0.25	600-1000	6	Yes	64-bit	2 X 64	100+	150+
2000	21364	0.18	1200+	N/A	N/A	64-bit	N/A	140+	200+

The third generation of Alpha implementations introduces out-of-order execution with its 6-issue 21264 processor (see *Gwennap* [115]). The 21264 was scheduled for volume production in late 1998 in 0.35 μm technology. It contains 15.2 million transistors; although most are in the large caches and the branch predictor, the CPU core contains about 6 million transistors. The Alpha 21264 is expected to operate at more than 1 000 MHz by the year 2000.

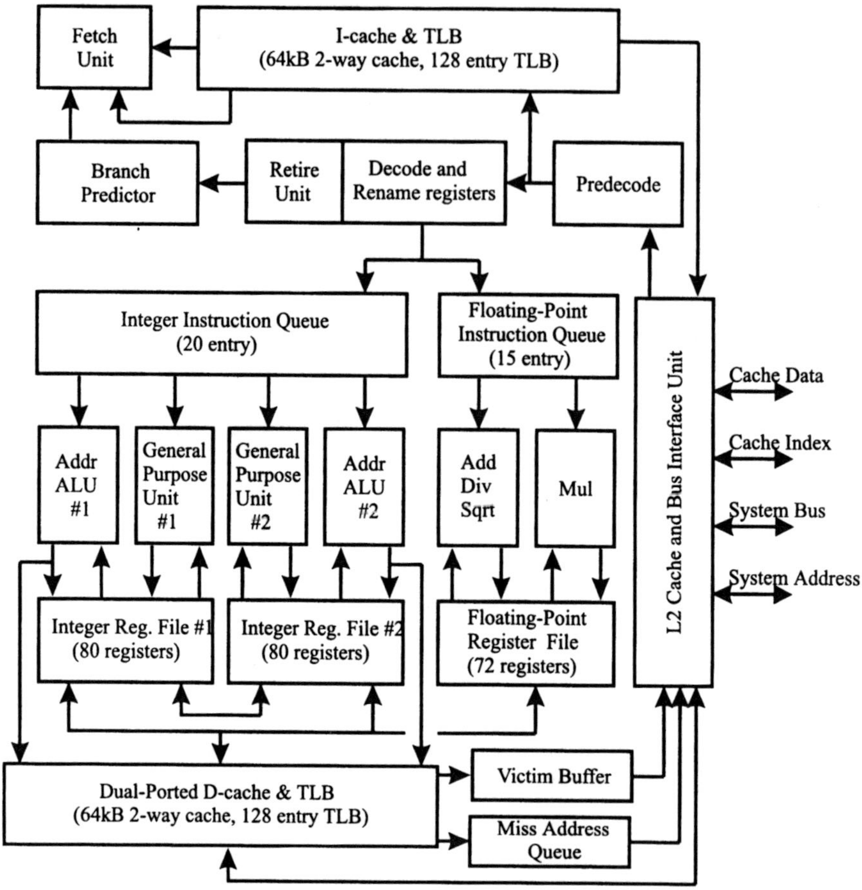

Fig. 4.25. The Alpha 21264 microprocessor

As Fig. 4.25 shows the 21264 processor contains a fetch unit, two general-purpose units, two address arithmetic logic units, two floating-point units, a retire unit, and a bus interface unit.

Simple instructions are processed in the following 7-stage pipeline (for load/store instructions 9 stages are necessary, floating-point operations need additional execute stages, see Fig. 4.26):

1. Fetch: fetch instructions using branch prediction,
2. Transit: transfer instructions to the decoder,
3. Map: rename registers,
4. Queue: place the instructions either in the integer or in the floating-point instruction queue,

5. Register: read operands and issue to FU,
6. Execute: execute integer (single cycle) or floating-point instructions (several cycles),
7. Write: write results.

The processor can keep up the peak execution rate of 6 IPC and a sustainable rate of 4 IPC on either integer or floating-point code. Up to 80 instructions can be in process at once in different pipeline stages. Let us describe this in more detail.

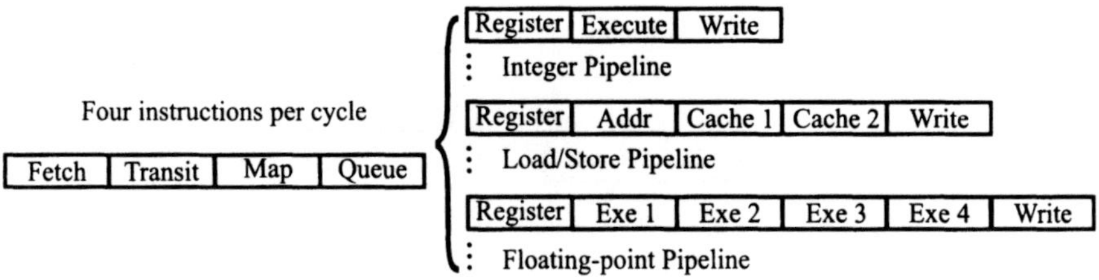

Fig. 4.26. Pipelines in Alpha 21264 microprocessor

The 21264 has on-chip I- and D-caches, which are 64 kB 2-way set-associative primary (L1) caches. Unfortunately, it generally takes two cycles to access such large primary caches: the cache access can be done in one cycle, but it takes nearly a full cycle, leaving no time to move the address/data any significant distance across the large die. This is especially the case with the D-cache. As a result, it takes two cycles for a load instruction to get an address to the D-cache, access the cache array, and return the data from the cache to the requester. Consequently, an instruction must wait if it requires data from an immediately preceding load instruction. Nevertheless, being an out-of-order execution processor, the 21264 may execute other independent instructions while waiting for the D-cache and thus compensate for the performance degradation caused by the additional cache access cycle.[24]

In the I-cache, each line holds four instructions along with a *next-line predictor* and a *set predictor*. Since the 21264 cache can produce a result in a single cycle, these two fields are immediately sent back to the cache inputs to start the next access (see Fig. 4.27). If the line contains a branch that is predicted taken, the predictor fields point to the cache line and set of the predicted target. Thus, assuming the prediction fields are correct, a taken branch has a zero-cycle delay, since the target group of instructions is fetched immediately after the branch. The predictors are initialized to point to the sequential address; they are updated with the branch target address when a branch is predicted taken.

[24] DEC estimates that adding a cycle of D-cache latency costs about 4 % in overall performance.

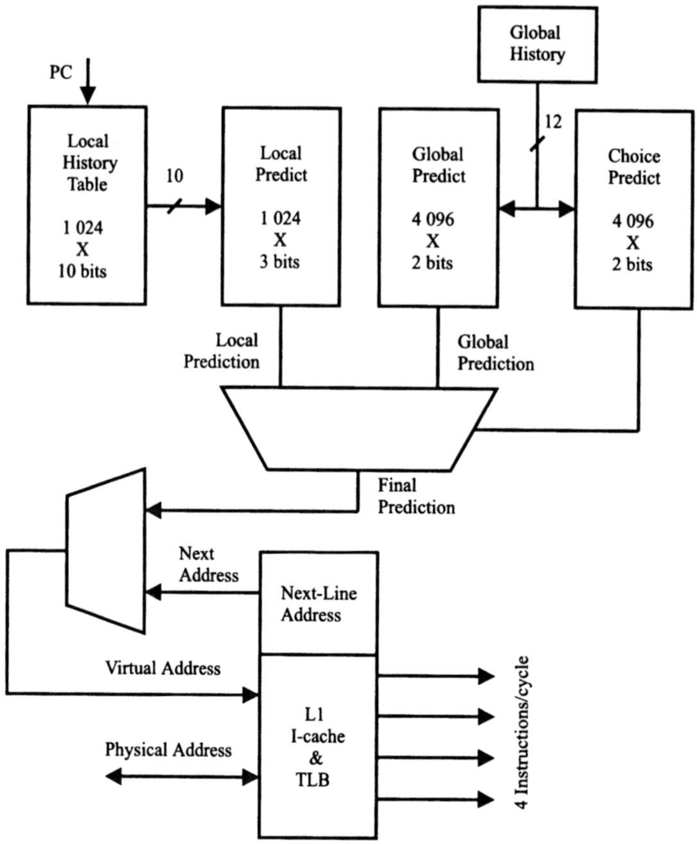

Fig. 4.27. Instruction fetch in the Alpha 21264 microprocessor

The prediction fields are controlled by the *branch prediction unit*, resulting in a high degree of accuracy (see Fig. 4.27). In the 21264, the average mispredicted branch penalty is more than 11 cycles; therefore, the extra I-cache cycle results in about a 10 % increase in the mispredicted branch penalty. The impact on overall performance is about 1 %. The processor uses a *hybrid* predictor developed by *McFarling* [196] (see p. 144). The cost of this branch prediction is about 35 kbits of storage for all the requisite branch history information, which consumes about 2 % of the processor's total die area. This figure does not include the predicted target addresses which are stored in the I-cache on a per-line basis, adding another 48 kbits.

Such a fetch stage allows feeding four instructions per cycle into the *decode unit* with little external intervention. Registers are renamed on-the-fly, with 41 extra integer registers (out of 80 in the integer register file) and 41 extra floating-point registers (out of 72 in the floating-point register file). After decode, instructions are assigned to either the *integer instruction queue* (20 entries) or the *floating-point instruction queue* (15 entries). Each cycle, all instructions that have their operands available arbitrate for access to the FUs, with instructions that have been in the queue longest having the highest priority. After arbitration, instructions can be issued to the FUs. Instructions that are data dependent and are waiting for data, are bypassed in favor of those that can execute right away, thus creating opportunities for out-of-order execution.

The 21264 includes four integer FUs: two *general-purpose units* and two *address ALUs*. The first pair execute arithmetic and logical operations, shifts, and branches; one general-purpose unit has a multiplier while the other handles the MVI instructions, which are also included in the 21264's instruction set (see Sect. 4.7). The address ALUs execute all load/store instructions (for either integer or floating-point) and can also perform simple arithmetic and logical operations. Hence, the integer FUs can execute four instructions per cycle for most instruction mixes.

A standard implementation of the above would require a register file with eight read ports and six write ports. The physical width of such a register file would cause the entire datapath to be distended, increasing the cycle time beyond the tight target. Instead, the *integer register file* was duplicated, with each copy having four read and six write ports, reducing the datapath width significantly. Each register file copy services one general-purpose unit and one address ALU in a grouping called a *cluster*. The register file copies are kept synchronized to ensure correct execution. However, due to the physical distance between the two register files and the minimal cycle time, it takes an extra cycle[25] to write data from one cluster's FU to the other's register file. The instruction queue understands the difference between the two clusters and issues instructions in an efficient[26] manner by issuing a stream of dependent instructions to the same cluster.

With only two floating-point FUs (*add/div/sqrt unit* and *multiply unit*) and a single physical register file, the floating-point part of the 21264 is organized more traditionally. Since floating-point load/store instructions are executed in address ALUs, the 21264 can sustain up to four floating-point instructions per cycle. Both floating-point FUs are fully pipelined, with a latency of 4 for add, multiply, and most other operations. A double-precision divide takes 16 cycles, while a double-precision square root requires 33 cycles; these operations are not pipelined.

The D-cache is dual ported, able to supply two independent 64-bit results per cycle by starting a new access on each half-clock cycle. In parallel to the D-cache access, virtual addresses are sent to the fully-associative 128-

[25] DEC's simulations showed a 1 % performance degradation from this penalty which is a small price to avoid degrading the cycle time.

[26] In addition, as simple integer instructions enter the queue, they are preassigned to either the general-purpose units or the address ALUs. This method speeds the arbitration logic since it reduces the number of instructions potentially arbitrating for each FU. DEC's simulations showed this simplification causes only a 1–2 % performance loss.

entry data TLB and the tag array. These structures, which have a slightly longer access time than the cache data array, are duplicated to support the dual porting. After translation, memory addresses are logged in the memory reorder buffer, where they are held until the associated instruction is retired. This structure checks for multiple accesses to the same physical address; if detected, such pairs of instructions must be executed in program order to avoid errors. Up to eight cache miss requests are held in the *miss address queue*, waiting for access to the external cache. Cache lines displaced from the D-cache are held in an 8-entry *victim buffer*. Both L1 and L2 caches are non-blocking, continuing to service other requests while these transactions are pending.

To increase the bandwidth of the system bus and at the same time keep the package size within limits, DEC decided to use a 64-bit system bus (instead of 128-bit as in the 21164 processor) but applied high-frequency design to run the bus at speeds up to 333 MHz. At this speed, the bus can sustain 2.0 GB/s. The external L3 cache is controlled directly by the processor and can be as large as 16 MB, although practical systems will probably implement 1 MB to 8 MB.

Compaq/DEC's future microprocessor, the 21364, is expected to enter volume production in mid-2000. The 21364 is scheduled to debut at 750 MHz and will eventually reach 1.2 GHz. It will be designed for use in SMP implementations, where up to 64 processors can be used in a single server. On-chip transistor count will jump to the 100 million range. The 21364 will also implement some of the advanced code-optimization techniques. The major difference between Merced and the 21364 in this respect is static vs dynamic: Merced is doing everything as statically as it possibly can, the 21364 is doing everything as dynamically as it possibly can.

4.9.5 Sun UltraSPARC family

UltraSPARC (*Tremblay and O'Connor* [301]) is Sun's new line in the SPARC microprocessor family and supports the SPARC ISA version 9, a 64-bit ISA with a multimedia extension called VIS. VIS instructions are used for specialized pixel operations that can operate in parallel on 8-bit, 16-bit, or 32-bit integer values packed in a 64-bit floating-point register. It also includes some 3D to 2D conversion, edge processing, data alignment, pixel distance, packing, etc. The UltraSPARC also added a block move instruction which bypasses the caches to avoid disrupting it.

Like all other SPARC members, the UltraSPARCs use register windows for integers (see Sect. 1.7.2). This makes register renaming extremely difficult to implement efficiently. Without register renaming, however, there is little to be gained from out-of-order execution. As a result, in UltraSPARC instructions are issued in order and retired in order.[27] To do this, the following multiple-issue 9-stage pipeline is used:

1. instructions are fetched from the I-cache,
2. instructions are decoded and placed in the instruction buffer,
3. up to four instructions are grouped and issued to FUs,
4. integer instructions are executed and virtual addresses are calculated,
5. the D-cache is accessed; cache hits and misses are determined, and branches are resolved,
6. if a cache miss was detected, the loaded miss enters the load buffer,
7. the integer pipe waits for the floating-point/graphics pipe to fill,
8. traps are resolved, and
9. all results are written to the register files and instructions are retired.

The latency of most instructions is 9 but some instructions may require more than that due to the nature of the instruction, a cache miss, or other resource contentions.

Figure 4.28 shows a functional block diagram of the UltraSPARC architecture with a *prefetch and issue unit* (PU), an *integer execution unit* (IEU), a *floating-point unit* (FPU), a *memory management unit* (MMU), a *load and store unit* (LSU), an *external cache unit* (ECU), a *graphics unit* (GRU), I-cache, and D-cache.

The PU ensures that all FUs remain busy by fetching instructions before they are needed in the pipeline. Instructions are retrieved from all levels of the memory hierarchy, predecoded and placed in the I-cache. A 12-entry *prefetch buffer* allows fetching to continue while the execution pipeline is stalled due to a dependency or cache miss. Conversely, FUs can continue to work even when the instruction fetch has a cache miss. I-cache is 16 kB 2-way associative and is physically indexed and tagged. UltraSPARC uses dynamic branch prediction based on a two-bit prediction scheme with a 2 048-entry BHT. UltraSPARC provides single-cycle branch following, i.e., it can rapidly fetch predicted branch targets.

Eight integer register windows and four sets of global registers are provided. Two ALUs form the main computational part of the IEU. An early-out multicycle integer multiplier and a multicycle integer divider are also part of the IEU. Loads, stores, and branches are also executed by IEU but outstanding load/store instructions are tracked by the LSU. The LSU is responsible for generating the virtual address of all loads and stores, for accessing the D-cache, for decoupling load misses from the pipeline through the 9-entry *load queue* and decoupling the stores through an 8-entry *store queue*. These queues allow overlapping of load/store instructions reducing the apparent memory latency. When a load instruction is issued, the IEU reads the required address-operands from the register file. If at least one address-operand

[27] The exceptions are some long-latency instructions – such as floating-point divide and square root, load/store, and multi-cycle integer instructions – which can be retired out of order.

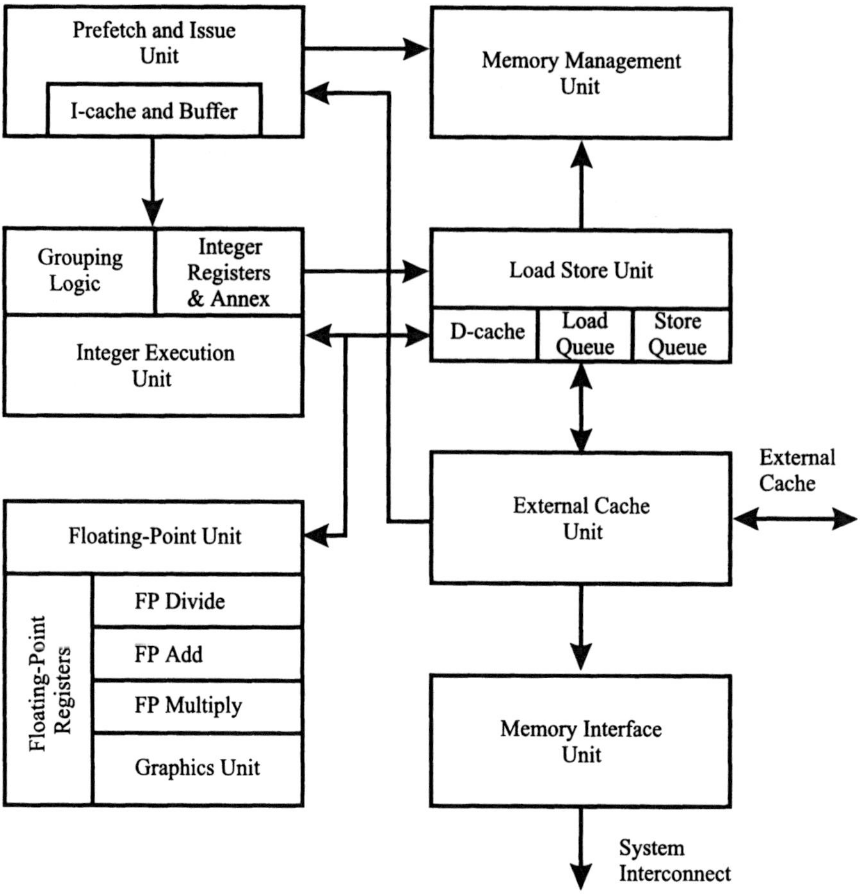

Fig. 4.28. The UltraSPARC microprocessor

is contained in the pipeline, it is extracted from the pipeline (bypassing the register file read), used to calculate the D-cache address of the item needed, and cache access is started. The accessed data is registered in the IEU and written to the register file. In the case of store instruction, the register file may be bypassed if the required operands can be obtained from the pipeline and used to calculated the virtual address. The data to be stored is then written to the D-cache.

There is a 32-entry floating-point register file where each entry can contain a 32-bit value or a 64-bit value. Three separate execution units (add, multiply, divide/sqrt) in the FPU allow UltraSPARC to issue and execute two floating-point instructions per cycle. Most instructions are fully pipelined. The divide and square-root instructions are not pipelined but they do not stall the processor: other instructions can be issued, executed, and retired to the register file before the divide/sqrt unit finishes. Two graphics execution units in the GRU provide VIS instruction execution. Execution latencies of some operations are given in Table 4.8.

Table 4.8. Some execution latencies in UltraSPARC-II and UltraSPARC-III

	UltraSPARC-II	UltraSPARC-III
64-bit Integer Mul	5-35	9
64-bit Integer Divide	64	64
FP Add (double-precision)*	3	4
FP Mul (double-precision)*	3	4
FP Divide	12	17
FP Square Root	12	24
FP Divide (double-precision)	22	20
FP Square Root (double-precision)	22	24

*Pipelined operations.

The MMU handles all memory operations, provides memory protection and performs the arbitration function between I/O, D-cache, I-cache, and TLB (64-entry, fully-associative). The MMU also implements virtual memory and translates virtual addresses to physical addresses in memory. The *memory interface unit* (MIU) handles I/O between local resources, including the processor, main memory, control space, and all external system resources.

The I-cache and D-cache are 16 kB, virtually-indexed and virtually-tagged caches. They are organized as 512 lines with 32 bytes per line in the case of I-cache and 2 × 16 bytes per line in the case of D-cache. Therefore, on a cache miss, 32 bytes are written from main memory in the case of I-cache and 16 bytes in the case of D-cache. I-cache is 2-way set-associative, while D-cache is direct-mapped. The ECU efficiently handles I-cache and D-cache misses, handling one access per cycle to the external cache. UltraSPARC supports a variety of external cache sizes ranging from 512 kB to 4 MB.

For scalable workstations and servers Sun plans to offer the following high-performance s-series of UltraSPARC processors: UltraSPARC-IIs (480 MHz, 0.25 μm, in 1999), UltraSPARC-IIIs (600–750 MHz, 0.18 μm, in 1999), UltraSPARC-IVs (1 GHz, 0.15 μm, in 2000), and UltraSPARC-Vs (1.5 GHz, 0.07 μm, in 2002). To lower the cost of single processor systems, Sun has integrated many system functions onto the i-series of UltraSPARC processors and will offer: UltraSPARC-IIi (400–480 MHz, 1999), UltraSPARC-IIIi (600–700 MHz, 2000), and UltraSPARC-IVi (1 GHz, 2001).

4.9.6 HAL SPARC64 family

HAL Computer Systems' first processor in the SPARC64 family was SPARC64-I. This processor was a six-chip design consisting of CPU, MMU, and four identical cache chips. It had nearly 22 million transistors, 87 % of which are in the MMU chip and four identical cache chips. Because of the way the MMU implements buffers (to cache page tables) and algorithms (for searching page tables in memory and reloading them into TLBs), HAL had to develop its own operating system and software development tools.

HAL's second processor, the SPARC64-II, combines similar MMU chip and cache chips with an improved CPU chip. That CPU chip has an 8 kB I-cache and a 2 k BHT (twice as large as in the SPARC64-I). The number of register windows was increased from four to five. Fabricated in Fujitsu's 0.34 μm process, the SPARC64-II operates at speeds up to 161 MHz.

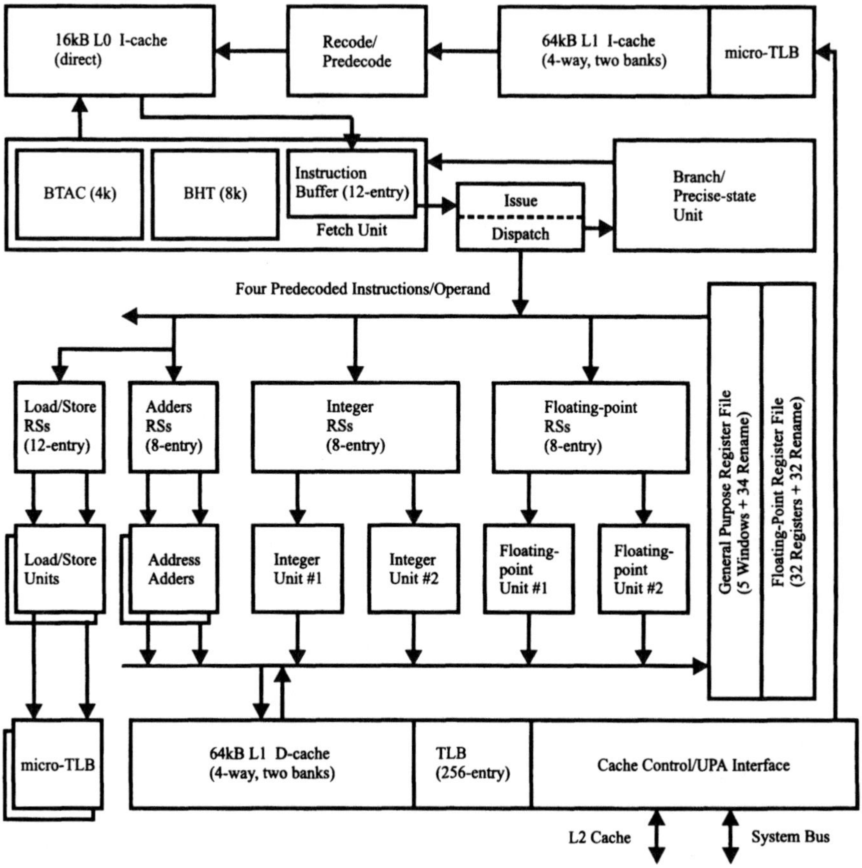

Fig. 4.29. The SPARC64-III microprocessor

HAL's latest design, the SPARC64-III, is designed to deliver high overall performance on server applications that use large data sets and intensive floating-point operations. It offers multiprocessor capability, which has become mandatory for today's server-class processors. The SPARC64-III integrates the SPARC64-II CPU chip, two cache chips with minor improvements, and a new memory subsystem onto a single die, as depicted in Fig. 4.29 (see

Song [275], 1997). The resulting I-cache and D-cache have 64 kB each with 4-way set-associative organization. The two-million transistor MMU chip from SPARC64-II is replaced with TLBs, an L2 cache interface, and a system bus that is compatible with Sun's UltraSPARC Port Architecture (UPA) bus. The new MMU design is compatible with Sun's Solaris operating system. In order to provide better support for huge applications, such as operating systems or database programs, the MMU supports page sizes up to 4 GB. The SPARC64-III has three 32-entry micro-TLBs to translate three addresses in each cycle – one for each instruction fetch and two for load/store accesses. An access that misses the micro-TLB takes four extra cycles to access the main 256-entry TLB. The processor has two 128-bit wide data buses for the L2 cache and system bus interfaces. The cache interface operates at either the full speed or half the speed of the processor, delivering 4 GB/s of peak bandwidth using 250 MHz SRAMs. The system bus can operate at 1/2, 1/3, 1/4, or 1/5 the speed of the processor.

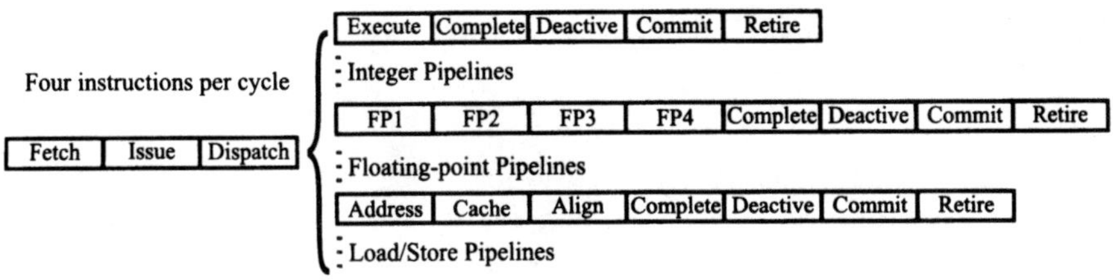

Fig. 4.30. Pipeline in the SPARC64-III

The SPARC64-III uses the pipeline as depicted in Fig. 4.30. In each cycle, up to four instructions are fetched from the L0 (level-zero) I-cache and placed in the 12-entry instruction buffer. The 16 kB direct-mapped L0 I-cache provides the recoded instructions and the predecoded bits. In the issue and dispatch stages, four instructions from the buffer are decoded, their destination registers are mapped to rename registers, and the rename registers for their source operands are identified. The source operands are read from the register file or the rename registers, or are forwarded from the FUs via result buses. The instructions and their operands are sent to the appropriate RSs by the end of the dispatch stage. RSs of a FU can accept two instructions per cycle. The load/store instructions are sent to both the address adders and load/store RSs.

The SPARC64-III uses a large BHT and a *gselect* prediction scheme (see p. 144). Its 8 k BHT is indexed using the 5-bit global history register concatenated with eight bits of the branch address, providing two bits of branch history. The global history register and the BHT entry are accessed during the fetch stage. Instead of waiting until the branch condition is known, the SPARC64-III speculatively updates the global history register and the BHT entry one cycle after the branch instruction is fetched.

The SPARC64-III has two integer units, two floating-point units, two address adders, and two load/store units. The first floating-point unit executes only add and subtract instructions, but occupies only half the area of the second, which is a multiply-add unit. Execution latencies of some operations are given in Table 4.9.

Table 4.9. Some execution latencies in the SPARC64-III

	SPARC64-III
64-bit Integer Mul	6
64-bit Integer Divide	2-37
FP Add (double-precision)*	3 or 4
FP Mul (double-precision)*	4
FP Divide	12
FP Square Root	12
FP Divide (double-precision)	22
FP Square Root (double-precision)	22

*Pipelined operations.

The SPARC64-III has a highly out-of-order execution core that can process 63 instructions at once. Each integer, address generation and floating point unit has eight RSs and dispatches up to two instructions per cycle. Since the operands use rename registers, there is generally no restriction on dispatching instructions out of order, provided that the oldest two are dispatched when more than two are ready. However, certain load/store instructions must be executed in program order, such as when the programming model requires an in-order execution or when the instructions access the same address. The D-cache has two banks, interleaved on eight byte boundaries, to support two accesses per cycle.

The SPARC64-III is built in Fujitsu's 0.24 µm process with 17.6 million transistors, of which 11.6 million are in the caches and TLBs and 6 million are used for the out-of-order execution engine. The processor is expected to operate at 250 MHz. At this frequency, HAL expects the chip to deliver 13 SPECint95 and 18 SPECfp95, using a 4 MB L2 cache and a 60 ns EDO DRAM. HAL is also working on a 0.18 µm derivative of the SPARC64-III design, hoping to reach a clock speed of 500 MHz.

4.9.7 HP PA-7000 family and PA-8000 family

In the early 1990s (see Table 4.10), Hewlett-Packard introduced the microprocessor PA-7100, which was HP's first superscalar implementation of its PA-RISC 1.1 ISA. At about the same time PA-7100LC appeared. This was the first processor to implement a multimedia instruction extension of an ISA (the extension was called MAX-1 by HP). The PA-7100 was a two-issue superscalar processor, with the restriction that only an integer and a floating-point instruction could be issued together for execution. A later version, the PA-7200 (1994), added a second integer unit. The original 48-bit addressing was expanded to 64 bits, using a segmented addressing scheme. The PA-7200 included a tightly integrated cache and memory management unit, a high-speed 64-bit bus. There was also a fast but complex fully-associative 2 kB on-chip assist cache between the simpler direct-mapped D-cache and main memory, which reduced thrashing (i.e., repeatedly loading the same cache line) when two memory addresses were aliased (i.e., mapped to the same cache line). Instructions were predecoded into a separate I-cache. The last 32-bit superscalar implementation of the PA-RISC was the PA-7300LC (*Blanchard and Tobin* [29]).

Table 4.10. The PA-RISC processor family

Year	Type	Technology (μm)	Clock (MHz)	Issue	Out-of-order issue	Word format	Internal caches (kB)	SPECint95	SPECfp95
1992	PA-7100	0.8	66-100	2	No	32-bit	none		
1992	PA-7100LC	0.8	100	2	No	32-bit	none		
1994	PA-7150	0.8	125	2	No	32-bit	none		
1994	PA-200	0.55	120	2	No	32-bit	none		
1995	PA-7300LC	0.5	160	2	No	32-bit	2 X 64		
1996	PA-8000	0.5	180	4	Yes	64-bit	none	11.8 @ 180 MHz	20.2
1997	PA-8200	0.5	240	4	Yes	64-bit	none	17.4 @ 240 MHz	28.5
1998	PA-8500	0.25	440	4	Yes	64-bit	512 + 1024	30+	50+

HP's next processors, named PA-8000, PA-8200, and PA-8500, implemented the 64-bit PA-RISC 2.0 ISA. The PA-8000 was introduced in 1996 (*Hunt* [141], *Kumar* [169]). The processor was a 4-issue superscalar and borrowed almost nothing from the older PA-7200, with the exception of the off-chip I-cache and D-cache. The PA-8000 microarchitecture served as the basis for the follow-up program which introduced the PA-8200 and PA-8500. The PA-8200 processor increased the number of TLB and BHT entries, and its operating frequency up to 300 MHz which resulted in 16.1 SPECint95 and 25.5 SPECfp95 performance (*Scott et al.* [257]).

The next PA-8500 processor also builds upon the foundation established in the PA-8000 but breaks with HP tradition by adding on-chip L1 caches – 0.5 MB I-cache and 1 MB D-cache (*Lesartre and Hunt* [179]). In two cycles, these L1 caches provide access to more data than many other processors provide in 10 or more cycles (via L1 and L2 cache). Thus, a challenge was to create such a large on-chip cache that could fit into the allocated die area and still keep up with the instruction reorder buffers whose responsibility it is to extract independent instructions and keep the FUs busy.

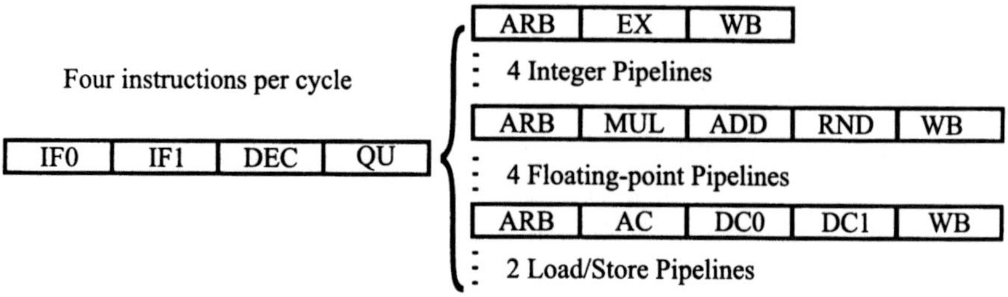

Fig. 4.31. Pipelines in the PA-8500 microprocessor

The length of the PA-8500 pipeline depends on the type of the instruction performed (Fig. 4.31). First, each instruction proceeds through a 4-stage pipeline (IF0, IF1, DEC, QU) performed by the fetch unit as follows: up to four instructions per cycle are fetched from I-cache in program order, and after being predecoded are inserted into the instruction reorder buffer. After arbitrating (ARB), up to four instructions can execute at once: two computation and two load/store instructions. Integer computations take one cycle (EX), while floating-point computation uses three cycles (MUL, ADD, RND). Similarly, load/store instructions take three cycles, one for address calculation (AC) and two for D-cache access (DC0, DC1). Once an instruction has executed, a temporary rename register holds its result and makes it available to subsequent instructions. As instructions retire (WB), the contents of the rename registers transfer to the architectural registers.

The PA-8500 supplies ten FUs: two integer ALUs, two integer shift/merge units, two floating-point multiply accumulate units (FMAC), two floating-point divide/square-root units, and two load/store units. See Fig. 4.32 for details. The FMAC units have a three-cycle latency and are fully pipelined to deliver up to four floating-point operations per cycle. The divide units have latency 17 and are not pipelined, but they run concurrently with the FMACs.

To keep most of these FUs busy, PA-8500 employs dynamic scheduling to extract the maximum parallelism from the instruction stream. Accordingly, the PA-8500 has a large decoupled instruction window called the *instruction reorder buffer* (IRB), which examines the 56 most current instructions to find four that can execute simultaneously.

The instruction *fetch unit* fetches blocks of four quadword-aligned instructions per cycle from the I-cache. The I-cache is a 0.5 MB four-way set-associative pipelined cache that provides 128 bits of instruction plus predecode bits per cycle to the instruction fetch unit. The fetch unit passes

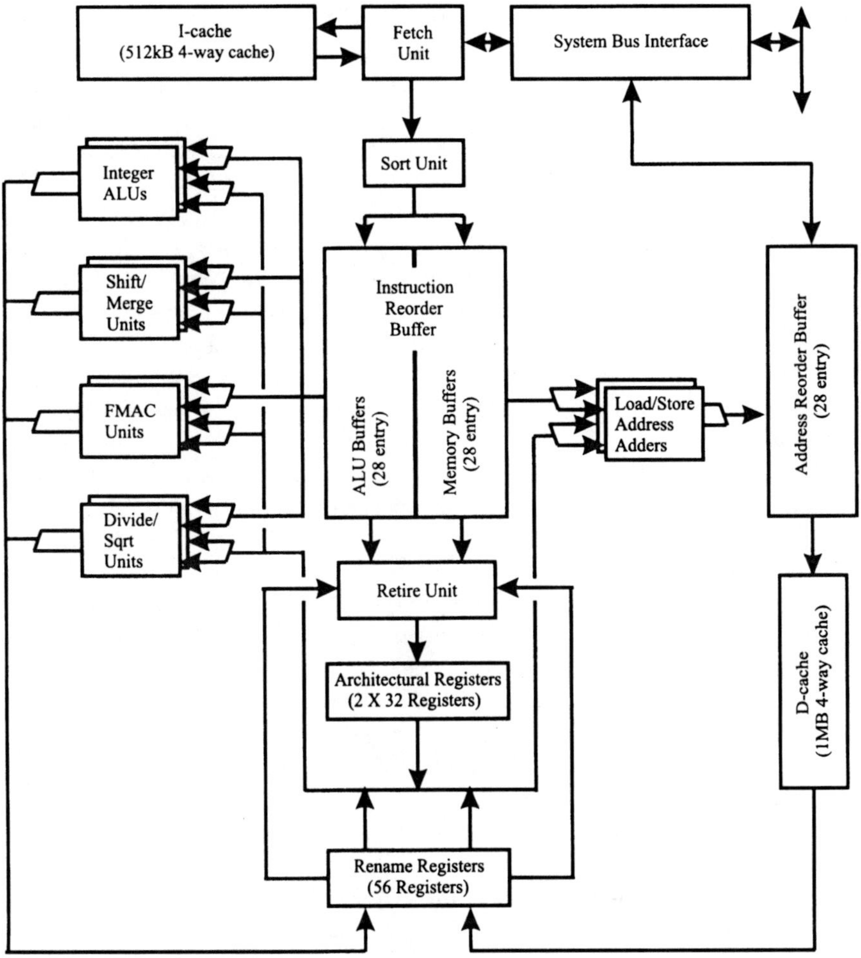

Fig. 4.32. The PA-8500 microprocessor

them to a *sort unit* which decodes and inserts four instructions into the IRB. The IRB consists of two 28-entry buffers: an *ALU buffer* that holds instructions destined for the integer units and floating-point units, and a *memory buffer* that holds load/store instructions. Certain instruction types enter both buffers; among these are branch instructions which in this way help recovery from misprediction. The IRB serves as the central control unit by supporting full register renaming for all instructions in the IRB and tracking instruction dependences to allow dataflow execution through the entire 56-instruction window. Registers are renamed via 56 *rename registers* (one for each IRB slot) and 64 *architectural registers* (32 integer and 32 floating-point). An instruction can be issued out of order but observing data dependences. Each of the ALU and memory buffers dispatches two instructions per cycle. When an instruction has been successfully executed it is sent to the *retire unit*. Up to four instructions per cycle can be retired.

When a load/store instruction in an IRB slot has received all its operands, it requests to be issued, just like an ALU instruction, but the destination is one of the *load/store address adder units*, to calculate its effective address. The calculated address is stored in a third 28-entry *address reorder buffer* (ARB), whose slots are associated one-to-one with the slots of the IRB's memory buffer. The effective address also goes to the TLB, which returns a physical address that is placed into the same ARB slot. With its address in the ARB, the load/store instruction starts arbitrating for access to the D-cache. The ARB hardware also checks data dependences and prevents RAW hazards. The ARB allows the memory control to track up to 28 loads, stores, and prefetches simultaneously and these memory operations can be completed in any order.

The D-cache supports two simultaneous memory operations while maintaining a two-cycle access to the dual-bank cache system (as in the PA-8000's off-chip D-cache). Access to the two banks is controlled by the ARB, which receives addresses that have been calculated by the load/store address adder units, gives priority to corresponding memory accesses, and then picks one access for each bank each cycle. A bypass path is provided to route an address directly from the address units if there is no outstanding access held in the ARB. This arrangement enables the cache ports to be utilized efficiently even when simultaneously calculated addresses happen to access the same half of the cache. Any delay in obtaining data in this scenario is hidden by the reorder buffers.

With on-chip L1 caches larger than most L2 caches, combined with the scheduling capabilities of its 56-entry IRB, the PA-8500 does not need to support a tightly coupled L2 cache directly from the CPU. This eliminates the need to include an integrated L2 cache controller.

Although priority is given to increasing the frequency of the PA-8500 by implementing it in 0.25 μm technology, some improvements have also been made by changing the branch prediction hardware which combines the advantages of static and dynamic two-bit branch prediction methods. The BHT in the PA-8500 is a standard array of 2-bit counters which record whether or not the branch went in the direction indicated by the static hint supplied by the compiler. If the static hint disagrees with the actual direction the branch followed, the counter is incremented; otherwise it is decremented. Each time a branch is fetched, the BHT is consulted and, if the counter is zero or one, the static hint encoded in the instruction is followed. If the counter is two or three, the hardware predicts that the branch will go in the direction opposite to the static hint.

HP pioneered the addition of multimedia instructions with the MAX-1 extensions in the 32-bit PA-7100LC and MAX-2 extensions in the 64-bit PA-8000, which allowed vector operations on two or four 16-bit subwords in 32-bit or 64-bit integer registers. This only required circuitry to slice[28] the integer ALU, adding only 0.1 % to the PA-8000 CPU area – using the floating-point registers like Sun's VIS and Intel's MMX would have required duplicating ALU functions. In favor of powerful "mix" and "permute" packing/unpacking operations, 8- and 32-bit support, multiplication, and complex instructions were also omitted.

Hewlett-Packard and Intel are working together on a new microprocessor architecture, IA-64, in order to provide breakthrough performance gains over current technology. The new processor will offer binary compatibility with PA-RISC software and Intel IA-32 software.

4.9.8 MIPS R10000 and descendants

Recall from Chap. 1 that between 1985 and 1994 MIPS Technologies (and its predecessor MIPS Computer Systems) introduced several scalar RISC processors and thus implemented three generations of their ISA (MIPS I to MIPS III, see p. 11). The fourth generation, MIPS IV ISA, was implemented in 1994 by the scalar processor R8000, and in 1995 by the first 4-issue superscalar processor R10000 (*Yeager* [330]). The 2-issue superscalar R5000 processor, which was introduced in 1996, is the third MIPS processor to implement the MIPS IV ISA.

In the following we discuss some of the R10000 features. The processor fetches and decodes four instructions per cycle and then appends them to one of three instruction queues. Each queue performs dynamic scheduling of instructions. The queues determine the execution order based on the availability of the required FUs. Though initially fetched and decoded in order, instructions can be issued, executed, and completed out of order, allowing the processor to have up to 32 instructions in various stages of execution. The high throughput is achieved through the use of wide, dedicated data paths, and large on-chip and off-chip caches. Running at 250 MHz, the R10000 delivers 14.7 SPECint95 and 24.5 SPECfp95. The processor is fabricated in 0.35 µm CMOS technology with 6.7 million transistors and operates at up to 275 MHz.

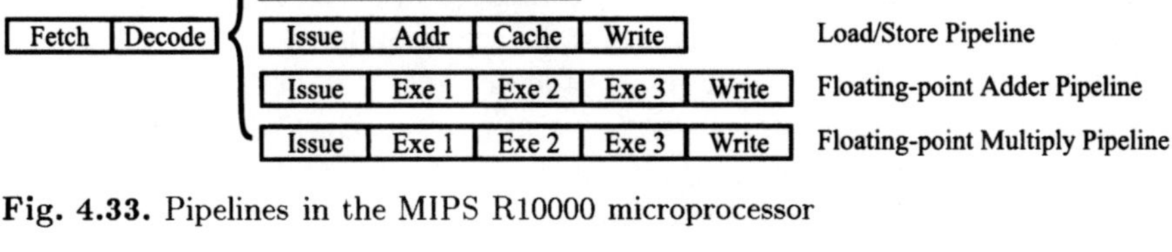

Fig. 4.33. Pipelines in the MIPS R10000 microprocessor

[28] Similar to bit-slice processors, such as the AMD 2901.

The R10000 is pipelined, with the pipeline depth depending on the type of the instruction (Fig. 4.33). Initially, instructions proceed through the instruction fetch pipeline which consists of fetch, decode, and issue stages:

1. in the fetch stage, four instructions are fetched and aligned,
2. in the decode stage, the instructions are decoded, register renaming is performed, and branch instructions are predicted,
3. in the issue stage (first half), the instructions are written to one of three 16-entry instruction queues; the validity of the operands is also determined.

Depending on its type, the instruction proceeds to one of the five execution pipelines. There are two integer and two floating-point pipelines, and one load/store execution pipeline. Each of these pipelines begins when a queue issues an instruction (issue stage) and continues as follows:

3. in the issue stage (second half), the processor reads operands from the register files,
4. the execution begins and takes
   a) one stage in the case of integer pipelines
   b) two stages in the case of the load/store pipeline
   c) three stages in the case of floating-point pipelines

5–7. in the last stage (depending on the type of the instruction) the result is written into the register file.

Figure 4.34 shows a block diagram of the R10000 microprocessor.

The D-cache is 32 kB in size and is arranged as two identical 16 kB banks. The cache is two-way interleaved. Each of the two banks is 2-way set-associative. Cache line size is 32 bytes. The virtual indexing allows the cache to be indexed in the same clock cycle in which the virtual address is generated. However, the cache is physically tagged as the L2 cache. The L2 cache interface provides a 128-bit data bus which can operate at a maximum of 200 MHz, yielding a peak data transfer rate of 3.2 GB/s. The L2 cache size is between 512 kB and 16 MB and the cache line size is programmable at either 64 or 128 bytes.

The I-cache is 32 kB with 64-byte line size and is 2-way set-associative. Instructions are predecoded before being placed in the I-cache. Four extra bits are appended to each instruction to identify the necessary FU.

The processor predicts the direction a branch will take by a 2-bit dynamic branch predictor and fetches instructions speculatively along the predicted path. To locate branch instructions in the instruction fetch pipeline, a branch bit is appended to each instruction in the decode stage. The path a branch will take is predicted using a 2-bit 512-entry BHT. The BHT is indexed by bits 11:3 of the branch instruction address. The MIPS architecture prescribes the delayed branch technique with one delay slot following a jump or branch instruction. The feature was retained for the R10000 for compatibility. The

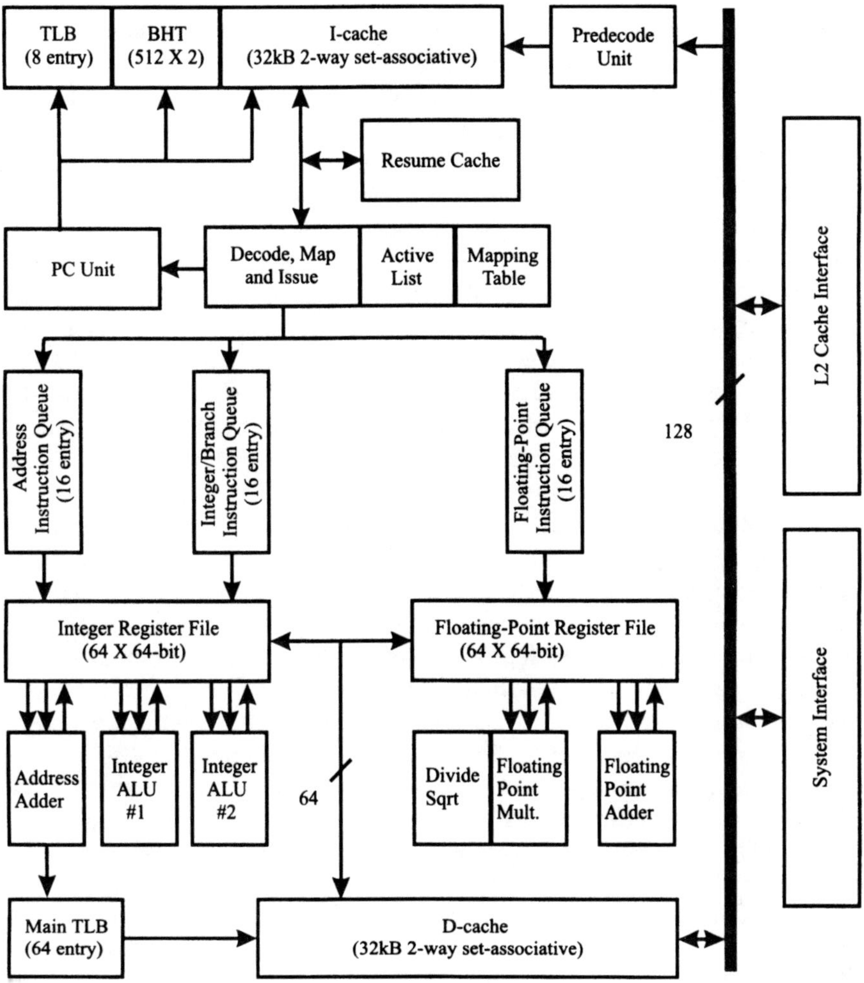

Fig. 4.34. The MIPS R10000 microprocessor

R10000 allows up to four outstanding branch predictions which can be resolved in any order. An on-chip 4-entry branch stack contains an entry for each branch instruction being speculatively executed.

Each entry contains the information needed to restore the processor's state if the speculative branch is incorrectly predicted (e.g., alternate branch addresses and corresponding register renaming). This allows the processor to restore the pipeline quickly when a branch misprediction occurs. When the branch stack is full, the processor continues decoding only until it encounters the next branch instruction. Decoding then stalls until resolution of one of the pending branches.

Let us briefly describe the register renaming. In the case of integer instructions 32 logical integer registers can be mapped to 64 physical integer registers; in the case of floating-point instructions 32 logical floating-point registers can be mapped to 64 physical floating-point registers. The renaming uses three structures: a *mapping table*, an *active list*, and a *free list*. Separate mapping tables and free lists are provided for integer and floating-point instructions while only one active list exists. The active and the free list can each contain a maximum of 32 values. After an instruction is fetched from the I-cache it is placed in the mapping table where logical registers are assigned to physical registers, i.e., register renaming is performed. Currently unassigned physical registers are kept in the free list. The active list maintains a listing of all instructions currently active within the processor. The instructions in the instruction queues can be executed out of order. Thus, before the result can be stored as final, it must be stored in order as, determined by the active list which is always kept in order. Once the result is stored it becomes obsolete, so the instruction can be retired and the logical destination can then be returned to the free list.

After being decoded, the instruction is placed into one of the three 16-entry *instruction queues* (integer/branch, floating-point, address) for scheduling. Theses queues supply instructions for five execution pipelines. An instruction is issued from the queue in out-of-order fashion, when the required operands and FU are available. The *integer queue* can get up to four integer or branch instructions per cycle and supplies two integer ALUs. Similarly, the *floating-point queue* can get up to four floating-point instructions per cycle and supplies either the floating-point adder or floating-point multiplier. The *address queue* is a circular FIFO that preserves the original program order of load/store instructions so that memory address dependences may be computed easily. An issued load/store instruction may fail to complete because of a memory dependency, a cache miss, or a resource conflict. In these cases the address queue must reissue the instruction until it is completed.

The R10000 contains five FUs which operate independently of one another: two integer ALUs, two floating-point units and a load/store unit. Both integer ALUs perform standard add, subtract, and logical operations. One integer ALU handles all branch and shift instructions, while the other handles all multiply and divide operations using iterative algorithms. The adder floating-point unit handles add operations and the multiply floating-point unit handles multiply operations. In addition, two secondary floating point units exist which handle long-latency operations such as divide and square root. The load/store unit consists of the address queue, address calculation unit, a 64-entry TLB, address stack, store buffer, and L1 D-cache. This unit performs load, store, prefetch, and cache instructions.

By scaling the R10000 design up to 300 MHz in clock speed, increasing the number of instructions that can be in various stages of execution from 32 to 48, adding a 32-entry two-way BTC, and quadrupling in size the BHT, the resulting R12000 microprocessor ensures a corresponding increase in applicaton performance. The R12000 was introduced in 1998. A faster version, operating at 400 MHz, and called R14000, has been scheduled for the second half of the 1999.

4.9.9 IBM POWER family

After the scalar RISC processors IBM 801 and ROMP (see p. 33), IBM in 1990 started designing superscalar processors based on IBM's POWER ISA. These processors were used in RS/6000 workstations and the RS/6000 SP multiprocessor (*Oehler and Groves* [217]).

The first processor POWER1 (initially called POWER CPU), which was one of the pioneering superscalars, was implemented on three chips for the CPU with two or four additional cache chips (8–32 kB I-cache and 32–64 kB D-cache). The CPU chips were *branch unit* (BU), *integer unit* (IU) and *floating-point unit* (FPU). In 1992, the RSC (RISC Single Chip) design integrated BU, IU, FPU, cache, memory controller, and I/O controller into one chip.

In 1993, IBM continued with a POWER ISA-based design and produced the POWER2 processor. This was also a multichip design with two floating-point load/store units, 256 kB D-cache, and added 128-bit floating-point support and a square root instruction. The processor could issue up to six instructions and four simultaneous load/store instructions. POWER2 was redesigned in 1996 into one chip called the POWER2 Super Chip (P2SC) which operated at up to 160 MHz and was made in 0.29 μm technology.

In 1998, the first unified POWER/PowerPC microprocessor POWER3 emerged. The 64-bit chip features eight FUs (three integer, two floating-point, two load/store, and branch) fed by a 6.4 GB/s memory subsystem. The POWER3 also has on-chip 64 kB D-cache and a 32 kB I-cache. The target operating frequency is 200 MHz with 0.25 μm technology. There are two high-bandwidth buses: a 128-bit 6XX architecture bus to main memory and a 256-bit bus to the L2 cache that runs at the processor speed. The POWER3 architecture provides the ability for thousands of microprocessors to be combined into a supercomputer.[29] Future versions of the POWER3 will be manufactured in a 0.20 μm process, that adopts copper wiring, and silicon on insulator technology, driving the POWER3 into the GHz range and beyond by 2001.

4.9.10 IBM/Motorola/Apple PowerPC family

Several 32-bit as well as 64-bit implementations of the PowerPC ISA appeared in the 1990s (see Table 4.11). The PowerPC ISA is based upon the IBM POWER architecture.

[29] The POWER3 processor is at the heart of a 4096-processor RS/6000 SP supercomputer called Deep Blue. When the system is upgraded with POWER3 processors, it is expected to deliver 3.0 TFLOPS.

Table 4.11. The PowerPC processor family

Year	Type	Technology (μm)	Clock (MHz)	Issue	Out-of-order execution	Word Format	Internal caches(kB)	SPECint95	SPECfp95
1993	PowerPC 601	0.6	66-80	3	Yes	32-bit	32		
1993	PowerPC 603	0.5	150-250	3	Yes	32-bit	2X16	7.1	4.2 @ 240 MHz
1995	PowerPC 602	0.5	66-80	2	Yes	32-bit	2X4		
1995	PowerPC 604	0.25	166-350	4	Yes	32-bit	2X32	14.6	9.0 @ 350 MHz
1998	PowerPC 740	0.25	200-266	4	Yes	32-bit	2X32	11.5	6.9 @ 266 MHz
1998	PowerPC 750	0.25	233-400	4	Yes	32-bit	2X32	17.6	12.2 @ 400 MHz
1994	PowerPC 620	0.5	133	4	Yes	64-bit	2X32		
1995	PowerPC RS64		165	3	Yes	64-bit	2X64		
1995	PowerPC RS64-II		262	4	Yes	64-bit	2X64		

In 1993, the PowerPC 601 processor reached the market (*Potter et al.* [235], *Becker et al.* [25]) as the first member of the first generation (G1) of PowerPC ISA implementations. Being a bridge, PowerPC 601 included both POWER and PowerPC features. The processor was based on the IBM POWER1 processor, except it had a single 32 kB cache rather than separate I-cache and D-cache.

In the same year, the PowerPC 603 processor (*Burgess et al.* [44] *Suessmith and Paap* [281]) was announced to be the first in the second PowerPC generation (G2). It further separated the main FUs by removing load/store operations from the integer unit and provided separate fetch, issue, and retire units. For speculative execution the PowerPC 603 used a rename buffer in the issue unit and renamed integer and floating-point registers, which were ordered properly by the retire unit, or discarded for mispredicted branches and exceptions. L1 cache was split into I-cache and D-cache with either 8 kB or 16 kB each.

The next processor, PowerPC 604, appeared in 1995 (*Ryan and Thompson* [250], *Song et al.* [276]). The 4-issue PowerPC 604 added dynamic branch prediction using a BHT, split the integer unit (IU) into three IUs (two for single-cycle operations and one complex IU for multicycle operations), and was capable of issuing four instructions at once.

The microarchitecture of the PowerPC 604 (see Fig. 4.35) extends the Tomasulo algorithm (see Sect. 3.3.2) to a four-issue scheme. Instructions are fetched from 16 kB I-cache in 4-instruction segments. During the fetch stage basic flow prediction is performed using a 256-entry BTAC. A second level of branch prediction during the decode/issue stage may override the flow prediction in the fetch stage. Branch prediction in the decode/issue stage uses a 2-bit prediction scheme. The bits are stored in the 512-entry BHT.

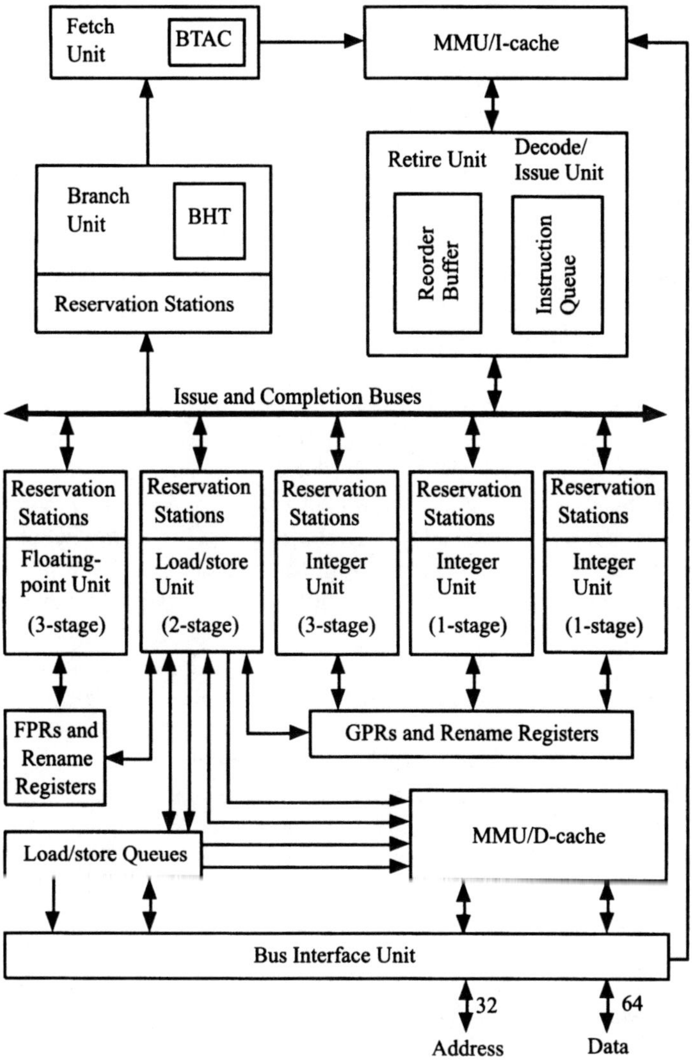

Fig. 4.35. The PowerPC 604 processor

Fetched instructions proceed into an 8-entry *instruction queue*. The bottom four entries of this queue decode the instructions, i.e., determine the resources that each instruction requires. Once decoded, up to four instructions per cycle can be issued in order to reservation stations in front of each FU. Results are moved into the appropriate architectural register after the instruction commits. When an instruction is decoded and issued, a logical rename register is allocated for the result of the operation. There are eight rename registers for 32 floating-point registers and twelve for 32 general-purpose registers. In addition, four loads and six stores can be buffered, further increasing the number of instructions issued.

The *decode/issue unit*[30] also assigns each instruction an entry in the 16-entry *reorder buffer*, which tracks the status of every instruction – including whether the instruction is executing speculatively – from issue to retirement. Thus, the PowerPC 604 can have no more than 16 instructions executing, speculatively or otherwise, at any one time. If the reorder buffer is full, issue stops until one or more entries become available.

Each FU is fronted by two *reservation stations* (RS). If the execution stage is busy, issued instructions wait in RSs until the first execution stage is clear. They also wait in RSs if any of their operands are not available. A data-forwarding mechanism feeds the RSs, allowing operands to be made available to data dependent instructions before the instructions that produce them commits. Note that if an instruction with all its operands available is issued to an idle FU whose RS is empty, the instruction dispatches immediately to the execution stage. With the branch, floating-point, and load/store units, instructions dispatch in order from the RS to the FU. With the integer units, an instruction can be dispatched out of order. Thus, the PowerPC 604 supports in-order issue; in-order dispatch within the branch, load/store, and floating-point units; out-of-order dispatch in the integer units; and out-of-order execution.

The *retire unit* uses the reorder buffer to retire instructions. It retires instructions in program order, up to four instructions per cycle. Recall that the retire unit knows the order of instructions because this information is supplied to the reorder buffer when the instructions are issued. It does not retire an instruction that is labeled speculative, nor an instruction that executed out of order unless all previous instructions have been retired.

In 1997, the PowerPC 604e operated at 350 MHz with 14.6 SPECint95 and 9.0 SPECfp95. The chip was fabricated in 0.25 μm technology with approximately 5.1 million transistors.

The PowerPC 620 is the first 64-bit implementation of the PowerPC architecture (*Gwennap* [113]). The PowerPC 620 is 4-issue superscalar with six FUs: three IUs, an FPU, a load/store unit, and a branch unit. The latter performs branch prediction and four-level speculative execution. The microarchitecture of the PowerPC 620 is similar to that of the PowerPC 604. It differs mainly in the 64-bit wide registers and data paths, and it has more reservation stations for speculative execution. It has the same 16-entry reorder buffer as the PowerPC 604, which tracks instructions from dispatch to completion to facilitate out-of-order execution. Since the PowerPC 620 releases

[30] In most of the papers on the PowerPC microprocessors cited in the references the terms *issue* and *dispatch* are used with reverse meaning and *completion* is also used with the meaning of *retire*. To avoid confusion, we use the terms as defined at the beginning of this chapter.

up to four instructions per cycle (compared to two in the PowerPC 604), it needs only 16 rename buffers, compared to 20 in the PowerPC 604.

Another two 64-bit processors, PowerPC RS64 and RS64-II, added decimal arithmetic and string instructions. The PowerPC RS64 (RS64-II) has a four (five) FUs, and can sustain a decode and execution rate of three (four) IPC. The PowerPC RS64 runs at 165 MHz (262 MHz in the case of RS64-II) and has large caches, high data path bandwidth and low latency. Both processors have separate 64 kB I-cache and D-cache. The PowerPC RS64 contains an L2 cache controller and a dedicated 16 byte (32 byte in the case of RS64-II) interface to a private external 2-way set-associative 4 MB (4-way, 8 MB in the RS64-II) L2 cache. The L2 interface runs at full processor speed and provides 2 GB/s (8 GB/s) of bandwidth. The PowerPC RS64-II processor also has a separate 16 byte system bus interface.

The 32-bit PowerPC 750 (G3, early 1998) refined the design and performance, adding a PowerPC 620-style backside cache bus, but made no other significant changes to the PowerPC 604. It is, however, the first microprocessor featuring the new copper CMOS technology. As a result, higher clock speeds and lower power consumption are achieved.

The PowerPC microprocessor roadmap expects the start of the G4 series of microprocessors in 1999 with new 32-bit and 64-bit microarchitecture, and advanced design methods with up to 50 million transistors in 0.25–0.18 μm technology.

4.9.11 Summary

Table 4.12 and Table 4.13 summarize selected features of some of the state-of-the-art superscalar processors. It can be seen that 64-bit processors as well as multimedia support have become standard.

4.10 VLIW and EPIC Processors

VLIW (very long instruction word) processors use a long instruction word that usually contains a fixed number of operations that are fetched, decoded, issued, and executed synchronously. All operations specified within a VLIW instruction must be independent of one another, otherwise the VLIW instruction would be similar to a sequential, multi-operation CISC instruction.

Some of the key issues of a VLIW processor that were described by *Colwell* [52] are:

- very long instruction word (128 to 1 024 bits per instruction),
- each instruction consists of multiple independent parallel operations,
- each operation requires a statically known number of cycles to complete,
- a central controller that issues a long instruction word every cycle,
- multiple FUs connected through a global shared register file.

Table 4.12. Superscalar processor features I

	Pentium II	AMD-K6	Alpha 21264	UltraSPARC-II	PA-8500	R10000	PowerPC620
Company	Intel	AMD	DEC	Sun	HP	MIPS	IBM/Motorola/Apple
Issue	3 (5 μops)	4 (6 RISC86ops)	4	4	4	4	4
Integer registers	8 x 32-bit	24 x 32-bit	32 x 64-bit	136 x 64-bit	32 x 64-bit	32 x 64-bit	32 x 64-bit
Integer rename registers	40 x 32-bit	24 x 32-bit	48 x 64-bit	--	56 x 64-bit	32 x 64-bit	8 x 64-bit
Floating-point registers	8 x 80-bit	8 x 80-bit	32 x 64-bit	32 x 64-bit	32 x 64-bit	32 x 64-bit	32 x 64-bit
Floating-point rename registers	--	--	40 x 64-bit	--	--	32 x 64-bit	8 x 64-bit
Load/store units	3	2	*	1	2	1	1
Integer and branch units	3	3	4	4	4	2	3
Floating-point units	1	1	2	3	4	2	1
Graphic/multimedia units	2	1	**	2	***	****	0
Multimedia support	MMX	MMX	MVI	VIS	MAX-2	MDMX	--

* Two of the integer units execute all load/store instructions

** One integer unit handles the MVI instructions

*** Integer units execute MAX-2 instructions

**** Integer units execute MDMX instructions

Table 4.13. Superscalar processor features II

	Pentium II	AMD-K6	Alpha 21264	UltraSPARC-II	PA-8500	R10000	PowerPC620
On-chip L1 I-cache (kB)	16	32	64	16	512	32	32
On-chip L1 D-cache (kB)	16	32	64	16	1024	32	32
On-chip L2 cache	No	No	No	No	No	No	No
On-chip controller for external L2 cache	Yes	Yes	Yes	Yes	No	Yes	Yes
I-cache TLB (entry)	32	64	128	64	160 unified	64	128
D-cache TLB (entry)	64	64	128	64		64	128
Virtual address (bit)	48	48	?	44	48	44	80
Physical address (bit)	36	32	44	36	40	40	40
External cache bus width (bit)	64	64	128	128	64	128	128
System bus width (bit)	64	64	64	128	64	64	128
Technology (μm)	0.25	0.25	0.35	0.35	0.25	0.35	0.35
Transistors (million)	7.5	8.8	15.2	5.4	140	6.7	7
Frequency (MHz)	450	233	575	250	440	275	200
Package	242 SEC	321 PGA	588 PGA	521 BGA	?	599 LGA	625 BGA

VLIW can best be explained as follows by setting it in contrast to superscalar:

- VLIW relies on a sequential stream of long instruction words, i.e., instruction tuples, in contrast to superscalar processors that issue from a sequential stream of "normal" instructions.
- The instructions are scheduled statically by the compiler (in contrast to superscalar processors which rely on dynamic scheduling by the hardware).
- More than one instruction can be issued each cycle, as in superscalar processors, but the number of simultaneously issued instructions is fixed during compile-time. Only in-order issue is possible.
- The instruction issue is, therefore, less complicated than the hardware scheduler of a superscalar processor. Complexity is moved from the hardware into the compiler by the VLIW technique. By explicitly encoding parallelism in the long instruction, a VLIW processor can eliminate the hardware needed to detect parallelism.
- Instruction parallelism and data movement in a VLIW processor are completely specified at compile-time. Run-time resource scheduling and synchronization are, therefore, completely eliminated. The disadvantage is that VLIW processors cannot react to dynamic events such as, for example, cache misses, with the same flexibility as superscalar processors.
- The number of instructions in a VLIW instruction word is usually fixed. Padding VLIW instructions with no-ops is needed in case the full issue bandwidth is not met. This increases code size. More recent VLIW architectures use a denser code format which allows removal of the no-ops.
- VLIW is an *architectural* technique, whereas superscalar is a *microarchitecture* technique (see p. 5). A 6-issue VLIW code cannot be executed on a 4-issue VLIW processor.
- VLIW processors take advantage of spatial parallelism, like superscalar processors, by using multiple FUs to execute several operations of a VLIW instruction concurrently.

The compiler groups independent instructions, executable in parallel, using optimization techniques such as software pipelining and loop unrolling, and schedules code blocks speculatively by predicting branch outcomes statically using a compiler technology called *trace scheduling*.

VLIW is not a recent concept; in fact, it originated as an extension of horizontal microcoding techniques[31] (*Fisher* [84]).

---

[31] CISC computers in the 1970s were microprogrammed machines. A complex instruction was decoded into a sequence of microinstructions that were executed serially under the control of a microcode sequencer. A microinstruction defined control points which activated connections between registers. *Horizontal* microinstructions partitioned control point actions into many different classes, any of which could be activated within any given cycle. In contrast, since the microprogram storage was expensive, *vertical* microinstructions encoded control points in some fashion, but required additional decoding to process the encoded control point information.

Due to the use of very long instruction words early VLIW processors were mostly implemented with microprogrammed control. Thus the clock rate is slow with the use of read-only memory (ROM). A large number of microcode access cycles may be needed for some instructions.

As stated by *Gwennap* [112], the objective of VLIW is to eliminate the complicated instruction scheduling and issue of superscalar microprocessors. Therefore, one can view a VLIW processor as an extreme instance of a superscalar processor in which all independent or unrelated operations are already synchronously compacted together in advance.

Older VLIW processors include the Trace Machine and the ELI-52 VLIW processor developed by *Fisher* [85]. VLIW is common in signal processors. As an example of this processor class, we will describe next the most sophisticated signal processor to date.

4.10.1 TI TMS320C6x VLIW Processors

In February 1997, Texas Instruments introduced the TMS320C6x (or 'C6x) generation of high-performance digital signal processors. Based on the so-called VelociTI VLIW architecture for a *digital signal processor* (DSP), the 'C6x processor family defines a family of high-performance DSPs as exemplified by the 'C6201 and the 'C6701 which use an 8-issue VLIW technique.

DSPs are distinguished from general-purpose microprocessors by the ability to reach extremely high fixed-point or floating-point performance on 32-bit values. Typical applications for DSPs are mass market applications in signal and image processing, image recognition, and data and voice communication.

The first processor in this new generation of DSPs is the 'C6201 fixed-point DSP [292] which features a maximum of eight instructions per cycle, 200 MHz clock rate, 1 600 MIPS.

The second DPS in the 'C6x family, the 'C67x, differs from the 'C6201 only in the floating-point functionality of six of its eight functional units and its wider bus structure. The 'C62x and 'C67x members of the 'C6x generation of DSPs are to date the industry's highest performing fixed-point and floating-point DSPs.

Figure 4.36 shows the overall structure of the TMS320C6201 processor.

Besides the processor core which is described below, the 'C6201 processor contains a 16-bit host port interface (HPI), four direct memory access (DMA) channels with bootloading capability, a 32-bit external memory interface (EMIF), two multichannel buffered serial ports (McBSPs) for simplified interface to telecommunication truncs and interprocessor communication, and two 32-bit timers.

The 32-bit EMIF supports a variety of synchronous and asynchronous memories including SDRAM, SBRAM, SRAM. Additionally, the host has direct access to the processor data memory by way of the separate 16-bit HPI. The 'C6201 can be booted from 8-, 16-, or 32-bit external ROM using one of the two DMA channels available on the chip.

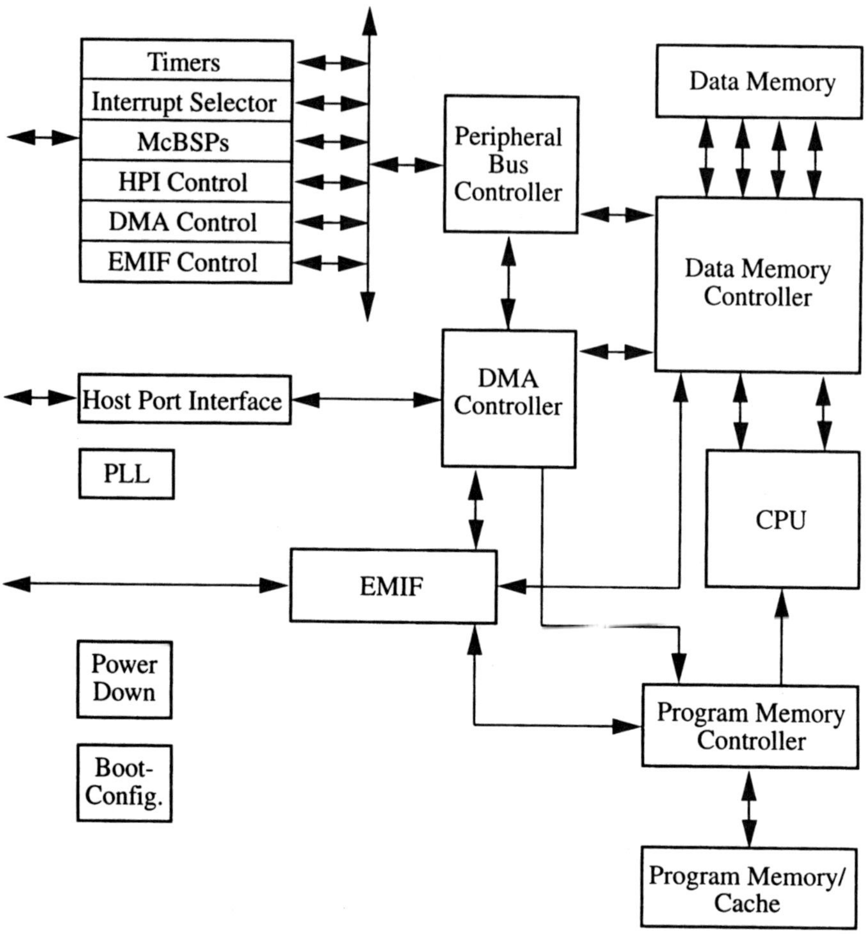

Fig. 4.36. TMS320C6201 processor overview

Figure 4.37 shows the structure of the TMS320C6201 processor core. The VLIW core contains eight functional units (six arithmetic-logic units and two load/store units) and thirty-two 32-bit general purpose registers.

The execution stage features two sets of functional units, each set with four units and a register set. One set contains functional units denoted .L1, .S1, .M1, and .D1, the functional units in the other set are denoted .D2, .M2., .S2, and .L2. The two .M units are dedicated for multiplies, the two .S and .L units perform a general set of arithmetic, logical, and branch operations with single-cycle latency, and the .L units are load/store units.

The data addressing units .D1 and .D2 are responsible for all data transfers between the register files and the memory.

The two register files contain sixteen 32-bit registers each. Each register file features its own connection to the memory.

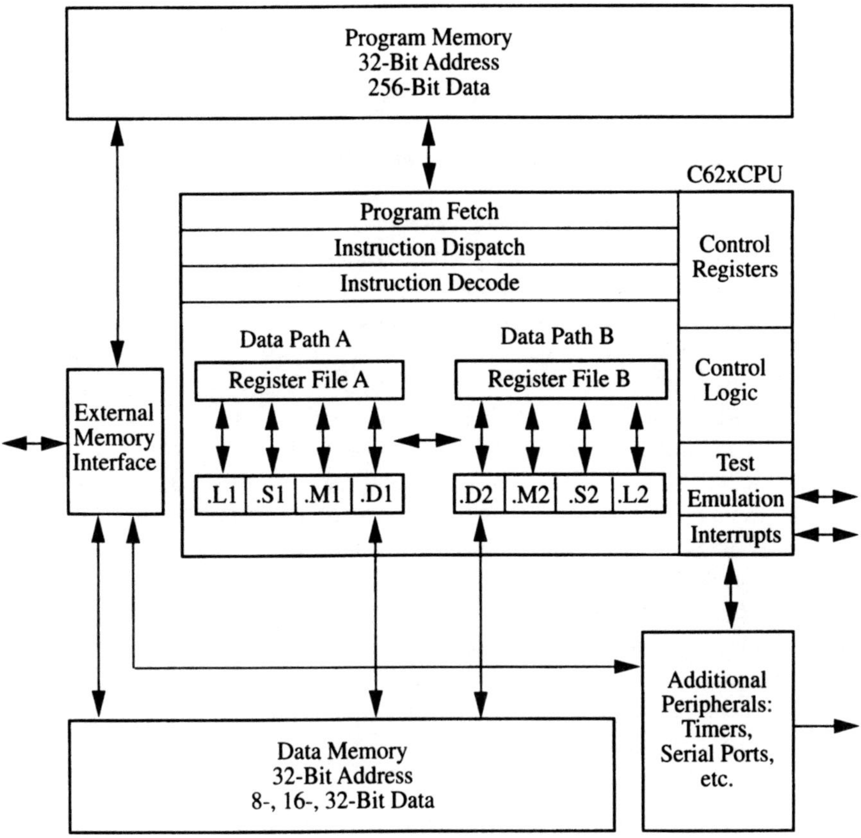

Fig. 4.37. TMS320C6201 processor core

The functional units in a set can freely share the sixteen registers belonging to the set. The register sets are connected by a data bus by which the two sets of functional units can access data from the other register file. The data bus supports only a single read and a single write access per cycle to the other register file.

128 kB on-chip RAM is split into 512 kbit internal data memory and 512 kbit internal program memory that can also be configured as cache memory.

The 'C6201 fetches a 256-bit wide VLIW packet with eight 32-bit instructions every clock cycle. The least significant bit of every 32-bit instruction chains the instructions together to execute packets. A "1" determines that the instruction belongs to the same execute packet as the previous instruction, while a "0" breaks the chain, determining that the next instruction should be executed in the following clock cycle as part of the next execute packet. Execute packets are dispatched to their respective functional units at the rate of one per cycle and the next 256-bit fetch packet is not fetched until all the execute packets from the current fetch packet have been dispatched. After decoding, the instructions in an execute packet are issued simultaneously to the functional units for a maximum execution rate of eight instructions every clock cycle.

The processor employs a load/store architecture; all instructions operate on registers.

All instructions are conditional (predicated instructions) and most can access any one of the 32 registers. Some registers support specific addressing modes only or act as condition registers for conditional instructions.

Figure 4.38 is taken from [337] and shows execution of a FIR filter code using the 8-issue VLIW format of the 'C6201 processor. The six instructions (`ADD`, `SUB`, `LDW`, `LDW`, `B`, and `MVK`) of the inner loop kernel of the FIR filter algorithm are condensed into a single VLIW instruction.

The 'C67x floating-point DSP [293] features 1 GFLOPS, 167 MHz (6 ns cycle time), eight 32-bit instructions per cycle of which six can be 32-bit IEEE floating-point instructions per cycle and reaches 1 GFLOPS single-precision and 420 MFLOPS double-precision performance. The 'C67x DSP adds floating point capability to six of the eight functional units available on the 'C6x VelociTI architecture. Therefore, the 'C67x instruction set is a superset of the 'C62x fixed point instruction set. All 'C62x instructions run unmodified on the 'C67x CPU.

The following is a list of the four functional units and their fixed and floating-point capabilities.

- The L-unit features
  - 32/40-bit fixed-point arithmetic and compare operations,
  - 32/64-bit floating-point arithmetic and compare operations (IEEE single and double precision),
  - 32-bit fixed-point logical operations,
  - fixed/floating-point conversions, and
  - 64 to 32-bit floating-point conversions.
- The S-unit executes
  - 32-bit fixed-point arithmetic operations,
  - 32/40-bit shifts and 32-bit bit-field operations,
  - branching and constant generation,
  - 32/64-bit floating-point reciprocal, absolute value, compares, and 1/sqrt operations, and
  - 32 to 64-bit floating-point conversions.
- The M-unit executes
  - 16 x 16-bit fixed-point multiplies,
  - 24 x 24-bit fixed-point multiplies,
  - 32 x 32-bit fixed-point multiplies,
  - 32 x 32-bit single-precision floating-point multiplies, and
  - 64 x 64-bit double-precision floating-point multiplies.
- The D-Unit performs

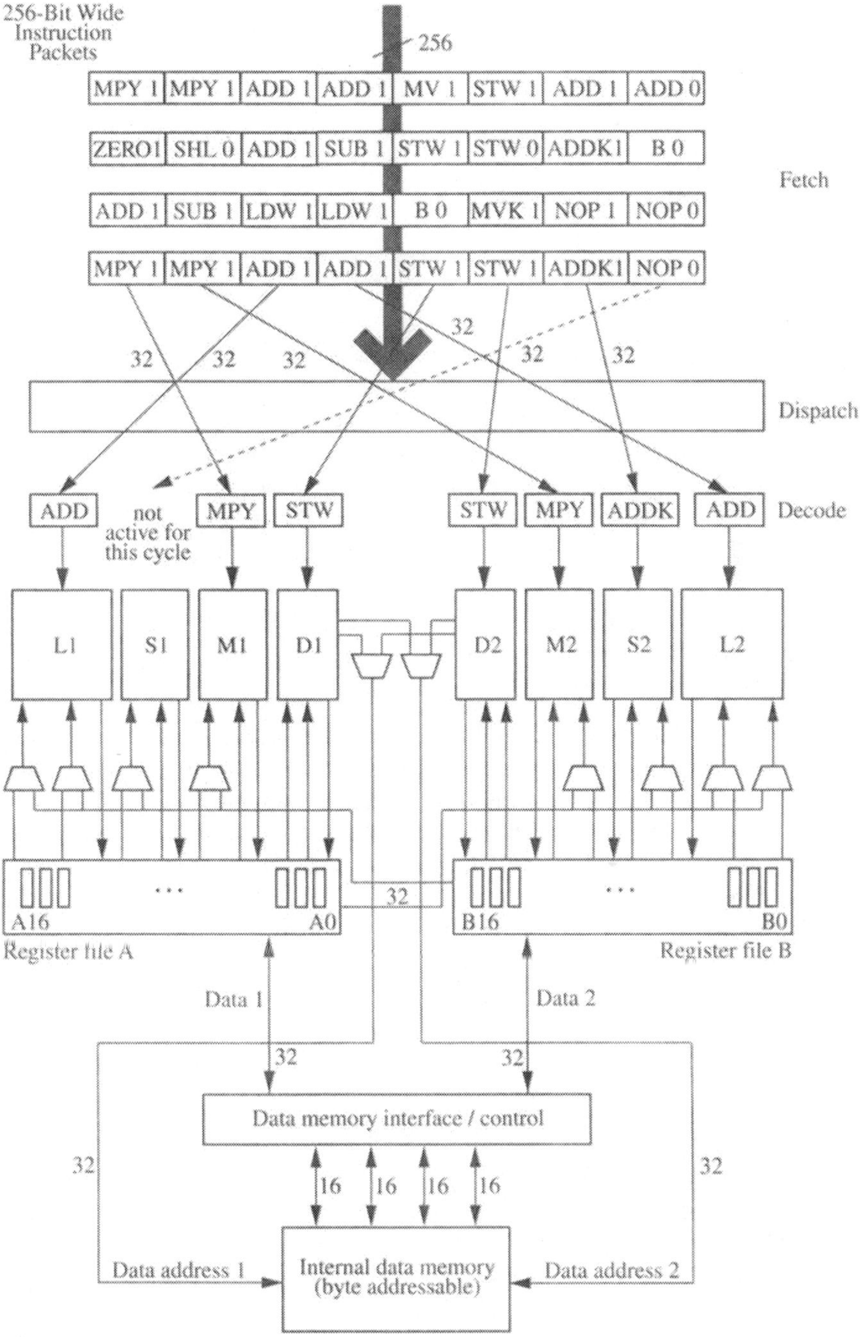

Fig. 4.38. Execution of a FIR filter code on the TMS320C6201 processor

- 32-bit add, subtract, linear, and circular address calculation,
- 8/16/32/64-bit loads, and
- 8/16/32-bit stores.

These highly orthogonal functional units provide code generation tools with many execution resources enabling these tools to maximize performance without extensive hand-coding of assembly instructions by software developers. The 'C67x's instruction-packing feature also allows these instructions to be executed in parallel, in serial or in parallel/serial combinations. This optimized scheme enables significant reductions in code size, program fetches and power consumption.

Just as with the 'C62x, the 'C67x core includes 8-byte, 16-byte, and 32-byte addressability; 8 bits of overflow protection; saturation; bit-field extract, set and clear; bit counting; normalization and two additional integer multiply functional units with 32-bit and 24-bit multiply support.

The 'C67x series of DSPs will begin sampling in the second half of 1998, with the first devices providing 1 GFLOPS of performance which will be fabricated by a 0.18 µm / 5-level metal process. By the year 2000 the 'C67x technology will allow devices that triple in performance to a full 3 GFLOPS.

4.10.2 EPIC Processors, Intel's IA-64 ISA and Merced Processor

Aimed at providing advanced technologies for workstation, server, and enterprise-computing products, Hewlett-Packard and Intel announced in June 1994 their joint R&D project, which includes development of the 64-bit instruction set and compiler optimization. This resulted in the *explicitly parallel instruction computing* (EPIC) design style, which includes several features and techniques, such as (*Gwennap* [117], 1997; and *Dulong* [71], 1998):

- specifying ILP explicitly in the machine code, that is, the parallelism is encoded directly into the instructions similarly to VLIW;
- a fully predicated instruction set;
- an inherently scalable instruction set (i.e., the ability to scale to many of FUs);
- many registers;
- speculative execution of load instructions,

which complement each other and can be effectively used by the compiler to enhance the performance of wide-issue processors (*August et al.* [19], 1998).

Also jointly developed by HP and Intel is the *Intel 64-bit Architecture*[32] (IA-64), which implements EPIC:

[32] In official HP-Intel announcements it was called 64-bit Instruction Set Architecture, 64-bit ISA. Although IA-64 is being developed jointly by HP, Intel is the only company authorized to manufacture and market microprocessors based on it.

- The IA-64 ISA specifies 128 64-bit general-purpose registers, 128 80-bit floating-point registers, and 64 1-bit predicate registers.[33]
- IA-64 instructions are 40-bit long and consist of op-code, predicate field (6 bits), two source register addresses (7 bits each), destination register address (7 bits), and special fields (including integer and floating-point arithmetic). The 6-bit predicate field in each IA-64 instruction refers to a set of 64 predicate registers.
- IA-64 instructions are packed by compiler into bundles. A *bundle* is a 128-bit long instruction word (LIW) containing three IA-64 instructions along with a so-called *template* that contains instruction grouping information. For example, 16 instructions could be packaged into six different bundles (three in each of the first five bundles, and one in the sixth), each bundle with its own template. Unlike some previous VLIW architectures, the IA-64 does not insert no-op instructions to fill slots in the bundles. The template explicitly indicates parallelism, that is, whether the instructions in the bundle can be executed in parallel or one or more must be executed serially (due to operand dependences) and whether the bundle can be executed in parallel with neighboring bundles. Thus, the bundled instructions do not have to be in their original program order, and they can even represent entirely different paths of a branch. The compiler can also mix dependent and independent instructions together in a bundle, because the template keeps track of which is which.

Scalability. A single bundle containing three instructions corresponds to a set of three FUs. If an IA-64 processor had n sets of three FUs each, then using the template information it would be possible to chain the bundles to create an instruction word of n bundles in length. Although not absolutely perfect, this is the way to provide scalability of IA-64 to any number of FUs.

Predication in IA-64 ISA. Predication was already introduced in Sect. 4.3.4. Its use is described here as applied to the IA-64 ISA. Normally, a compiler turns a source code branch statement into alternate blocks of machine code arranged in a sequential stream (Fig. 4.39a). Depending on the outcome of the branch, the processor will execute one of those basic blocks by jumping over the others. Today's superscalar processors predict the branch outcome and speculatively execute the target block, paying a heavy penalty in lost cycles if they mispredict. IA-64 compilers will use predication to remove the penalties caused by mispredicted branches and by the need to fetch from noncontiguous target addresses by jumping over blocks of code beyond branches. When the compiler finds a branch statement it marks all the instructions that represent each path of the branch with a unique identifier called a *predicate* (Fig. 4.39b). IA-64 defines a 6-bit field (predicate register

[33] 128 registers are a considerable number when compared with the 8 general-purpose registers of the x86 family.

address) in each instruction to store this predicate. Thus, there are 64 unique predicates available at any one time. Any instructions that share a particular branch path will share the same predicate. After marking the instructions with predicates, the compiler determines which instructions the processor can execute in parallel.[34] Specifically, the main idea of predication is that when the processor encounters a predicated branch at runtime, it will begin executing the code along all destinations of the branch, exploiting as much parallelism as possible (see *Mahlke et al.* [188], 1995). Now the compiler can start assembling the machine code instructions into 128-bit bundles of three instructions each.

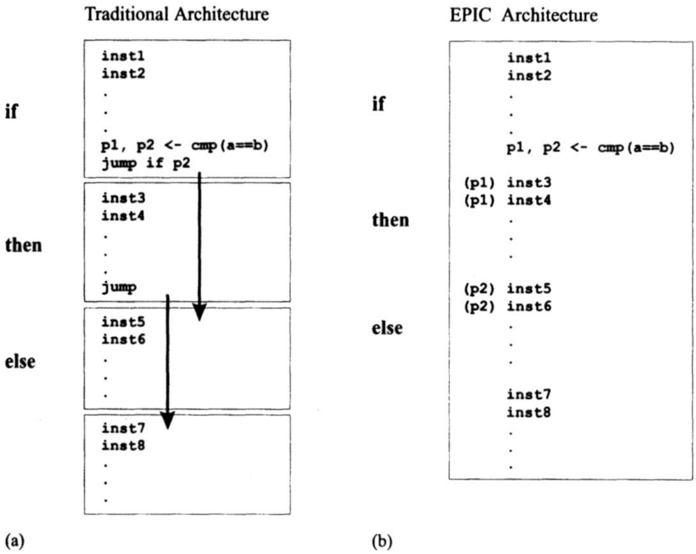

Fig. 4.39. An if-then-else statement (a) four basic blocks in traditional architecture (b) one basic block in EPIC architecture

At run-time, the CPU scans the templates, picks out the independent instructions, and issues them in parallel to the FUs. When the processor finds a predicated branch, it begins executing the code for every possible branch outcome. Suppose that predicate register **PR1** has been associated to the instructions in the THEN path, and the predicate register **PR2** to those in the ELSE path of the branch. Eventually, the processor will evaluate the compare operation of the branch and the branch outcome will become known. If the condition is (say) TRUE, the processor will store a 1 in **PR1** to represent TRUE, and it will store a 0 in **PR2** to represent FALSE. In spite of the fact that the processor has probably executed some instructions from both

[34] This requires the compiler to know the processor's microarchitecture well, because different IA-64 chips will have different numbers and types of FUs.

possible paths, none of the (possible) results has been stored yet. To do this, the processor checks the predicate register of each of these instructions. If the predicate register contains a 1 (in our example that would be **PR1**), the instruction is on the TRUE path (i.e., valid path), so the processor retires the instruction and stores the result. If the register contains a 0 (**PR2** in our example), the instruction is invalid, so the processor discards the result.

Simulations indicate that predication can eliminate more than half of the branches in a typical program and, therefore, reduce by half the number of potential mispredictions. This has several benefits:

- code fragmentation is reduced because the compiler can merge small basic blocks into larger blocks that branches do not chop up,
- the compiler has more freedom to rearrange instructions for parallel execution (due to larger blocks),
- the number of mispredicted branches is drastically reduced because many branches do not require the processor to predict their outcome,
- the FUs are kept busy because more instructions can be issued in parallel.

The downside of predication is that whether the predicated condition evaluates TRUE or FALSE, the processor does perform redundant work.

Speculative loading. Another key feature of IA-64 is speculative loading. This technique allows IA-64 processors to load data from memory well before the program needs it, and thus to effectively minimize the impact of memory latency. Like predication, speculative loading is a combination of compile-time and run-time optimizations.

The compiler is looking for any instructions that will need data from memory and, whenever possible, hoists a load at an earlier point in the instruction stream, well ahead of the instruction that will actually use the data. In most of today's superscalar processors the load can be hoisted up to the first branch instruction which represents a barrier (see Fig. 4.40a). Since branches typically occur about every seven instructions, it seems that they also severely inhibit the IA-64's ability to load data from memory long before needed. However, speculative loading combined with predication gives the compiler more flexibility to reorder instructions and to hoist loads above branches. We call such a shifted load a *speculative load* instruction and denote it by `ld.s`. The compiler also inserts a matching *speculative check* instruction, denoted by `check.s`, immediately before the particular instruction that will use the data (see Fig. 4.40b). At the same time, the compiler rearranges the surrounding instructions so that the processor can issue them in parallel.

At run-time, the processor encounters the `ld.s` instruction first and tries to retrieve the data from the memory. More precisely, `ld.s` performs memory fetch and exception *detection* (e.g., checks the validity of the address). If an exception is detected, `ld.s` does not deliver the exception (i.e., it does not call the operating system routine for handling the exception). Instead, `ld.s`

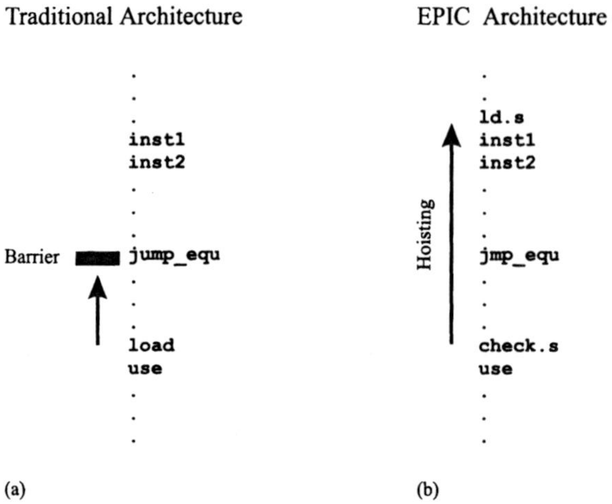

Fig. 4.40. Control speculation **(a)** in traditional architecture a branch is the barrier for hoisting a load instruction **(b)** in EPIC a load can be initiated even before the branch

only marks the target register (by setting a token bit). The exception *delivery* is the responsibility of the matching `check.s` instruction. When encountered, `check.s` calls the operating system routine if the target register is marked (i.e, if the corresponding token bit is set), and does nothing otherwise. Of course, whether the `check.s` instruction will be encountered depends on the outcome of the branch instruction. Thus, it may happen that an exception detected by this `ld.s` is never delivered (see *Dulong* [71], 1998).

Many EPIC ideas are not new. Predication has already appeared in the ARM architecture (see p. 43), which has included a form of predication since its inception in the 1980's; and in TI's VelociTI architecture which is the foundation of the TMS320Cx processors, whose instructions included a conditional field for similar purposes. Speculative loading with `ld.s/check.s` machine level instructions resembles the `TRY/CATCH` statements in some high-level programming languages (e.g., Java).

Merced. Merced[35] is the code name of Intel's general-purpose 64-bit microprocessor, which should be the first member of the IA-64 family. It is currently under development and is scheduled for production in mid-2000 in 0.18 μm technology. Not much is publicly known about the Merced at the time of writing. The processor is targeted at servers with moderate to large numbers of processors, and should provide full compatibility with Intel's x86 family. In x86 mode, it will probably match the performance of a 500 MHz Pentium II. Merced, which is otherwise expected to operate at speeds of around 800 MHz, will deliver 50 SPECint95 and 100 SPECfp95.

[35] The processor is named after Merced city, located near San Jose, CA.

Because of the comparable estimated performances of other competitors, such as Sun's UltraSPARC-III, DEC's Alpha 21264, and IBM's POWER3, Intel and Hewlett-Packard are now pinning their performance hopes on the second generation IA-64 processor, code-named McKinley (*Gwennap* [118], 1998).

4.11 Conclusions on Multiple-Issue Processors

Current microprocessors utilize ILP by a deep processor pipeline and by the superscalar instruction issue technique. A superscalar processor is able to issue multiple instructions each clock cycle from a conventional linear instruction stream. DEC Alpha 21164, IBM/Motorola/Apple PowerPC 604 and 620, MIPS R10000, Sun UltraSPARC-II or IIi and HP PA-8000 issue up to four instructions per cycle from a single thread. DEC Alpha 21264 already issues six instructions per cycle. VLSI technology will soon allow future microprocessors to exploit instruction-level parallelism (ILP) of eight or more instructions per cycle.

As the issue rate of future microprocessors increases, the compiler or the hardware will have to extract more ILP by analyzing a larger instruction window. However, ILP found in a conventional instruction stream is limited (*Butler et al.* [45], *Wall* [317], *Lam and Wilson* [170]). In general, integer-dominated programs feature a rather low ILP, while a high ILP can be extracted from floating-point programs by transforming loop-level parallelism into ILP using compiler techniques like loop unrolling or software pipelining.

Recent studies have shown the limits of processor utilization even of today's superscalar microprocessors. Using the SPEC92 benchmark suite, the PowerPC 620 showed an execution of 0.96 to 1.77 IPC (see *Diep et al.* [68]), and even an 8-issue Alpha processor will fail to sustain 1.5 IPC (see *Tullsen et al.* [303]). The out-of-order superscalar processors MIPS R10000 and HP PA-8000 reach a slightly higher IPC due to their larger instruction windows and the out-of-order issue facility. All these measurements were usually based on the SPEC92 benchmark suite. Similar CPI values between 0.5 and 1.5 have been reported by *Bhandarkar and Ding* [27] for SPEC95 benchmark programs on the Pentium Pro.

Newer measurements by *Keeton et al.* ([160], 1998) of commercial online transaction processing (OLTP) database workloads on a quad Pentium Pro symmetric multiprocessor show an overall CPI of 3.39. Similar measurements by *Barroso et al.* ([23], 1998) show an even less favorable CPI of 7.0 (IPC of only 0.14 !) for an OLTP and a CPI of 1.3–1.9 for a decision support system (DSS) workload on a four-issue Alpha 21164 processor. The low IPC stems from the 75 % of time that is spent stalling for memory accesses due to the large working set, common in database transaction processing, that leads to a high cache miss ratio even of a large L2 cache.

An unsolved problem in today's microprocessors is the *memory latency* caused by cache misses. As reported by *Loudon and Lenoski* [173] for the SGI Origin 2000 distributed shared memory system, these latencies are 11 processor cycles for a L1 cache miss, 60 cycles for a L2 cache miss, and can be up to 180 cycles for a remote memory access (see also p. 257 for latencies on an Alpha Server 4100 SMP). In many missed instruction slots the latencies should be multiplied by the issue bandwidth. Only a small part of the memory latency can be removed or hidden in modern microprocessors, even when advanced techniques are employed, such as out-of-order execution, write buffer, cache preload hardware, lockup-free caches, and a pipelined system bus. Thus microprocessors often idle and are unable to exploit the high degree of internal parallelism provided by a wide superscalar approach. The rapid context switching of dataflow and multithreaded architectures shows a superior way out. Idling is avoided by switching execution to another context.

An 8-issue (or even higher) superscalar processor will be possible in the next generation of microprocessors. Finding enough fine-grain parallelism to fully exploit the processor will be the main problem. One solution is to *enlarge the instruction window* to several hundred instruction slots with hopefully more simultaneously executable instructions present. However, there are two drawbacks to this approach. First, considering that all instructions stem from a single instruction stream and that on average every seventh instruction is a branch instruction, most of the instructions in the window will be speculatively assigned with a very deep speculation level (today's depth is normally four at maximum). Thereby most of the instruction execution will be speculative. The principal problem here arises from the single instruction stream that feeds the instruction window. Second, if the instruction window is enlarged, the updating of the instruction states in the slots and matching of executable instructions lead to more complex hardware logic in the issue stage of the pipeline, thus limiting the cycle rate increase which is essential for future generations of microprocessors. Solutions include the *decoupling of the instruction window* with respect to different instruction classes as in the HP PA-8000, the *partitioning of the issue stage* into several pipeline stages, and *alternative instruction window organizations*. One alternative instruction window organization, proposed by *Palacharla et al.* [222], is the multiple FIFO-based organization in the dependence-based microprocessor. Only the instructions at the heads of a number of FIFO buffers can be issued to the execution units in the next cycle. The total parallelism in the instruction window is restricted in favor of a less costly issue that does not slow down processor cycle rate. The potential fine-grained parallelism is thereby limited – a technique somewhat similar to the threaded dataflow approaches described in Sect. 2.3.1.

Another problem of today's superscalar microprocessors may not be found in the instruction window organization. It is the necessity of instruction commitment due to the serial semantics of the instruction stream. For example, if a load instruction causes an L2 cache miss, the whole reorder buffer may soon be clogged by succeeding instructions (succeeding in sequential program order) that are already executed. Because of the sequential program order that should be restored by the retire stage, these instructions cannot be retired and removed from the reorder buffer even if the instructions are independent of the load instruction that caused the cache miss. The commitment may be the main obstacle for further performance increase in microprocessors.

In principle, an algorithm defines a partial ordering of instructions due to control and data dependences. The total ordering in an instruction stream for today's microprocessors stems from von Neumann languages. But why should

- a *programmer*
  - design a *partially* ordered algorithm,
  - and then code the algorithm in *total* ordering because of the use of a sequential von Neumann language,
- the *compiler*
  - regenerate the *partial* order in a dependence graph,
  - and then generate a reordered "optimized" sequential machine code,
- the *microprocessor*
  - dynamically regenerate the partial order in its out-of-order section, execute due to a *micro dataflow* principle,
  - and then re-establish the unnatural serial program order for in-order commitment in the retire stage ?

Ideally, an algorithm should be coded in an appropriate higher-order language (e.g., dataflow-like languages might be appropriate). Next, the compiler should generate machine code that still reflects the parallelism and not an unnecessary serialization. Here, a dataflow graph viewed as machine language might show the right direction. A parallelizing compiler may generate this kind of machine code even from a program written in a sequential von Neumann language. The compiler could use compiler optimization and coding to simplify the dynamic analysis and issue out of the instruction window. The processor dismisses the serial reordering in the completion stage in favor of only a partial reordering. The retire unit retires instructions not in a single serial order but in two or more series (as in the simultaneous multithreaded processors). Clogging of the reorder buffer is avoided since clogging of one thread does not restrict retirement of instructions of another thread.

5. Future Processors to use Fine-Grain Parallelism

Everything that can be invented has been invented.
US Commissioner of Patents, 1899

I think there is a world market for about five computers.
Thomas J. Watson Sr., IBM founder, 1943.

5.1 Trends and Principles in the Giga Chip Era

Forecasting the implications of technology is even harder than forecasting technological trends, although even the latter can prove wildly inaccurate. This and the next two chapters describe some ideas about future microarchitectures based on technological forecasts. Principal problems and possible application domains are stated. All architectures described are still in survey stage. Up to now only simulation studies exist and very few prototype implementations are available. So most of this and the following two chapters is highly speculative.

This section is organized into subsections separately covering technology-related aspects such as technological forecasts, future impacts of wire delays, and limits in miniaturization; application- and economy-related aspects like the growing fabrication costs, the increasing importance of multimedia workloads, and consumer and mobile product; and future architecture proposals.

5.1.1 Technology Trends

State-of-the-art microprocessors in 1999 feature up to 15-million-transistor chips (in the case of the IBM RISC System/6000SP processor) and up to 600 MHz clock rate (in the case of an Alpha 21164 processor). The most advanced processors use a 0.25 μm CMOS technology. Achieving clock rates of 1 GHz has already been demonstrated for a relative simple, scalar RISC processor (*Hofstee et al.* [139]).

In its 1994 road map, the Semiconductor Industry Association (SIA) predicted the course of semiconductor technology over the next 15 years. The

SIA predicted that by 2007, industry would be manufacturing 350-million-transistor processors, and by 2010, it would be manufacturing 800-million-transistor processors with thousands of pins, a 1 000-bit bus, and clock speeds over 2 GHz. Such "giga" chips would produce a predicted maximum of 180 W (*Burger and Goodman* [39]). These large numbers of transistors result from greatly reduced feature sizes and lead to higher wiring densities. Thus, a major challenge is to use these transistors effectively and to accommodate the dramatic shifts in design constraints that will result from these changes (*Smith and Vajapeyam* [271]).

The 1997 National Technology Roadmap for Semiconductors published by the SIA [336] is similar optimistic (see Table 5.1).

Table 5.1. The 1997 National Technology Roadmap for Semiconductors

Year of 1st shipment	1997	1999	2001	2003	2006	2009	2012
Local clock (GHz)	0.75	1.25	1.5	2.1	3.5	6	10
Across chip (GHz)	0.75	1.2	1.4	1.6	2	2.5	3
Chip size (mm^2)	300	340	385	430	520	620	750
Feature size (nm)	250	180	150	130	100	70	50
Number of chip I/O	1450	2000	2400	3000	4000	5400	7300
Transistors/chip	11M	21M	40M	76M	200M	520M	1.4G

It foresees that the most powerful processors in 2012 will run at 10 GHz, contain 1 400 million transistors and feature a 0.05 µm technology. DRAMs will grow to 4 Gbits in 2003. An Alpha 21164 processor today contains about 9.3 million transistors. If 1 000 million transistors could be integrated in a single chip in 2012, then 100 copies of an Alpha processor could be arranged on such a chip. However, to approach feature sizes of 0.1 µm or less, new technologies have to be developed [335].

According to *Tremblay* [300], the design challenges for future microprocessors focus on three issues: increasing clock speed, the amount of work that can be performed per cycle, and the number of instructions needed to perform a task. While these issues can be competing goals, most ideas usually just improve one factor, thereby making the others worse. For instance, the original RISC idea successfully supported higher clock speeds and a smaller CPI count through pipelining, but increased the number of instructions needed to perform a task compared to contemporary CISC processors.

Today's general trend in processor design again goes toward more complex designs and is opposed by the wiring delay within the processor chip as the main technological problem. At increasingly higher clock rates with sub-quarter-micron designs, on-chip interconnecting wires cause a significant portion of the delay time in circuits (*Burgess* [43]). In particular global interconnects within a processor chip cause problems with higher clock rates. Maintaining the integrity of a signal as it moves from one end of the chip to the other becomes more difficult. Copper metallization is worth a 20–30 %

reduction in wiring delay, but will not in general solve the problem for future processor generations (*Killian* [161]). While in the past designers focused on gate delay, in the future wire delay will be an increasingly important issue. To compensate for wire delays, a design must be partitioned in a way that maximizes the local communication of data values. The architecture, floor planning, and circuit design will be more tightly interwoven.

The study of *Palacharla et al.* [222] estimated the impacts of delays of key pipeline components of superscalar processors, when technology moves to smaller features. The study shows that for a given feature size, instruction-issue logic moderately increases delays for higher degrees of instruction-level parallelism (ILP). Reducing feature size decreases the delays, but they are likely to remain an important design consideration because of faster clock cycles. What really matters are bypass delays: the complexity of bypass paths (long wires) grows quadratically with the number of functional units required to support higher ILP. This is significant because wire delays are not likely to scale well for smaller features, so bypass delays will become critical (*Smith and Vajapeyam* [271]).

5.1.2 Application- and Economy-Related Trends

While the physical limits of how far one can shrink a CMOS processor may not be reached in the near future (if we keep shrinking transistors as in the past, we will run into the limits of physics within the next 25 years), cost is a possibly limiting factor. Economy of scale may hinder the manufacture of densely integrated chips as proposed by SIA. Fabrication plants now cost about 2 000 million USD, a factor of ten more than a decade ago. Manufacturers can only sustain such development costs if larger markets with greater economies of scale emerge. This problem may be tightened if consumers begin to prefer cheaper computers instead of faster ones.

As *Colwell* [51] from Intel remarks the real threat for processor designers is shipping 30 million CPUs only to discover they are imperfect in some way that causes a recall, threatening the bankruptcy of a firm. This leads to the challenge of chip debug, validation, and testing. More than 20 % of Pentium Pro design efforts was associated with validation. Formal methods prove useful; they are already mature enough that they could have avoided the Pentium divide unit flaw. Chip verification techniques potentially render testing unnecessary. DEC has already been able to verify functionally the Alpha 21264 processor (see *Rubinfeld* [249]). But the performance of today's chip design tools is still rather limited when the task is as complex as designing a whole state-of-the-art microprocessor.

The large design teams necessary for complex processor designs are another obstacle, using much manpower and prolonging the design period. A microprocessor design takes about two-and-a-half years, but the lifetime of a microprocessor is only one-and-a-half years. It follows that several different designs have to be developed at once by the same engineers or by different design teams (*Grohoski* [107]). Less complex processors will need less time for debugging and smaller design teams.

There is a general agreement among processor designers that future microprocessors will support user interactions like video, audio, voice recognition, speech processing, and 3D graphics. Multimedia features, like video and audio, may prove minor challenges for the future, since there is a limit to what the human visual and audio system can perceive. A performance level that is sufficient for human perception will be reached within one of the next generations of microprocessors. Long-term user interaction issues are applications that are scalable in their computation requirements. Such scalable applications that profit from increasingly higher performance microprocessors are outstanding 3D graphics, handwriting and speech recognition, object recognition and classification from pictures or videos. Moreover, future computers will primarily be communication devices.

Furthermore, special application challenges will come from large data sets and huge databases, large data-mining applications, transaction processing, huge EDA applications like CAD/CAM software, virtual reality computer games, signal processing, and real-time control.

5.1.3 Architectural Challenges and Implications

Further architectural challenges besides the economic and application demands are:

- Preserving object code compatibility, to deal with legacy binaries, as noted in 1998 by *Tremblay* [300] and *Colwell* [51]. This may be avoided by a virtual machine that targets run-time ISAs, as the Java Virtual Machine or *Fisher*'s approach of walk-time techniques [86].
- It is necessary to find ways of expressing and exposing more parallelism to the processor. It is doubtful whether enough ILP is available. The language structures of von Neumann languages used today limit the amount of extractable parallelism. We need to look at parallelism at higher than ILP and run more than one thread in parallel. Higher-level parallelism allows smaller computational units to work in parallel on these threads and thereby favors a modular design approach (*Tremblay* [300], 1998).
- Buses may probably scale. Area-array flip-chip packaging will allow the implementation of much wider buses in future (*Rubinfeld* [249], 1998). Slow buses are a low-end vendor problem. SGI is already shipping 400 MHz chip-to-chip communication by point-to-point unidirectional communication over five meters of cable (*Killian* [161], 1998).
- A related issue is the memory bottleneck. Memory latency may be solved by a combination of technological improvements in memory chip technology (*Katayama* [156], 1997) like SLRAM, SLDRAM (*Gillingham and Vogley* [100], 1997) and Rambus (*Crisp* [55], 1997), and by applying advanced memory hierarchy techniques such as larger caches, more elaborate

cache hierarchies, prefetching, compiler tricks, streaming buffers, intelligent memory controllers, and bank interleaving [51, 249]. Many other authors disagree with this assumption and center their designs on the memory processor bottleneck.
- Power consumption is of specific importance for the increasing market for mobile computers and appliances.
- Soft errors by cosmic rays of gamma radiation may be met by fault-tolerant design through the chip [249].

A general-purpose processor will not be able to serve all these differing requirements adequately. Possible solutions are:

- a focus of processor chips on particular market segments (*Killian* [161], 1998);
- multimedia pushes desktop personal computers while high-end microprocessors will serve specialized applications due to different performance/cost tradeoffs (*Burgess* [43], 1998);
- integrate functionalities to systems on a chip, for example, AGP on-chip (*Colwell* [51], 1998);
- a partition of the microprocessor into a client chip part that focuses on general user interaction enhanced by server chip parts that are tailored for special applications (*Tremblay* [300], 1998);
- a CPU core that works like a large ASIC block and that allows system developers to instantiate various devices on a chip with a simple CPU core (*Grohoski* [107], 1998); and
- reconfigurable on-chip parts that adapt to application program requirements.

An architectural implication of the long interconnect wire delay problem is a strict functional partitioning within the microarchitecture and a floor planning that avoids long interconnects. Designers can probably best accomplish this by dividing the microarchitecture into multiple processing elements, each no larger than today's superscalar processors. Coordinating these processing elements to act as a single, unified processor will require, as noted by *Smith and Vajapeyam* [271], an additional level of microarchitecture hierarchy for both control (distribution of instructions) and data (for communicating values among the processing elements).

We will see in future whether a modular design is possible for a very complex single instruction stream general-purpose processor, or whether the pendulum swings back to less complex processors. Several simple processors could be combined into a multiprocessor chip or a simple processor could be integrated on a DRAM for tighter processor–memory integration. A 256 Mbit memory chip together, with memory compression, can afford to devote some reasonable percentage of that die to a simple CPU (*Grohoski* [107], 1998).

Very complex uniprocessor approaches that retain the result serialization of the von Neumann architecture exhibit the following microarchitecture design principles:

- Advanced superscalar processors will scale up from current designs to issue 16 or 32 instructions per cycle (see Sect. 5.2).
- Superspeculative processors enhance wide-issue superscalar performance by speculating aggressively at every point in the processor pipeline (see Sect. 5.3).
- Multiscalar processors divide a program into a collection of tasks that are distributed to a number of parallel processing units under the control of a single hardware sequencer (see Sect. 5.4).
- Trace processors facilitate high ILP and a fast clock rate by breaking up the processor into multiple, distinct cores similar to multiscalar, and breaking up the program into traces (dynamic sequences of instructions). One core executes the current trace while the other cores execute future traces speculatively (see Sect. 5.5).
- A DataScalar processor runs the same sequential program redundantly across multiple processors using distributed data sets. Loads and stores are performed only locally on the processor memory that owns the data, but a local load broadcasts the loaded value to all other processors (see Sect. 5.6).

Multiprocessor alternatives that optimize the throughput of a multiprogramming workload while each thread or process retains the result serialization of the von Neumann architecture are:

- Chip multiprocessors which place a small number of distinct processors (4–16) on a single chip and run parallel programs and/or multiple independent tasks on these processors (see Sect. 6.2.3).
- Simultaneous multithreaded processors which share an aggressive pipeline between multiple tasks when there is insufficient ILP in any one task to use the pipeline fully (see Sect. 6.4).

Highly parallel chip architectures that deviate from the von Neumann architecture model are made available by two related approaches that are covered in some detail in Chap. 7:

- The processor-in-memory (PIM) or intelligent RAM (IRAM) approach integrates processor and memory on the same chip to increase memory bandwidth (see Sect. 7.1).
- Reconfigurable processors allow the hardware to adapt dynamically at runtime to the needs of an application (see Sect. 7.2).

5.2 Advanced Superscalar Processors

One end of the spectrum of architectural design choices is to enhance today's superscalar multiple issue processors but retain result serialization as defined by the von Neumann architecture. To reach highest execution of a single instruction stream involves delivering the maximum possible instruction bandwidth each cycle to the execution core and consuming the delivered bandwidth within the execution core. Delivering optimal instruction bandwidth requires a high number of instructions fetched each cycle, a minimal number of cycles in which instructions that are fetched for a wrongly predicted path are subsequently discarded, and a very wide, full instruction issue each cycle. Consuming this instruction bandwidth requires sufficient data supply so that instructions are not unnecessarily inhibited from executing, and sufficient functional units to handle the instruction bandwidth.

Patt et al. [229] suggest for advanced superscalar processors for one giga transistor chips in year 2005 (see Fig. 5.1):

- an I-cache that provides for out-of-order fetch in the presence of I-cache misses;
- a large sophisticated trace cache (*Rotenberg et al.* [246], 1996) for providing a contiguous instruction stream;
- an aggressive multi-hybrid branch predictor using multiple, separate branch predictors, each tuned to a different class of branches with support for context switching, indirect jumps, and interference handling;
- a very wide-issue superscalar processing with an issue width of 16 or 32 IPC;
- a large number of reservation stations to accommodate approximately 2 000 instructions;
- 24 to 48 highly optimized, pipelined functional units;
- sufficient on-chip D-cache (more than half of the transistors on chip are allocated to data and on-chip L2 caches), and
- sufficient resolution and forwarding logic.

The *trace cache* is a new paradigm for caching instructions (see also Sect. 5.5). The trace cache is accessed like an I-cache by using the starting address of the next block of instructions. Unlike the I-cache, a trace cache stores contiguously fetched and executed instructions in physically contiguous storage. A trace cache line stores a segment of the dynamic instruction trace across multiple, potentially taken branches. As a group of instructions is processed, it is latched into the so-called *fill unit* that maximizes the size of the trace segment, and finalizes a segment when the segment can be expanded no further. Finalized segments are written to the trace cache. Instructions can be sent from the trace cache to the reservation stations without having to undergo a large amount of processing and rerouting. Because the fill unit is outside the critical path, it may take several cycles to analyze and build a trace. To measure the performance of a trace cache, three applications from

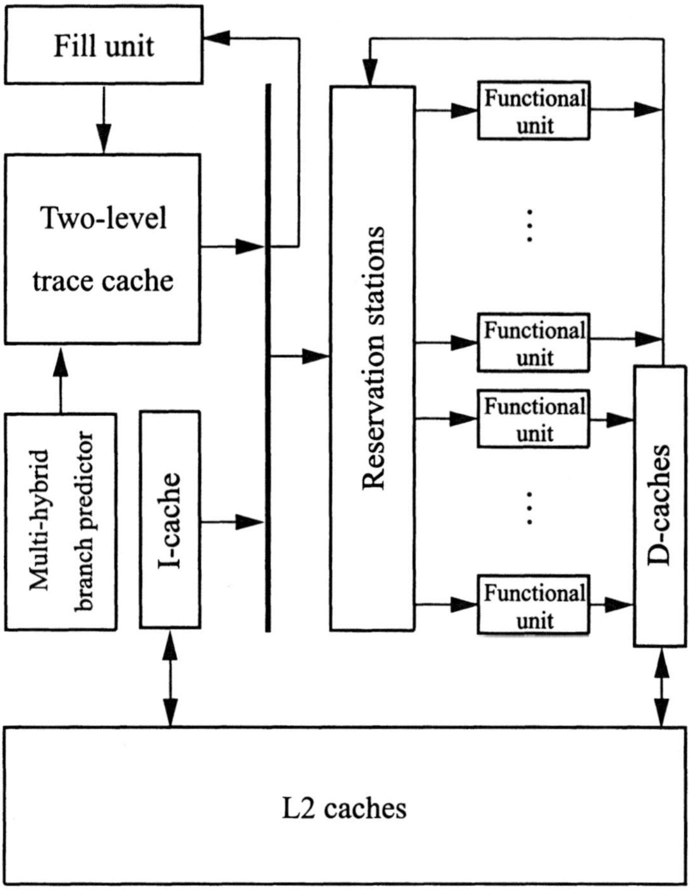

Fig. 5.1. Advanced superscalar architecture

the SPECint95 benchmarks were simulated by *Patt et al.* [229] on a 16-wide issue machine assuming perfect branch prediction. The simulations showed the trace cache as effective in delivering more than one basic block per cycle, and the trace cache continued to add performance as the size of the storage structure increased.

As already emphasized in Sect. 4.3 on branch prediction, a fast and accurate branch prediction is essential for advanced superscalar processors with hundreds of in-flight instructions. Branch prediction itself is already a well-developed part of microarchitecture design. One observation is that many branches display different characteristics that cannot be optimally predicted by a single-scheme branch predictor. *Evers et al.* [80] propose hybrid branch predictors, a technique that was previously proposed by combining two predictors (*McFarling* [196], 1993) and that is already implemented in the PowerPC 620. Hybrid predictors comprise several predictors, each targeting different classes of branches. The principal idea is that each predictor scheme works best for a particular branch type.

As predictor tables increase in size, they often take more time to react to changes in a program (warm-up time). A hybrid predictor with several components can solve this problem by using component predictors with shorter warm-up times while the larger predictors are warming up. Examples of predictors with shorter warm-up times are two-level predictors with shorter histories as well as smaller dynamic predictors [229].

The *multi-hybrid branch predictor* of *Evers et al.* [80] and *Patt et al.* [229] uses a set of selection counters for each entry in the branch target buffer, in the trace cache, or in a similar structure, keeping track of the predictor currently most accurate for each branch, and then using the prediction from that predictor for that branch. The multi-hybrid predictor performs better than regular hybrid predictors. It reaches a prediction rate of 95 % for 16 kB predictor size and up to near 97 % for 256 kB predictors using programs of the SPECint95 benchmark suite [229]. Despite this high prediction rate, the remaining mispredictions still incur a large performance penalty. Other branch techniques must be combined with branch prediction. Such techniques are predication (see Sect. 4.3.4) to enlarge the number of instructions between two speculative predictions or both-path execution (as in the PolyPath architecture by *Klauser et al.* [163], 1998) in the case of low branch prediction confidence (*Grunwald et al.* [110], 1998).

To increase instruction supply in case of trace cache misses, an out-of-order fetch should be employed. An in-order fetch processor, upon encountering a trace cache miss, waits until the miss is serviced before fetching any new segments. An out-of-order fetch processor temporarily ignores the segment associated with the miss, attempting to fetch, decode, and issue the segments that follow it. After the miss has been serviced, the processor decodes and issues the ignored segment. A related technique is to fetch instructions that appear after a mispredicted branch, but are not control-dependent upon that branch. Out-of-order fetch provides a way to fetch such control-independent instructions by skipping the block that follows a hard-to-predict branch until either an accurate prediction can be made or the branch is resolved. The processor fetches, decodes, and issues instructions that begin at the merge point of the alternative paths that follow the branch. These instructions are guaranteed to be on the program's correct path. As soon as a prediction can be made or the branch is resolved, the fetch unit will return to the branch and restart fetching there. Upon reaching the merge point, the processor will jump past the instructions that it has already fetched [229].

A 16-wide-issue processor will need to execute about eight loads/stores per cycle. The primary design goal of the D-cache hierarchy is to provide the necessary bandwidth to support eight loads/stores per cycle. The size of a single, monolithic, multiported, L1 D-cache would probably be so large that it would jeopardize the cycle time. Because of this, *Patt et al.* expect the L1 D-cache to be replicated to provide the required ports – a technique already proposed for the register file of the Alpha 21264 processor. Further features of the data supply system are a larger, L2 D-cache with fewer port requirements and less data prefetching.

In addition to control prediction, data prediction (see Sect. 5.3) will also be widely used in future superscalar processors. Whenever the full issue bandwidth cannot be filled by enough instructions that are executed in parallel, control and data prediction can be applied for better processor utilization and a potential performance increase. *Patt et al.* expect prefetching and set prediction in caches to become the norm in processor design. Future processors will also predict the addresses of loads, allowing loads to be executed before the computation of operands needed for their address calculation. Processors will predict dependences between loads and stores, allowing them to predict that a load is always dependent on some older store (a more extensive description of data speculation follows in the next section).

The execution core must consume 16 to 32 IPC to be as quick as the fetch engine is providing the instructions. As in today's superscalar processors logical registers must be renamed to avoid unnecessary delays due to false dependences, and instructions must be executed out of order to compensate for the delays imposed by the data dependences that are not predicted. *Patt et al.* envision an execution core comprising 24 to 48 functional units supplied with instructions from large reservation station units and having a total storage capacity of 2 000 or more instructions. For better functional partitioning and shorted signal propagation on the processor die, the execution units will be partitioned into clusters of three to five units. Each cluster will maintain an individual register file. Each functional unit has its own reservation station unit. Data forwarding within a cluster will take one cycle, while data forwarding between different clusters will require multiple cycles. Instruction scheduling will be done in stages to solve the difficulty of scheduling instructions from a centralized large instruction window.

Despite all extensive speculation mechanisms, such highly parallel uniprocessors only make sense if enough ILP can be supplied by the application programs. *Patt et al.*'s [229] simulations of the SPECint95 with an instruction window having 2 048 instructions, perfect caches and perfect branch prediction show an IPC rate of about 10 for an issue/execution width of 16 IPC, increasing to 12 for an issue rate of 24, and approximately 13 for an issue rate of 32. The simulation results show the high potential for IPC improvements over contemporary superscalar processors by applying aggressive superscalar techniques. Further improvements may be attained by the value prediction and data speculation technique that is still in its infancy (see next section).

5.3 Superspeculative Processors

The basis for the superspeculative approach is the astounding observation that instructions generate many highly predictable result values in real programs. Consumer instructions can thus frequently and successfully speculate on their source operand values and begin execution without results from the producer instructions. Consequently, a superspeculative processor can remove the serialization constraints between producer and consumer instructions, enabling program performance potentially to exceed the classical dataflow limit which states that a program cannot execute faster than the longest execution path set by the program's data dependences.

The reasons for the existence of value locality are manifold (*Lipasti et al.* [184]). Some reasons are:

- Due to register spill code the re-use distance of many shared values is very short in processor cycles. Many stores do not even make it out of the store queue before their values are needed again.
- Input sets often contain data with little variation (e.g., sparse matrices or text files with white spaces).
- A compiler often generates run-time constants due to error-checking, switch statement evaluation, and virtual function calls.
- The compiler also often loads program constants from memory rather than using immediate operands.

Superspeculative processors enhance wide-issue superscalar performance by speculating aggressively at every point in the processor pipeline (*Lipasti and Shen* [182, 183, 184]). Superspeculative processors speculate on data dependences in addition to branch prediction.

Conventional superscalar processors employ the *strong-dependence model* for program execution, which implies a total instruction ordering of a sequential program. In the strong-dependence model two instructions are identified as either dependent or independent, and when in doubt, dependences are pessimistically assumed to exist. Dependences are never allowed to be violated and are enforced during instruction processing. To date, most machines enforce such dependences in a rigorous fashion. This traditional model is overly rigorous and unnecessarily restricts available parallelism.

Instead, the *weak-dependence model* is applied for superspeculative processors, specifying that dependences can be temporarily violated during instruction execution as long as recovery can be performed prior to affecting the permanent machine state. The weak-dependence model's advantage is that the machine can now speculate aggressively and temporarily violate the dependences as long as corrective measures are in place to recover from misspeculation. If a significant percentage of speculations are correct, the machine can effectively exceed the performance limit imposed by the traditional, strong-dependence model.

Similar in concept to branch prediction's implementation in current processors, superspeculation uses two interacting engines:

- The *front-end engine* assumes the weak-dependence model and is highly speculative, predicting instructions to speculate aggressively past them. When predictions are correct, these speculative instructions will effectively have skipped over certain stages of instruction execution.
- The *back-end engine* still uses the strong-dependence model to validate the speculations, recover from misspeculation, and provide history and guidance information to the speculative engine.

Figure 5.2 identifies the three key parameters that a superspeculative microarchitecture must maximize:

- *instruction flow*, the rate at which useful instructions are fetched, decoded, and dispatched to the execution core;
- *register dataflow*, the rate at which results are produced and register values become available; and
- *memory dataflow*, the rate at which data values are stored and retrieved from data memory.

These three flows roughly correspond to the processing of branch, arithmetic/logical, and load/store instructions, respectively. In a superspeculative microarchitecture, aggressive speculative techniques are employed to accelerate the processing of all these instruction types.

The superspeculative processor proposed in 1997 by *Lipasti and Shen* [183] speculates on the instruction flow, the register dataflow, and the memory dataflow.

Speculation on the instruction flow uses a two-phase branch predictor with local and global branch history, combined with a trace cache to execute more than one taken branch per cycle, which is similar to *Patt et al.*'s advanced superscalar architecture in the previous section. The misprediction latency is reduced by data speculation.

Speculation in the register dataflow comprises source operand value prediction, and value stride prediction. *Source operand value prediction* eliminates data dependences by use of a dynamic value prediction table per static instruction. *Value stride prediction* speculates on constant, incremental increases in operand values to increase the accuracy of value prediction. In value stride prediction, a dynamic hardware mechanism detects constants, incremental increases in operand values (strides), and uses them to predict future values. *Dependence prediction* is applied to predict the inter-instruction dependences. Instructions that are data ready are allowed to execute in parallel with the dependence checking for these instructions. Dependence prediction is used when the dynamic history shows that value prediction cannot be successfully applied. It can be implemented by a dependence prediction table with entries that are indexed by hashing together the instruction address bits,

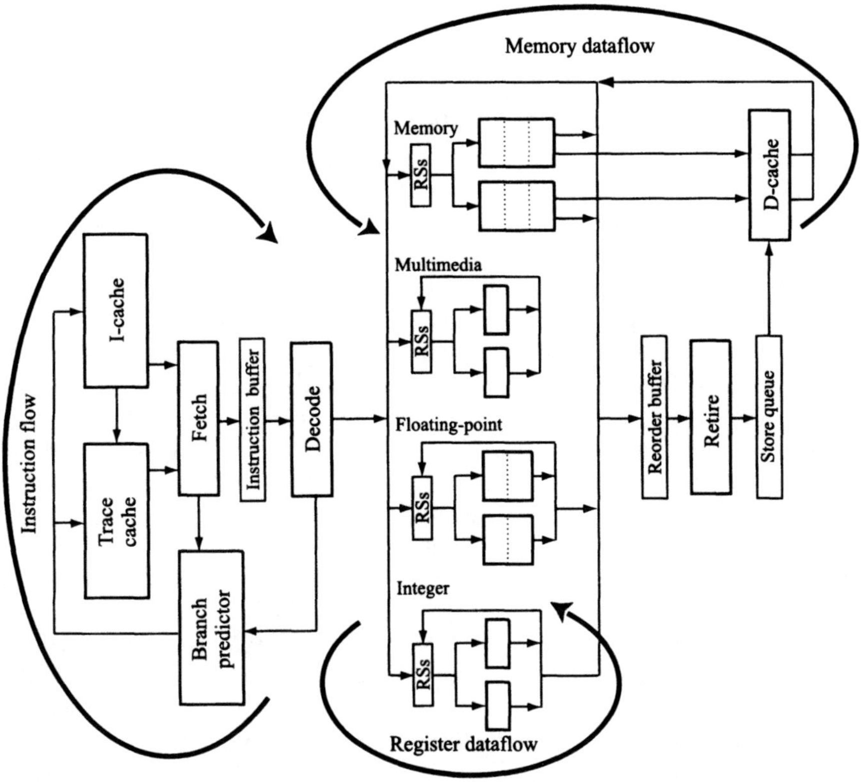

Fig. 5.2. Superspeculative architecture

the *gshare* branch predictor's BHR, and the relative position of the operand being looked up [182].

Deeper pipelining often results in dependence checking and dispatch in multiple pipelined stages. With dependence and value prediction a three-cycle dispatch nearly matches the performance of single-cycle dispatch.

The memory dataflow is used to predict *load values* to bridge the latency of accessing the storage device, *load addresses* to eliminate address generation interlock (i.e., the delay of a load the address of which is not yet known), and *aliases* with earlier outstanding stores. *Load value prediction* [184] predicts the results of load instructions at the time of dispatch by exploiting the affinity between load instruction addresses and the values the loads produce. The predictions are implemented by prediction tables. Memory loads are predicted by a *load value prediction unit*, which consists of a load value prediction table for generating value predictions, a load classification table for deciding which predictions are likely to be correct, and a constant verification unit that replaces accessing the conventional memory hierarchy for verifying highly predictable loads.

Alias prediction is related to dependence prediction. Rather than predict the dependence distance to a preceding register write, the alias prediction predicts the distance to a preceding store to memory. The predicted distance is then used to obtain the load value from that offset in the processor's store queue, which hold outstanding stores. For this speculative forwarding to occur, neither the load nor the store need to have their addresses available. A related approach is called *memory dependence prediction* which identifies stores upon which a load depends. The processor uses a load's store set, i.e., the set of stores upon which the load has ever depended, to predict which stores a load must wait for before executing (*Chrysos and Emer* [49]).

To evaluate superspeculation's performance potential, *Lipasti and Shen* [183] simulated the performance of a superspeculative processor called *Superflow* with a fetch width of 32, a 128-entry reorder buffer, 64 kB 4-way set-associative D-cache and I-cache with 10-cycle miss delay to a perfect, pipelined 16 MB unified L2 cache, and a 128-entry, fully associative store queue. Additional resources were a value load table for generating value predictions, a classification table to decide which predictions are likely to be correct, a dependence prediction table, an alias prediction table, and additional pattern history tables.

Superflow simulations yielded 7.3 IPC for SPECint95 benchmark suite on a realizable processor configuration and up to 9 IPC with an issue bandwidth of 32 instructions per cycle (the higher IPC value by *Patt et al.* mentioned in the previous section is reached because of the assumption of perfect caches and perfect branch prediction, in contrast to the Superflow simulations). The results demonstrate the enormous performance potential of superspeculative techniques, although such techniques are far from mature yet. Recent research (*Gabbay and Mendelson* [96, 97], *Rychlik et al.* [251], *Chrysos and Emer* [49]) in the area of value prediction surveys the impacts of such techniques in superscalar, multiscalar, or trace processors, and how to combine value prediction techniques in hybrid value predictors.

5.4 Multiscalar Processors

The multiscalar model of execution (*Franklin* [91], *Sohi et al.* [274]) represents another paradigm to extract a large amount of inherent parallelism from a sequential instruction flow. A program is divided into a collection of tasks by a combination of hardware and software. The tasks are distributed to a number of parallel PEs within a processor. Each PE fetches and executes instructions belonging to its assigned task. A functional decomposition of the processor chip as required in order to reach short wire delays in future generation high-density processor chips is thus naturally realized.

The name *multiscalar* was chosen because of the structure of the processor which can be viewed as a collection of sequential (or scalar) processors that cooperate in executing a sequential program (*Sohi* [273]). The difference to a single chip multiprocessor (CMP) is the denser coupling of the PEs in the multiscalar processor. While a CMP executes different threads of control that are statically determined by the programmer or by a parallelizing compiler, the multiscalar processor executes a sequential program that is enriched by sequencing information.

A static program is represented as a control flow graph (CFG), where basic blocks are nodes, and arcs represent the flow of control from one basic block to another. Dynamic program execution can be viewed as walking through the CFG, generating a dynamic sequence of basic blocks which have to be executed for a particular run of the program. To achieve high performance, the multiscalar processor must walk through the CFG with a high level of parallelism. The primary constraint of any parallel walk is that it must preserve the sequential semantics assumed in the program. A program is statically partitioned into tasks which are demarcated by annotations of the CFG. The left side of Fig. 5.3 shows a static program represented as a CFG, where each task is a collection of instructions (e.g., part of a (large) basic block, a basic block, a collection of basic blocks, a single loop iteration, an entire loop, a function call, etc.). A task sequencer (speculatively) sequences through the program a task at a time, assigning the task to a PE, which in turn unravels the task to determine the dynamic instructions to be executed, and executes them.

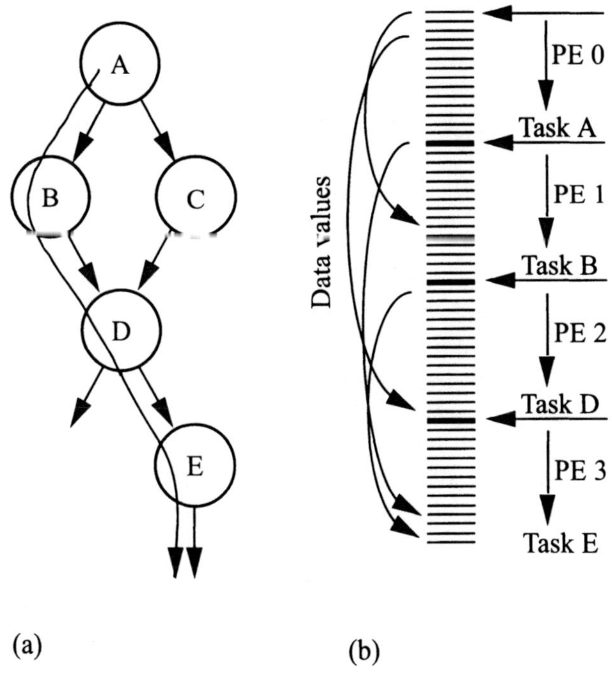

Fig. 5.3. Multiscalar mode of execution

The right side of Fig. 5.3 shows the dynamic instruction stream divided into task-sized steps. The four tasks A, B, D, and E are assigned to PE 0...PE 3.

A multiscalar processor walks through the CFG speculatively, taking task-sized steps, without pausing to inspect any of the instructions within a task. A task is assigned to one of a collection of PEs for execution by passing the initial program counter of the task to the PE. For each step of its walk, a multiscalar processor assigns a task to a PE for execution, without concern for the actual content of the task, and continues from this point to the next point in the CFG.

A possible microarchitecture for a multiscalar processor is shown in Fig. 5.4. A multiscalar processor can be considered as a collection of PEs with a sequencer that assigns tasks to the PEs. Once a task is assigned to a PE, the PE fetches and executes the instructions of the task until it is complete. Multiple PEs, each with its own internal instruction sequencing mechanism, support the execution of multiple tasks, and thereby multiple instructions, in any given step. Multiple tasks then execute in parallel on the PEs, resulting in an aggregate execution rate of multiple IPC [274].

The concept proposed by *Sohi et al.* [274] in 1995 sounds simple, but the key point is the proper resolution of inter-task data dependences. That concerns, in particular, data that is passed between instructions via registers and memory. It is in this area of inter-task data communication that the multiscalar approach differs significantly from the more traditional multiprocessing methods.

To maintain a sequential appearance, a twofold strategy is employed. First, each processing element adheres to sequential execution semantics for the task assigned to it. Second, a loose sequential order is enforced over the collection of processing elements, which in turn imposes a sequential order of the tasks. The sequential order on the processing elements is maintained by organizing the elements into a circular queue. Head and tail pointers indicate respectively the elements that are executing the earliest and the latest of the current tasks.

Because a sequential execution model views storage as a single set of registers and memory locations, multiscalar execution must maintain this view as well. In order to provide this behavior, communication between tasks is synchronized.

The appearance of a single logical register file is maintained, although copies are distributed to each parallel PE. Register results are dynamically routed among the many parallel processing elements with the help of compiler-generated masks.

In the case of registers, the control logic synchronizes the production of register values in predecessor tasks with the consumption of these values in successor tasks via reservation on the registers. The register values a task may produce can be determined statically and maintained in a create mask.

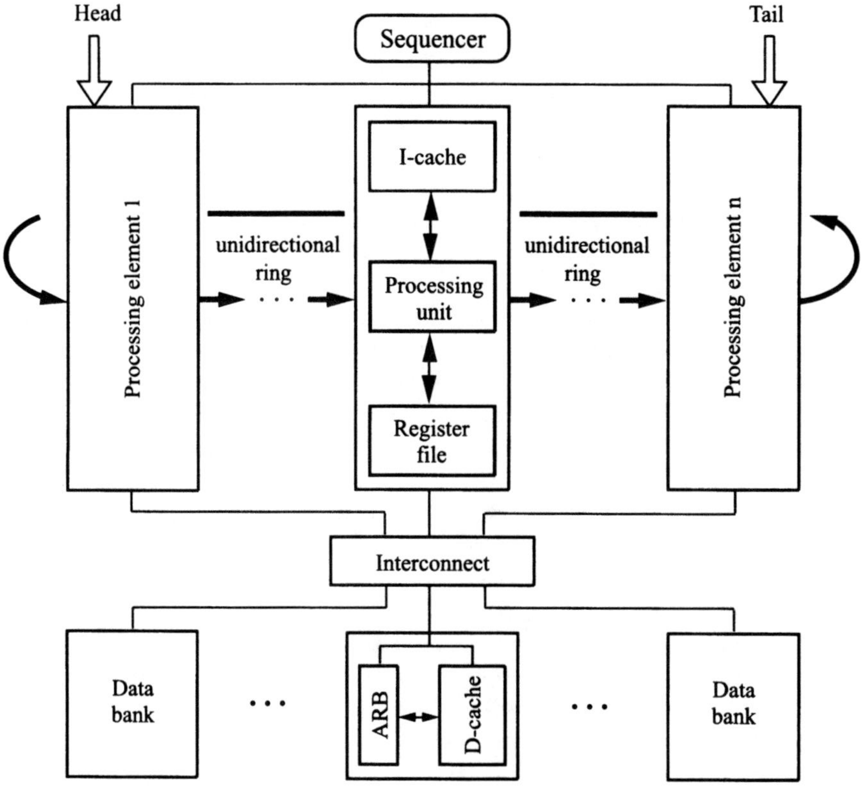

Fig. 5.4. A multiscalar processor

Bits in the create mask correspond to each of the logical registers: a bit is set to one if the register is potentially written by the task. At the time a register value in the create mask is produced, it is forwarded via a circular unidirectional ring to later tasks, i.e., to PEs which are logical successors. The reservations on registers for a successor task are given in the accum mask, which is the union of the create masks of currently active predecessor tasks. As values arrive from the predecessor PEs, reservations are cleared in the successor PEs. If a task uses one of these values, the consuming instruction can proceed only if the value has been received; otherwise, it waits for the value to arrive.

For memory operations, the situation is more complicated. When a PE is ready to execute a load, it does not even know whether previous tasks have stores, let alone stores to a given memory location. Here multiscalar processing employs data dependence speculation – speculating that a load does not depend on instructions executing in predecessor tasks. Memory access may occur speculatively without knowledge of preceding loads and stores. Addresses are disambiguated dynamically, many in parallel, and processing waits only for data dependences. An *address resolution buffer* (ARB) is provided to hold speculative memory operations and to detect violations of memory dependences. The ARB checks that the speculation was correct, squashing instructions if it was not.

Thus the multiscalar paradigm has at least two forms of speculation (*Sohi* [273]): *control speculation*, which is used by the task sequencer, and *data dependence speculation*, which is performed by each PE. It could also use other forms of speculation, such as data value speculation, to alleviate inter-task dependences.

Multiscalar processors use multiple internal sequencers (PCs) to sequence through a sequential program. The internal sequencers require information about which tasks are possible successors of any given task in the CFG. Such information can be determined statically and placed in a task descriptor. Each internal sequencer may also speculatively sequence through a task. The task descriptors may be dispersed within the program text – for instance, before the code of the task – or placed in a single location beside the program text. A multiscalar program may be generated from existing binaries by augmenting the binary with task descriptors and tag bits.

Compared to superscalar processors each task is equivalent to a subwindow of the instruction window; collectively the multiple sequencers capture a portion of the dynamic instruction stream. Interoperation communication can be carried out more efficiently if the total instruction window is broken into subwindows, with more frequent intrawindow and less frequent interwindow communication. Likewise, instruction scheduling becomes more efficient if the overall schedule is treated as an ensemble of (several) smaller schedules, where the smaller schedule is the schedule in a subwindow, as is achieved by the multiscalar model of execution.

The *Superthreaded Architecture* (*Tsai and Yew* [302], 1996; *Li et al.* [181], 1996) is a related approach using a thread pipelining execution model that allows threads with data and control dependences to be executed in parallel. As in the multiscalar approach a compiler statically partitions the control-flow graph of a program into tasks from which it generates threads to be executed on the thread processing units of the superthreaded architecture. In this architectural model the thread processing units are connected by a unidirectional bus to exchange data that arise from loop-carried dependences, similar to the multiscalar proposal. The superthreading technique is designed for the exploitation of loop-level parallelism where each interaction is executed within another thread. The superthreaded architecture, like the multiscalar approach, is to very closely related the CMP and multithreaded architectures that are covered in the next chapter. However, both the superthreaded and the multiscalar approaches use more closely coupled processing elements and are designed to increase single thread performance using a compiler-supported task partitioning.

5.5 Trace Processors

The focus of a trace processor is the trace cache which has already been mentioned in Sect. 5.2 on advanced superscalar and in the Sect. 5.3 on super-speculative processors. A *trace* is a sequence of instructions that potentially covers several basic blocks starting at any point in the dynamic instruction stream. The number of instructions within a trace is limited by the trace cache line size. The number of basic blocks covered is called the *branch predictor throughput*. In principle, a trace is fully specified by its starting address and a sequence of branch outcomes which describes the path followed.

A *trace cache* can be constructed in two ways. The first implementation is to store the trace start addresses and only the branch target addresses of the branches within a trace. This approach reduces storage space and allows the prediction of multiple branches per cycle. However, taken branches introduce the problem of noncontiguous instruction fetching: the dynamic sequence exists in the I-cache, but the instructions are not in contiguous I-cache locations. Therefore, the second and preferred implementation stores the whole instruction sequence that makes up a trace.

A trace cache is a special I-cache that captures dynamic instruction sequences in contrast to the I-cache that contains static instruction sequences. Each line in the trace cache stores a dynamic code sequence, which may contain one or more taken branches. Each line stores a snapshot, or trace, of the dynamic instruction stream. Dynamic instruction sequences are built as the program executes. The trace construction is off the critical path; it does not lengthen the pipeline. Figure 5.5 shows that a trace cache stores an instruction sequence contiguously, while the same instruction sequence is stored in the I-cache in noncontiguous areas because of branch or jump instructions. Moreover, since several predicted branches may be captured within a trace, trace prediction automatically leads to multiple predicted branches per cycle within the portion of a trace fetched for a wide-issue processor from the trace cache. A single entry in the trace cache holds an entire trace. The trace cache

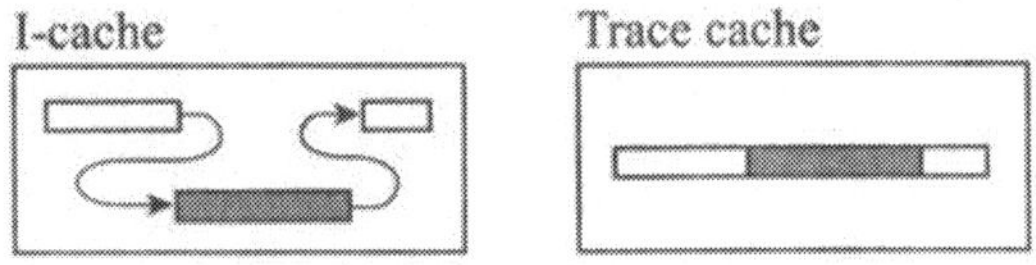

Fig. 5.5. Trace cache principle

is indexed using the next address field of the previously fetched trace combined with prediction information for next trace prediction. Thus, an entire trace consisting of multiple basic blocks is fetched in one clock cycle, without the need for multiple cache lookups, multiported caches, or complicated and time-consuming mask, alignment, and concatenation operations on multiple cache blocks. Such logic is moved off the critical path to the trace construction hardware.

The main ideas of the trace processor were presented in 1997 by *Rotenberg et al.* [247], *Smith and Vajapeyam* [271], and *Vajapeyam and Mitra* [310], who proposed to create subsystems similar in complexity to today's superscalar processors and combine replicated subsystems into a full processor. A *trace processor* (see Fig. 5.6) is partitioned into multiple distinct PEs (similar to multiscalar). The code is broken up into traces that are captured and stored by hardware. One PE executes the current trace while the others execute future traces speculatively.

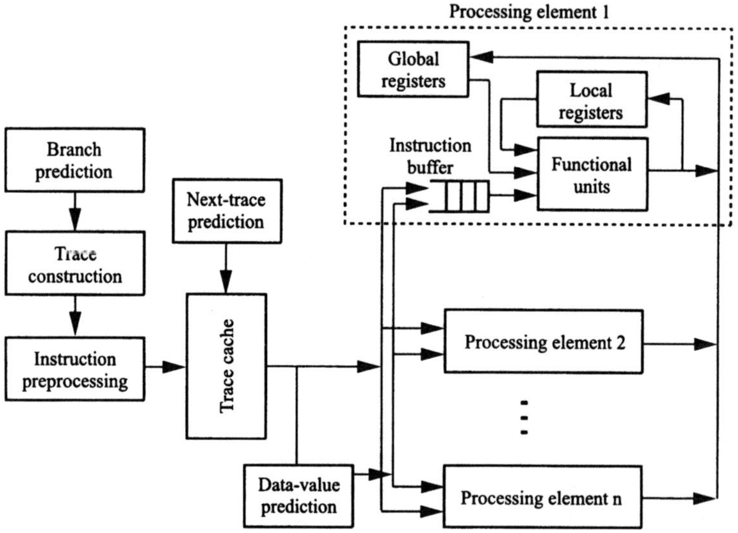

Fig. 5.6. A trace processor

Instruction fetch hardware fetches instructions from the I-cache and simultaneously generates traces of 8–32 instructions including predicted conditional branches. Traces are built as the program executes and they are stored in a trace cache. A trace fetch unit reads traces from the trace cache and parcels them out to the parallel PEs.

A trace cache miss causes a trace to be built through conventional instruction fetching with branch prediction. Blocks of instructions are preprocessed before being put in the trace cache, which greatly simplifies processing after they are fetched. Preprocessing can include capturing data dependence relationships, combining and reordering instructions, or determining instruction resource requirements – all of which can be re-used. To support precise interrupts, information about the original instruction order must also be saved with the trace.

During the dispatch phase, instructions move from the trace cache to the instruction buffers in the PEs. Only inter-trace dependence checking and register renaming are required.

Because traces are the basic units for fetching and execution, control-flow prediction is moved up to the trace level. The unit of control prediction should be a trace, not individual branches. That suggests a next-trace predictor. Next-trace prediction predicts multiple branches per cycle.

Trace processors also employ data value prediction (see Sect. 5.3). Data value prediction speculates on the input data values of a trace and is combined with next-trace prediction. Successfully predicting a trace's input data values makes the trace independent of data availability, and leads to a further decoupling of traces, allowing the trace to execute immediately and in parallel with any other trace. Data value prediction and speculation is restricted to inter-trace dependences.

The expected parallelism within a single trace is suitable for execution in a modest superscalar unit that can be chosen to implement the PEs. As multiple PEs issue instructions in parallel, both intra-trace and inter-trace parallelism are exploited.

Because the PEs and register files are distributed, so is the communication of register data. The relatively simple bypass paths within a unit allow local result forwarding in a single cycle. Global paths are used for communicating global register results between PEs. The global bypass paths are likely to require multiple clock cycles.

The trace processor uses a conventional set of logical registers. Physical registers are divided into local and global sets. The hierarchical organization of registers allows small register files with fast access times and fewer ports per file. The trace dispatcher remaps the trace's source and destination registers to the global registers without the need for intra-trace dependence checking. The dispatcher maps local registers with re-usable mappings based on the intra-trace dependences detected during instruction preprocessing. Dispatch logic can remap a 16-instruction trace line using register rename logic as complex as that used by a conventional four-way superscalar processor.

Memory systems for the trace processor will have to provide very high bandwidth to supply enough data to the processor's multiple PEs. Distributed, multiported caches can be employed, provided that coherence among distributed caches is maintained. A large, interleaved cache system is also possible, although designers will have to deal with the additional latency in such systems.

Each PE in the trace processor generates a stream of load and store requests to memory. Moreover, these address streams are generated speculatively and out of order. The hardware to sort out the address streams and make sure that all memory locations are accessed in the correct order will have to be fairly sophisticated. The ARB as proposed for multiscalar processors (see Sect. 5.4) solves the problem of parallel resolution of memory addressing hazards. However, the ARB is a centralized device, separate from D-caches, and must be developed further, using distributed mechanisms that merge address resolution and data caching.

5.6 DataScalar Processors

The DataScalar model (*Burger et al.* [42], 1997) of execution runs the same sequential program redundantly across multiple processors. The data set is distributed across physical memories that are tightly coupled to their distinct processors. Each processor broadcasts operands that it loads from its local memory to all other processors. Instead of explicitly accessing a remote memory, processors wait until the requested value is broadcasted. Stores are completed only by the processor that owns the operand, and are dropped by the others.

The most heavily accessed data is statically replicated by duplicating whole memory pages that are stored in each processor's local memory. Access to a replicated page requires no data communication. The address space is divided into a *replicated* and a *communicated* section. The latter holds values that only exist in single copies and are owned by the respective processor. Replicated pages are mapped into each processor's local memory, and the communicated section of the address space is distributed among the processors.

Figure 5.7 demonstrates the execution of load and store operations for replicated and communicated memory. Assume that both processors execute a sequence of `load-1`, `store-1`, `load-2`, and `store-2`. Operations `load-1` and `store-1` are issued to the replicated memory and can therefore complete locally on both processors. Operations `load-2` and `store-2` are issued to the communicated memory of `Processor 1`. The `load-2` of `Processor 1` is deferred until the value is broadcasted from `Processor 1`, which owns the value. Since all processors are running the same program, they all generate the same store value, which is stored only in the communicated memory of the processor that owns the value. Therefore, `store-2` is completed at `Processor 1`, but is dropped at `Processor 2`.

The main goal of the DataScalar model of execution is the improvement of memory system performance by introducing redundancy in execution by replicating processors and part of the data storage. Since all physical memory is local to at least one processor, a request for a remote operand is never sent, thus reducing memory access latency and bus traffic. All communication is one-way. Writes never appear on the global bus.

The processors execute the same program in slightly different time steps, due to asynchronous memory accesses and the ability to perform out-of-order execution. One processor, the lead processor, runs slightly ahead of the others, especially when it is broadcasting while the others wait for the broadcasted value. When the program execution accesses an operand that is not owned

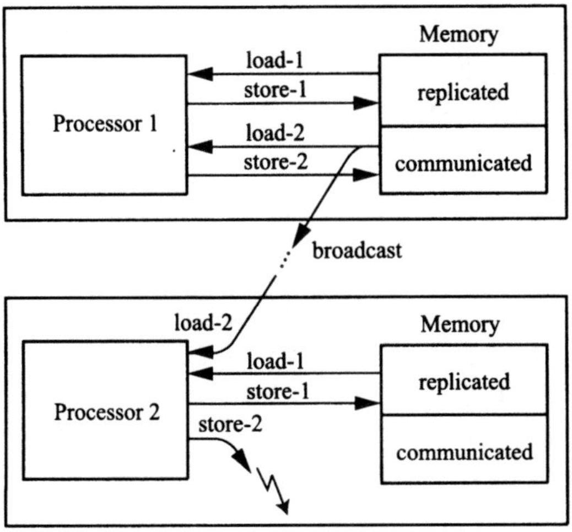

Fig. 5.7. Accesses of DataScalar processors to replicated and communicated memory

by the lead processor, a lead change occurs. All processors stall until the new lead processor catches up and broadcasts its operands. The capability for each processor to run ahead on computation that involves operands owned by the processor is called *datathreading* and was described by *Burger et al.* [42] in 1997.

The DataScalar model creates opportunities for new optimizations. Because each processor executes the instructions in a different order due to the out-of-order execution facility, it is possible for a processor to execute a private computation, broadcasting only the result and not the operands to the other processors. This technique, called *result communication*, deviates from the strict single-program-single-data stream (SPSD) model of DataScalar computation. Speculation is another optimization opportunity. However, the additional bus traffic that might be caused by speculation must be weighed against the possible performance advantage. The broadcast of data may be a critical limitation of the DataScalar model, and frequent superfluous broadcasts would greatly hinder performance. Possible solutions are to hold onto speculative broadcasts until the speculative condition is resolved or to send the broadcast immediately upon issue, followed by a corresponding squash message if the load that generated the broadcast is squashed.

The DataScalar model is primarily a memory system optimization intended for codes that are limited in performance by the memory system and difficult to parallelize. However, every DataScalar machine is a *de facto* multiprocessor. When codes contain coarse-grain parallelism, the DataScalar machine can also run like a traditional multiprocessor. DataScalar and multiprocessing can be viewed as two endpoints on a spectrum, where the DataScalar model is restricted to run sequential code and makes no attempt to exploit coarse-grained parallelism in the code, while multiprocessing requires compiler and/or programmer support to generate parallel code, and its main focus is explicit exploitation of coarse-grain parallelism.

Simulation results of *Burger et al.* [42] suggest that the DataScalar model of execution works best with codes for which traditional parallelization techniques fail. Six unmodified SPEC95 binaries ran from 7 % slower to 50 % faster on two nodes, and from 9 % to 100 % faster on four nodes, than on a system with a comparable, more traditional memory system.

However, current technological parameters do not make DataScalar systems a cost-effective alternative to today's microprocessors. For a DataScalar system to be more cost effective than the alternatives, the following three conditions must hold:

- Processing power must be cheap, and the dominant cost of each node should be memory.
- Remote memory accesses should be slower than local memory accesses.
- Broadcasts should not be prohibitively expensive.

The DataScalar model can be applied to speed up sequential execution wherever multiprocessor hardware is available without a sufficient load of parallel tasks. Three possible candidates for DataScalar systems were proposed:

- The concept of merging processor and memory on a chip as proposed by the IRAM approach (see Sect. 7.1). IRAM-based systems connected by a bus or a point-to-point ring would exhibit the parameters needed for a cost-effective DataScalar implementation, because remote memory accesses to other IRAM chips would certainly be more expensive than on-chip memory accesses.
- The DataScalar model of execution may also be applied within a single chip to alleviate wiring delays – for example, extending the concept of a CMP (see Sect. 6.2.3). CMPs access operands from a processor-local memory faster than requesting an operand from a remote processor memory across the chip, due to wiring delays.
- Networks of workstations (NOWs), where DataScalar could be an alternative to paging, provided that broadcasts were sufficiently inexpensive. Some network topologies like fat trees support efficient broadcasts. Alternatively, optical interconnects, especially free-space optical interconnects, provide extremely cheap broadcasts. However, the data threads of the DataScalar model may prove too fine-grained regarding the communication latencies of today's NOWs.

At first glance the DataScalar model looks like an immense waste of processing power. However, conclusions are that the DataScalar model of execution may be advantageous, when remote memory accesses are significantly slower than local memory accesses, when global broadcasts are relatively inexpensive, and when the cost of additional processors is a only small addition to the total system cost. The communication bandwidth itself may be limited.

The DataScalar model only produces broadcasts, but never remote stores and remote load requests.

5.7 Conclusions

It is difficult to look ahead to the next generation of microprocessors early in the lifetime of the current generation. However, as a microprocessor generation matures and future hardware technologies become better defined, the next generation starts to become visible. We are at that stage now. The last time we were in a similar position was almost 10 years ago – when the superscalar processors emerged. Once again, it is time to lay the ground-work for many more years of high-performance processor development.

In this chapter we have surveyed the uniprocessor alternatives, such as advanced superscalar, superspeculative, multiscalar, trace, and DataScalar processor, which are all examples of new microarchitectures suitable for the next generation. Developing superspeculative techniques prepare the way for exceeding the classical dataflow limit imposed by data dependences. The multiscalar processor proposed the basics for a functional distribution of processing elements working simultaneously on tasks generated from sequential machine programs. The trace processor is similar to the multiscalar processor except for its use of hardware-generated dynamic traces rather than compiler-generated static tasks. The DataScalar approach reduces interprocessor data traffic in favor of broadcasting. A real uniprocessor chip of the future will most likely combine some of these execution modes within a single microarchitecture. For instance, *Patt et al.*'s advanced superscalar processor proposal already uses superspeculative techniques and a trace cache.

All proposed microarchitecture techniques increase memory bandwidth requirements compared to today's superscalar microprocessors. Therefore, all proposed microarchitecture techniques may be combined in future with the processor-in-memory (PIM) or intelligent RAM (IRAM) approaches that combine processing elements of various complexity with DRAM memory on the same chip to solve the memory latency bottleneck.

All architecture proposals described so far retain result serialization – the serial instruction flow as seen by the programmer and forced by the von Neumann architecture. However, the microarchitectures strive to press as much fine-grained or even coarse-grained parallelism from the sequential program flow as can be done by hardware. Unfortunately, a large portion of the exploited parallelism is speculative parallelism which in case of incorrect speculation leads to an expensive reroll mechanism and to a waste of instruction slots. Therefore, the result serialization of the von Neumann architecture poses a severe bottleneck.

6. Future Processors to use Coarse-Grain Parallelism

There are strong indications that multithreading will be utilized in future processor generations to hide the latency of local memory access ... ... It also establishes an architectural direction that may yield much greater latency tolerance in the long term.

David Culler, Jaswinder Pal Sing, Anoop Gupta
Parallel Computer Architecture: A Hardware / Software Approach
(Morgan Kaufmann Publishers, 1999)

6.1 Utilization of more Coarse-Grain Parallelism

Current superscalar microprocessors are able to issue up to six multiple instructions each clock cycle from a conventional linear instruction stream. VLSI technology will allow future microprocessors with an issue bandwidth of 8–32 instructions per cycle.

However, ILP found in a conventional instruction stream is limited. In general, integer-dominated programs feature a rather low ILP, while a high ILP can be extracted from floating-point programs. One set of solutions which was pursued by the architectural techniques described in the last chapter, is to apply an even higher degree of speculation in combination with a functional partitioning of the processor.

The solution surveyed in this chapter is the additional utilization of more coarse-grained parallelism. The main approaches are the *(single) chip multiprocessor* (CMP) and the *multithreaded processor*. The chip multiprocessor (sometimes called the *multiprocessor chip*) integrates two or more complete processors on a single chip. Therefore, every unit of a processor is duplicated and used independently of its copies on the chip.

In contrast, the multithreaded processor interleaves the execution of instructions of different threads of control in the same pipeline. Therefore, multiple program counters are available in the fetch unit and the multiple contexts are often stored in different register sets on the chip. The functional units are multiplexed between the thread contexts that are loaded in the register sets. Depending on the specific multithreaded processor design, only a single-issue instruction pipeline (as in scalar RISC processors) is used, or a single issue unit issues instructions from different instruction streams simultaneously.

Supported by multiple register sets, context switching is very fast. Multithreaded processors tolerate memory latencies by overlapping the long-latency operations of one thread with the execution of other threads – in contrast to the chip multiprocessor approach. While the chip multiprocessor is easier to implement, the use of multithreading in addition to a wide-issue bandwidth is a promising approach.

6.2 Chip Multiprocessors

6.2.1 Principal Chip Multiprocessor Alternatives

Today the most common organizational principles for multiprocessors (see Fig. 6.1) are the symmetric multiprocessor (SMP), the distributed shared memory multiprocessor (DSM), and the message-passing shared-nothing multiprocessor.

The SMP and the DSM multiprocessors feature a common address space, which is implemented in the SMP as a single global memory where each memory word can be accessed in uniform access time by all processors (UMA – uniform memory access). In the DSM multiprocessor a common address space is maintained despite physically distributed memory modules. A processor in a DSM may access data in its local memory faster than in the remote memory (the memory module local to another processor). DSM multiprocessors are therefore nonuniform memory access (NUMA) systems. Shared-nothing multiprocessors feature physically distributed memory modules and no common address space. Therefore, communication can only be performed by passing messages between processors. Shared-nothing multiprocessors are highly scalable but harder to program than shared-memory multiprocessors. They are beyond the scope of today's reasoning about chip multiprocessors, which, by their tight physical coupling on a single chip, may also feature a very tight coupling of instruction streams, usually expressed by a common memory organization.

The principal organizational forms of multiprocessors, as expressed in Fig. 6.1, do not regard cache organization. Commodity microprocessors, which are usually used today as building blocks for multiprocessors, contain on-chip caches, often coupled with off-chip L2 cache memories. Shared-memory multiprocessors maintain cache coherence by a cache coherence protocol which is a bus-snooping coherence protocol like M.E.S.I. for SMPs or a directory-based coherence protocol for DSMs[1]. SMPs consist of a moder-

[1] DSM multiprocessors can be further classified into cache-coherent NUMA (CC-NUMA) systems that maintain cache-coherence over the whole system as in the Cray/SGI Origin, the HP/Convex Exemplar, and the Sequent NUMA-Q, and non-cache-coherent NUMA (NCC-NUMA) such as the Cray T3D and T3E.

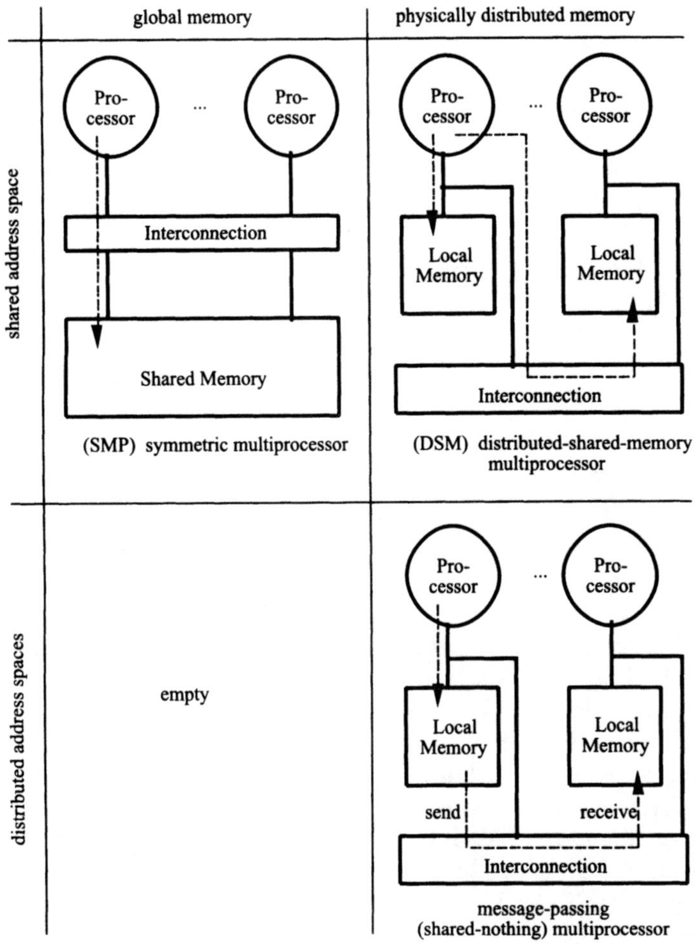

Fig. 6.1. Organizational principles of multiprocessors

ate number of commodity microprocessors with cache memories coupled by a fast memory bus with the global memory. So, the typical symmetric multiprocessor is organized as in Fig. 6.2. In the latest SMPs the memory bus is replaced by an address bus (necessary for the bus-snooping) and a data crossbar switch for faster transfer of cache lines. SMPs are the starting point for chip multiprocessors.

Thread-level parallelism is more coarse-grained than ILP. However, there are at least three more levels to distinguish that influence the microarchitecture of a potential chip multiprocessor (and a multithreaded processor, too).

- The most common coarse-grained thread-level parallelism is to execute multiple processes in parallel. This implies for CMPs that different logical

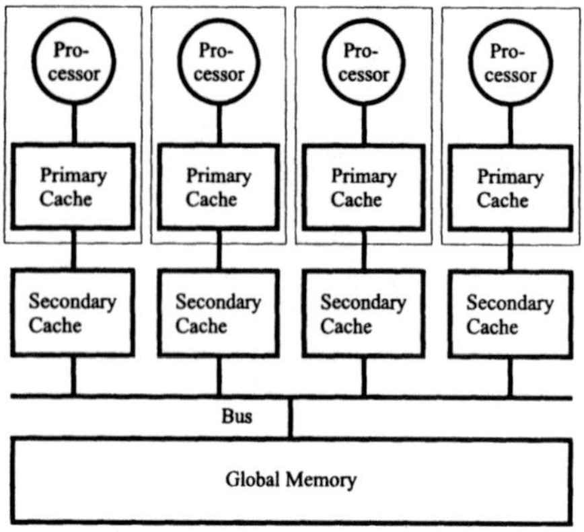

Fig. 6.2. Typical SMP

address spaces have to be maintained for the different instruction streams that are in execution by the processors within the CMP.

- The parallel execution of multiple threads from a single application usually implies a common address space for all threads. Here threads of control are identical with the threads (light-weighted processes) of a multithreaded operating system such as Sun Solaris, IBM AIX, and Windows NT, used by today's symmetric multiprocessor workstations and servers. This approach has several architectural advantages for chip multiprocessors (and for multithreaded processors as well). Cache organization is simplified, when a single logical address space is shared. Moreover, thread synchronization, as well as exchange of global variables between threads, can be made very efficient by providing common on-chip memory structures (shared caches or even shared registers). If the threads shared a SPMD program structure, a single multiported I-cache might be taken into consideration. However, although most desktop applications like Acrobat, Netscape, Photoshop, Powerpoint, and Winword today use 3–8 threads, most thread activity in these program systems is restricted to a single main thread (*Lee* [174], 1998). This drawback may be alleviated by parallelizing compilers in conjunction with regularly structured application problems such as, for example, numerical problems or multimedia applications.
- Execution of a sequential application may be accelerated by extracting threads of control dynamically from a single instruction stream (as exemplified by the architectures described in the previous chapter).

From the applications perspective, whether a CMP works best depends on the amount and the characteristics of the parallelism in the applications. These fall into three broad classes depending on the degree of interprocessor communication, which can be low, moderate, or high. From the architectural perspective, the performance of a CMP will depend on the level of the memory hierarchy at which the CPUs of the CMP are interconnected.

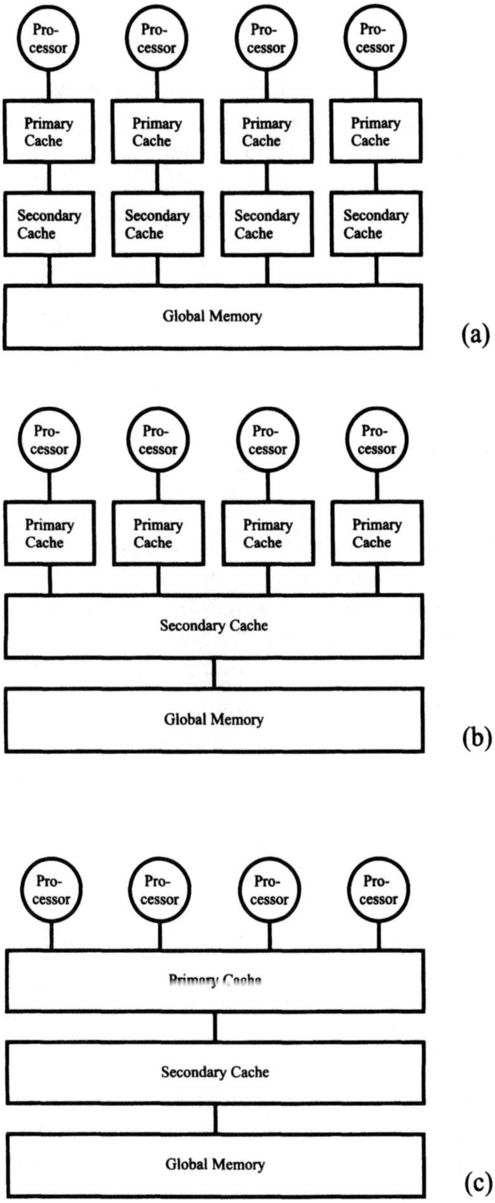

Fig. 6.3. Some shared memory candidates for CMPs (**a**) shared-main memory (**b**) shared-secondary cache (**c**) shared-primary cache

In order to develop insight about the most appropriate memory hierarchy level for connecting the CPUs in a CMP, *Nayfeh et al.* [211] in 1996 compared three alternatives: a shared-main memory multiprocessor (i.e., the typical symmetric multiprocessor today), a shared-secondary cache multiprocessor, and a shared-primary cache multiprocessor (Fig. 6.3). They found that, when applications have a high or moderate degree of interprocessor communication, both shared-primary cache and shared-secondary cache architectures perform similarly and outperform the shared-main memory architecture substantially. There are two reasons for this. First, the shared cache was assumed large enough to accommodate most of the working sets of independent threads running on different CPUs, so that the cache miss rate is low. Second, when there is interprocessor communication, it is handled very efficiently in the shared (primary or secondary) cache. Even for applications with little or no interprocessor communication, the performance of the shared-primary cache architecture is still slightly better than shared-main memory architecture.

To maintain the performance growth of microprocessors, *Olukotun et al.* [219], in 1996, discussed the details of implementing a *single chip multiprocessor* (CMP). We will describe Hydra, a research vehicle currently being designed at Stanford University in an effort to evaluate a variant of the shared-secondary cache CMP, as an alternative for future microprocessor development (*Hammond and Olukotun* [121], 1998). This is not the first project on CMP design, though. In 1994, Texas Instruments introduced the TMS320C80 multimedia video processor (MVP) [291], a variant of the shared-primary cache CMP which contained five processors on a single chip.

6.2.2 TI TMS320C8x Multimedia Video Processors

The Texas Instruments TMS320C8x (or 'C8x) family of processors are single chip multiprocessors (CMPs) suitable for system-level and embedded implementations [291]. Applications include image processing, 2D and 3D graphics, audio/video digital compression and playback, real-time encryption/decryption and digital telecommunications. The processor is dubbed MVP, the multimedia video processor. A single MVP replaces several system components by integrating multiple processors, memory control logic, I-cache and internal memory, an advanced DMA controller, and video timing generation logic ('C80 only) onto a single chip. They provided an order of magnitude increase in computational power over existing digital signal processors (DSPs) and general-purpose processors in 1994.

Two types of processors are combined in the MVP architecture: A single RISC *master processor* (MP) and a number of VLIW DSP-like *parallel processors* (PP) (Fig. 6.4). Moreover, the chip contains a programmable DMA *transfer controller* (TC), a video controller (VC), and a boundary-scan test access port (TAP). All processors are interconnected by a crossbar with I-caches, and data RAM and parameter RAM areas.

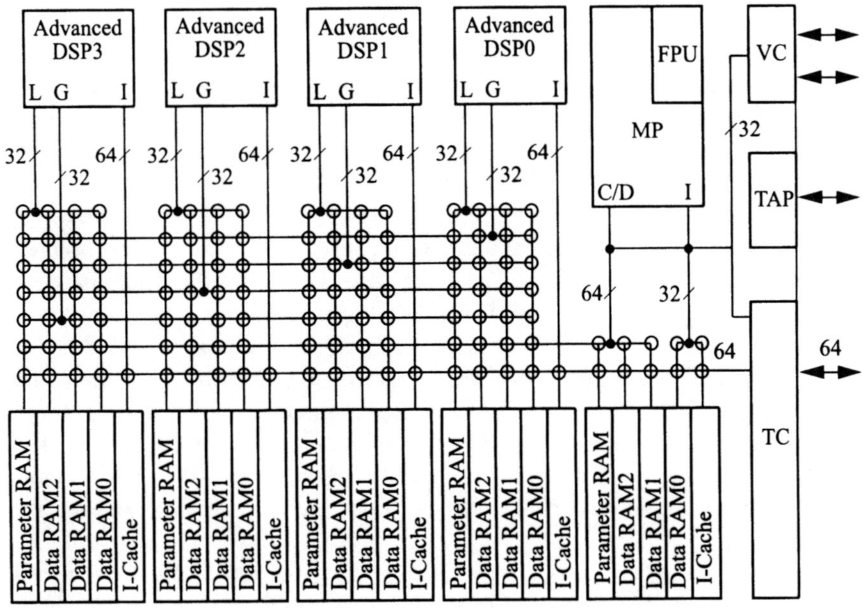

Fig. 6.4. TI's Multimedia Video Multiprocessor

The 'C8x family consists of two members, the 'C80 [291], which features four PPs, and the 'C82 (*Golston* [104]) with only two on-chip PPs.

The MP functions as a 32-bit RISC master control processor, and is intended to handle interrupts and external requests, and control the operation of the system as a whole. MP has an ANSI/IEEE 754-1985 compliant floating-point unit. Although the MP ISA is classified as a RISC, it has the capability of issuing explicit parallel operations in VLIW fashion. Up to three such operations can be issued: a multiply, an arithmetic/logical, and a load/store operation.

The PPs are designed to perform 64-bit multiply-intensive, multiple pixel, and bit-field manipulation operations on data in registers and internal RAM. The PPs are a collection of 32-bit fixed-point subprocessors connected by data path multiplexers. This allows a wide variety of operations to issue simultaneously. A local and global address unit, a three-input ALU, and a multiplier unit are integrated along with a barrel rotator[2], bit-detection hardware, mask generator, and a bitfield expander. In this way, an algorithm can be mapped onto the PP in far fewer overall instructions than traditional DSP and superscalar RISC processors. The multiplier supports signed/signed or unsigned/unsigned (but not signed/unsigned) multiplies when 16-bit input operands are used, and signed/unsigned or unsigned/unsigned (but not

[2] An FU that is similar to a barrel shifter but performs rotation by an arbitrary number of bits in one instruction cycle, as opposed to shifting by an arbitrary number of bits.

signed/signed) multiplies when 8-bit input operands are used. The ALU is capable of executing all 256 combinations of logical operations on three variables. This allows operations that may take several instructions on most DSPs to be performed in a single ALU operation.

The TC is designed to handle all off-chip data transfer operations required by the MP and PPs. The TC is programmed by a 64-byte data structure called a *packet transfer request* (PTREQ) describing the organization of the transfer desired, and by internal processor interrupts. The TC completely controls the bandwidth utilization of the chip. Hardware for prioritizing requests such as cache-service interrupts, individual PTREQ, and direct external access requests (DEA) is built into the TC. Each PTREQ can be linked to another PTREQ structure, providing an unbroken transition from one transfer to another, while giving the TC the opportunity to service other transfer types. The MP and each PP may make independent PTREQs; the TC services them according to an internal priority scheme. The TC thus provides the DSP programmer with the necessary tool to interleave data transfer and computation.

The same MVP's memory space is used for program and data accesses, and is shared by all of the processors on the chip. The MVP crossbar allows each PP to perform two independent parallel data accesses to the on-chip shared RAMs and one instruction fetch every cycle. Each PP has three crossbar ports. The *global port* connects to any of the shared RAMs. If an access is attempted over this port to an address not in the shared RAMs, a DEA request is sent to the TC. The *local port* connects to any of its local RAMs. When a PP attempts a memory access over this port to an address not in local RAMs, the access is diverted to the global port and tried on the following cycle. Finally, the *instruction port* accesses instructions from the PP's I-cache.

The MVP is not a general-purpose microprocessor. Rather it is designed and used as an extremely fast digital signal processor. Due to its complex architectural structure – a chip multiprocessor structure with two different kinds of processors, one of which even features a VLIW ISA – it is difficult to develop a compiler that generates efficient code. Therefore, most programs have to be hand-coded in assembly language, which is not at all easy.

6.2.3 Hydra Chip Multiprocessor

While the TI MVP is an existing commercial microprocessor, the Hydra Chip Multiprocessor is simulated in software to evaluate the CMP alternatives for future 10^9-transistor chips. The Hydra proposal (*Hammond and Olukotun* [121], 1998) is composed of four 2-issue superscalar CPUs on a single chip. Each of the CPUs is similar to a small MIPS R10000 processor and is attached to its own on-chip primary (L1) I-cache and D-cache. In addition, a single, unified secondary (L2) cache is included on the chip (Fig. 6.5).

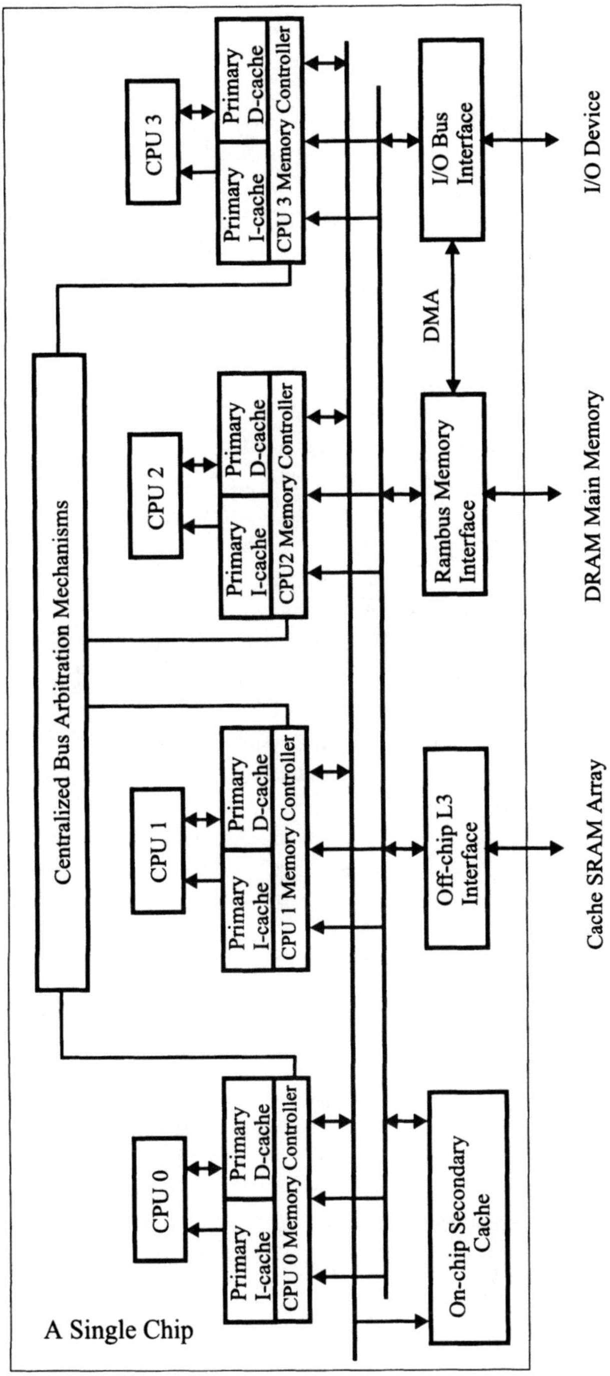

Fig. 6.5. A schematic overview of Hydra

The Hydra memory system uses a four-level arrangement, that is, an L1, L2, L3 SRAM cache and a DRAM main memory. It is a shared-secondary cache single-chip multiprocessor.

The individual L1 I-cache and D-caches are designed to supply each CPU in a single cycle while supporting the very high memory bandwidth needed to sustain processor performance. Each L1 cache has 16 kB capacity and is organized in 32-byte lines. The connection to its CPU is provided by a 64-bit bus. In order to maintain cache coherence, each D-cache must snoop the write bus and invalidate any lines to which other CPUs write. Measurements have shown that, in this way, in typical applications more than 90 % of loads hit in L1 caches and thus save the progress further down the memory hierarchy. The large 512 kB L2 cache acts as a large on-chip cache to back up the L1 caches with a nearby memory which is five or more cycles slower than L1. The L2 cache is a kind of write buffer between the CPUs and the outside world. On the other hand, L2 also acts as a communication medium through which the four CPUs can communicate using shared-memory mechanisms. The 8 MB off-chip L3 cache has an access time of 10 cycles to the first word, and can be accessed through a 128-bit port that operates at half the processor speed. Even with a large L3 cache, applications exist with large enough working sets to miss in all of the caches frequently. For this reason, up to 128 MB DRAM with at least 50 cycles access time is attached to the Hydra chip via a Rambus memory interface.

The *read bus* and *write bus* are the principal paths of communication across the Hydra chip. The read and the write bus are 256-bit and 64-bit wide, respectively. *Hammond and Olukotun* found that the contention for both the read bus and the write bus slows performance by only a few percent over a perfect crossbar, even in the worst cases. Furthermore, with the write bus-based architecture, no sophisticated coherence protocols are necessary to keep the on-chip caches coherent.

The 1998 Hydra CMP [121] addresses an expensive, high-end design, with many high-speed SRAM and DRAM chips, directly attached to the Hydra chip. Alternative designs are possible, however, in systems with different constraints. One interesting alternative is a design with no off-chip L3 at all. In this way the system cost can be reduced dramatically since expensive SRAM chips are eliminated and the number of I/Os on Hydra chip is halved. Another alternative is that the L2 cache is replaced with on-chip DRAM, thus making the L3 superfluous. In 1997, *Yamauchio et al.* [328] evaluated the performance of a Hydra CMP integrated with 256 MB DRAM. On floating-point applications with large working sets, the on-chip DRAM Hydra performed on average 52 % faster than the L2-on-chip/L3-off-chip Hydra.

6.3 Multithreaded Processors

6.3.1 Multithreading Approach for Tolerating Latencies

The memory access latency problem arises for each memory access after a cache miss – in particular, when a processor of a shared-memory multiprocessor accesses a shared-memory variable located in a remote-memory module. To perform such a remote-memory access in a DSM multiprocessor, the processor issues a request message to the communication network that couples the processor – memory nodes. The request message traverses the network to the memory module. The memory module reads the value, respectively the cache line, and sends a result message back to the requesting processor. Depending on the coherence scheme, further actions may be necessary to guarantee memory consistency or cache coherence before the requested value or cache line is sent back. The interval between the sending of the request message until the return of the result message is called *(remote) memory access latency*, or often just the *latency*[3]. The latency becomes a problem if the processor spends a large fraction of its time sitting idle and waiting for remote accesses to complete.

Load access latencies measured on an Alpha Server 4100 SMP with four 300 MHz Alpha 21164 processors are (*Barroso et al.* [23], 1998):

- 7 cycles for a primary cache miss which hits in the on-chip L2 cache of the 21164 processor,
- 21 cycles for a L2 cache miss which hits in the L3 (board-level) cache,
- 80 cycles for a miss that is served by the memory, and
- 125 cycles for a dirty miss, i.e., a miss that has to be served from another processor's cache memory.

For DSM multiprocessors supporting up to 1 024 processors we can expect latencies of 200 cycles or more (see also p. 218 for the SGI Origin 2000 latencies). Furthermore, memory access latencies are expected to increase over time as off-chip speeds are reduced more quickly than on-chip speeds.

In a *single-threaded* architecture the computation conceptually moves forward one step at a time through a sequence of states, each step corresponding to the execution of one enabled instruction. The state of a single-threaded machine consists of the *memory state* (program memory, data memory, stack) and the *processor state* which consists of the *continuation* or *activity specifier* (program counter, stack pointer) and the *register context* (a set of register contents). The activity specifier and the register context make up what is also called the *context* of a thread. Today most processors are of a single-threaded processor architecture.

[3] Latencies that arise in a pipeline are defined with a wider scope – for example, covering also long-latency operations like `div` or latencies due to branch interlocking.

According to *Dennis and Gao* [63], a *multithreaded* architecture differs from a single-threaded architecture in that there may be several enabled instructions from different threads which all are candidates for execution. Similar to the single-threaded machine, the state of the multithreaded machine consists of the memory state and the processor state; the latter, however, consists of a *collection* of activity specifiers and a *collection* of register contexts. A thread is a sequentially ordered block of instructions with a grain-size greater than one (to distinguish multithreaded architectures from fine-grained dataflow architectures).

Another notion is the distinction between *blocking* and *non-blocking* threads. A *non-blocking thread* is formed such that its evaluation proceeds without blocking the processor pipeline (for instance by remote memory accesses, cache misses or synchronization waits). Evaluation of a non-blocking thread starts as soon as all input operands are available, which is usually detected by some kind of dataflow principle. Thread switching is controlled by the compiler harnessing the idea of rescheduling, rather than blocking, when waiting for data. Access to remote data is organized in a split-phase manner by one thread sending the access request to memory and another thread activating when its data are available. Thus a program is compiled into many, very small threads activating each other when data become available. The same hardware mechanisms may also be used to synchronize interprocess communications to awaiting threads, thereby alleviating operating systems overhead. In contrast, a *blocking thread* might be blocked during execution by remote memory accesses, cache misses or synchronization needs. The waiting time, during which the pipeline is blocked, is lost when using a von Neumann processor, but can be efficiently bridged by a fast context switch to another thread in a multithreaded processor. Switching to another thread in a single-threaded processor usually inhibits too much context switching overhead to mask the latency efficiently. The original thread can be resumed when the reason for blocking is removed.

Use of non-blocking threads typically leads to many small threads that are appropriate for execution by a hybrid dataflow computer or by a multithreaded architecture that is closely related to hybrid dataflow. Blocking threads may just be the threads (e.g., P(OSIX)threads or Solaris threads) or whole UNIX processes of a multithreaded UNIX-based operating system, but also even smaller microthreads generated by a compiler to utilize the potentials of a multithreaded processor.

Note that we exclude in this chapter dataflow hybrid architectures that are designed for the execution of non-blocking threads. Although these architectures are often called *multithreaded*, we have categorized them as threaded dataflow (Sect. 2.3.1) or large-grain dataflow (Sect. 2.3.2) because a dataflow principle is applied to start the execution of non-blocking threads. Thus, multithreaded architectures (in the more narrow sense applied here) stem from the modification of scalar RISC, VLIW, or even superscalar RISC processors.

Multithreading techniques use coarse-grain parallelism to speedup computation of a multithreaded workload by better utilization of the resources of a single processor. The latencies that arise in the computation of a single instruction stream are filled by computations of another thread. This ability is in contrast to RISC processors or today's superscalar processors, which use busy waiting or a time-consuming, operating system-based thread switch.

The minimal requirement for a multithreaded processor is the ability to pursue two or more threads of control in parallel within the processor pipeline and a mechanism that triggers a thread switch. Thread-switch overhead must be very low, from zero to only a few cycles. A fast context switch is supported by multiple program counters and often by multiple register sets on the processor chip.

The following principal approaches to multithreaded processors exist:

- *Cycle-by-cycle interleaving technique*: An instruction of another thread is fetched and fed into the execution pipeline at each processor cycle (see Sect. 6.3.3).
- *Block interleaving technique*: The instructions of a thread are executed successively until an event occurs that may cause latency. This event induces a context switch (see Sect. 6.3.4).
- *Simultaneous multithreading*: The wide superscalar instruction issue is combined with the multiple-context approach. Instructions are simultaneously issued from multiple threads to the FUs of a superscalar processor (see Sect. 6.4).

Research on multithreaded architectures has been motivated by two concerns: tolerating latency and bridging of synchronization waits by rapid context switches. Older multithreaded processor approaches from the 1980s usually extend scalar RISC processors by a multithreading technique and focus at effectively bridging very long remote memory access latencies. Such processors will only be useful as processor nodes in DSM multiprocessors. However, developing a processor that is specifically designed for DSM multiprocessors is commonly regarded as too expensive. Multiprocessors today comprise standard off-the-shelf microprocessors and almost never specifically designed processors (with the exception of Tera MTA). Therefore, newer multithreaded processor approaches also strive for tolerating smaller latencies that arise from primary cache misses that hit in L2 cache, from long-latency operations, or even from unpredictable branches.

Note that multithreaded processors aim at a low execution time of a multithreaded workload, while a superscalar processor aims at a low execution time of a single program. Depending on the implemented multithreading technique, a multithreaded processor running only a single thread does not reach the same efficiency as a comparable single-threaded processor. The penalty may be only slight in the case of a block-interleaving processor or be several times as long as the run-time on a single-threaded processor in the case of a cycle-by-cycle interleaving processor.

6.3.2 Comparison of Multithreading and Non-Multithreading Approaches

Before we present the different multithreading approaches in detail, we briefly review the main principles of architectural approaches that exploit instruction-level parallelism and thread-level parallelism.

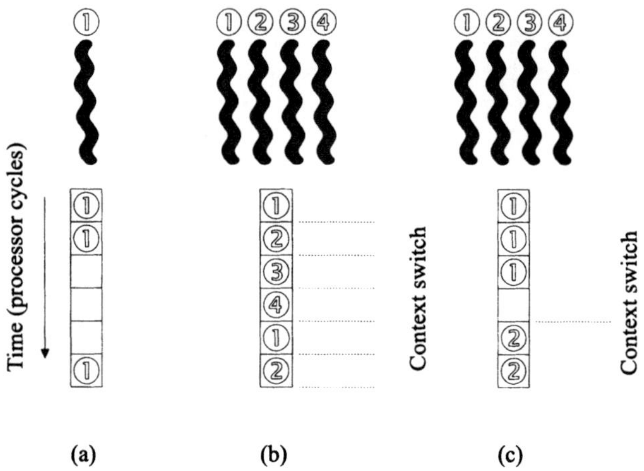

Fig. 6.6. Different approaches possible with scalar processors **(a)** single-threaded scalar **(b)** cycle-by-cycle interleaving multithreaded scalar **(c)** block-interleaving multithreaded scalar

Figure 6.6 demonstrates the different approaches possible with scalar (i.e., single-issue) processors: single-threaded (Fig. 6.6a), multithreaded with cycle-by-cycle-interleaving (Fig. 6.6b), and multithreaded with block interleaving (Fig. 6.6c).

Another way to look at latencies that arise in a pipelined execution is the *opportunity cost* in terms of the instructions that might be processed while the pipeline is interlocked, e.g., waiting for a remote reference to return. The opportunity cost of single-issue processors is the number of cycles lost by latencies. Multiple-issue processors (e.g., superscalar, VLIW, etc.) potentially execute more than one IPC, and thus the opportunity cost is the product of the latency cycles by the issue bandwidth plus the number of issue slots not fully filled. We expect that future single-threaded processors will continue to exploit further superscalar or other multiple-issue techniques, and thus further increase the opportunity cost of remote-memory accesses.

Figure 6.7 demonstrates the different approaches possible with four-issue processors: single-threaded superscalar (Fig. 6.7a), single-threaded VLIW (Fig. 6.7b), superscalar with cycle-by-cycle interleaving (Fig. 6.7c), and VLIW with cycle-by-cycle interleaving (Fig. 6.7d). Each row represents the issue slots for a single execution cycle. An empty box represents an unused slot; `N` stands for a no-op operation in the VLIW case.

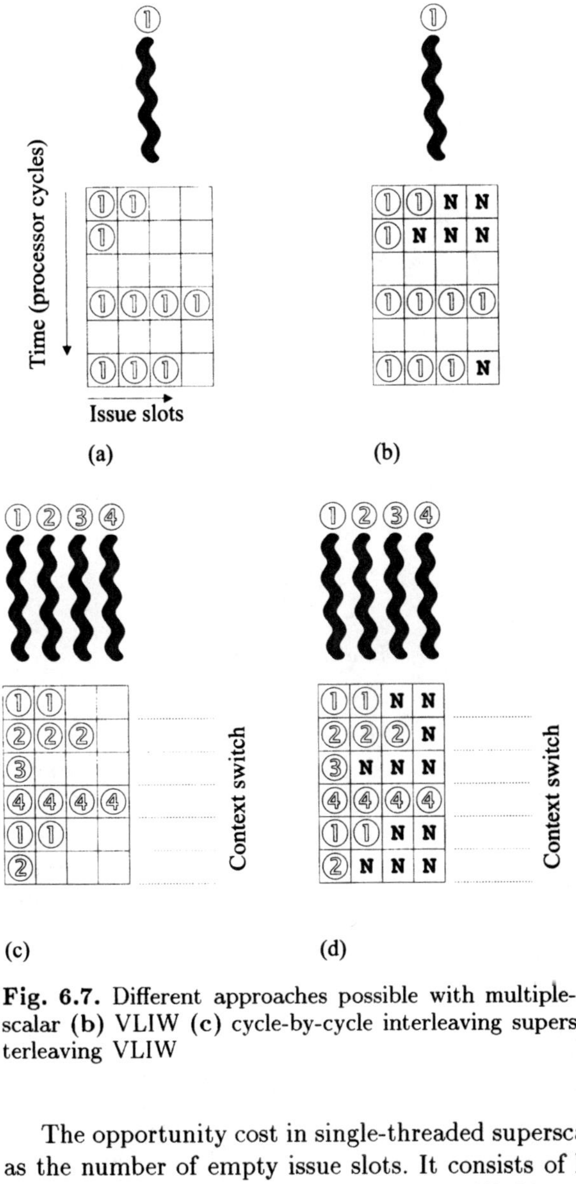

Fig. 6.7. Different approaches possible with multiple-issue processors **(a)** superscalar **(b)** VLIW **(c)** cycle-by-cycle interleaving superscalar **(d)** cycle-by-cycle interleaving VLIW

The opportunity cost in single-threaded superscalar can be easily counted as the number of empty issue slots. It consists of horizontal losses (unfilled issue slots when not all issue slots can be filled in a cycle) and the even more harmful vertical losses (cycles where no instructions can be issued). In VLIW processors, horizontal losses appear as a no-op operation. The opportunity cost of single-threaded VLIW is about the same as single-threaded superscalar. Cycle-by-cycle interleaving superscalar (or VLIW) is able to fill the vertical losses of the single-threaded models by instructions of other threads, but not the horizontal losses. Further design opportunities, block-interleaving superscalar or VLIW models (not shown in the figures) would fill several succeeding cycles with instructions of the same thread before context switching. The switching event is more difficult to determine and a context-switching overhead of one to several cycles might arise.

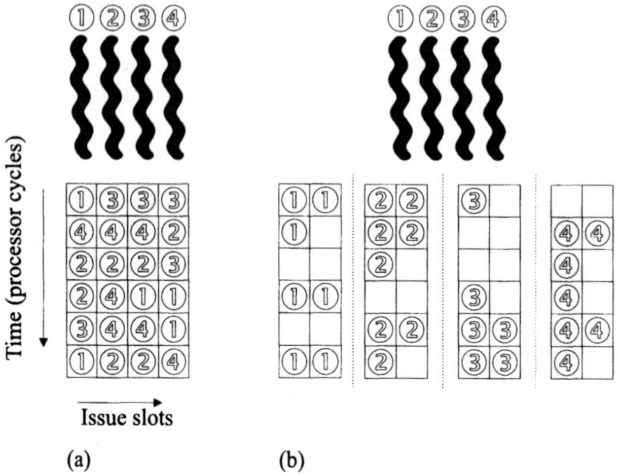

Fig. 6.8. Issue slot partitioning in **(a)** simultaneous multithreading (SMT) **(b)** chip multiprocessor (CMP)

Figure 6.8 demonstrates a four-threaded four-issue simultaneous multithreading (SMT) processor (Fig. 6.8a) and a chip multiprocessor (CMP) with four two-issue processors (Fig. 6.8b). The processor model in Fig. 6.8a exploits ILP by selecting instructions from any thread (four in this case) that can potentially issue. If one thread has high ILP, it may fill all horizontal slots depending on the issue strategy of the SMT processor. If multiple threads each have low ILP, instructions of several threads can be issued and executed simultaneously. In the CMP with four two-issue CPUs on a single chip that is represented in Fig. 6.8b, each CPU is assigned a thread from which it can issue up to two instructions each cycle. Thus, each CPU has the same opportunity cost as in a two-issue superscalar model. The CMP is not able to hide latencies by issuing instructions of other threads. However, because horizontal losses will be smaller for two-issue than for high-bandwidth superscalars, a CMP of four two-issue processors will reach a higher utilization that an eight-issue superscalar processor.

6.3.3 Cycle-by-Cycle Interleaving

In the *cycle-by-cycle interleaving* model (sometimes also called *fine-grain multithreading*) the processor switches to a different thread after each instruction fetch. In principle, an instruction of a thread is fed into the pipeline after the retirement of the previous instruction of that thread. Since cycle-by-cycle interleaving eliminates control and data dependences between instructions in the pipeline, pipeline hazards cannot arise and the processor pipeline can be easily built without the necessity of complex forwarding paths. This leads to a very simple and therefore potentially very fast pipeline – no hardware interlocking is necessary. Moreover, the context-switching overhead is zero cycles. Memory latency is tolerated by not scheduling a thread until the memory transaction has completed. This model requires at least as many threads as pipeline stages in the processor. Interleaving the instructions from many threads limits the processing power accessible to a single thread, thereby degrading the single-thread performance. There are two possibilities to overcome this deficiency:

- The *dependence lookahead technique* adds several bits to each instruction format in the ISA. The additional opcode bits allow the compiler to state the number of instructions directly following in program order that are not data- or control-dependent on the instruction being executed. This allows the instruction scheduler in the cycle-by-cycle interleaving processor to feed non-data- or control-dependent instructions of the same thread successively into the pipeline. The dependence lookahead technique may be applied to speed up single-thread performance or in the case where a workload does not comprise enough threads.
- The *interleaving technique* proposed by *Laudon et al.* [172], adds caching and full pipeline interlocks to the cycle-by-cycle interleaving approach. Contexts are interleaved on a cycle-by-cycle basis, yet a single-thread context is also efficiently supported.

The most well-known examples of cycle-by-cycle interleaving processors are the HEP (*Smith* [267]), the Horizon (*Thistle and Smith* [294]) and the Tera MTA (*Alverson et al.* [9]). All three employ the dependence-lookahead technique; the Horizon and Tera MTA feature a VLIW ISA. The HEP processor applies a cycle-by-cycle interleaving of instructions of eight threads within an 8-stage pipeline. To tolerate even longer memory access latencies, a large number of threads and non-blocking memory accesses are necessary. The Horizon and the Tera MTA each contain 128 thread contexts and register sets per processor node to mask remote memory access latencies effectively.

Further cycle-by-cycle interleaving processor proposals include the MASA (*Halstead and Fujita* [119]), the SB-PRAM/HPP (*Formella et al.* [90]), and MicroUnity's MediaProcessor (*Hansen* [122]), as an example of a multithreaded signal processor. The SB-PRAM is 32-threaded, and the MediaProcessor interleaves five contexts. In principle, cycle-by-cycle interleaving can also be combined with a superscalar instruction issue, but simulations of *Eggers et al.* [76] confirm the intuition that simultaneous multithreading is the more efficient technique (see Sect. 6.5).

Some machines that use cycle-by-cycle interleaving are described in more detail below.

Tera MTA Processor

The Tera Multi-Threaded Architecture (MTA) computer (*Alverson et al.* [9]) features a powerful VLIW instruction set, uniform access time from any processor to any memory location, and zero-cost synchronization and swapping between threads of control. It was designed by the Tera Computer Company (Seattle, WA) research team led by *Burton J. Smith*, who was also the principal architect of the Horizon and HEP computers.

MTA systems are constructed from resource modules. Each resource module contains up to six resources which can be a *computational processor* (CP), an *I/O processor* (IOP), an *I/O cache* (IOC) units, and either two or four *memory units* (MUs) (Fig. 6.9). Each resource is individually connected to a

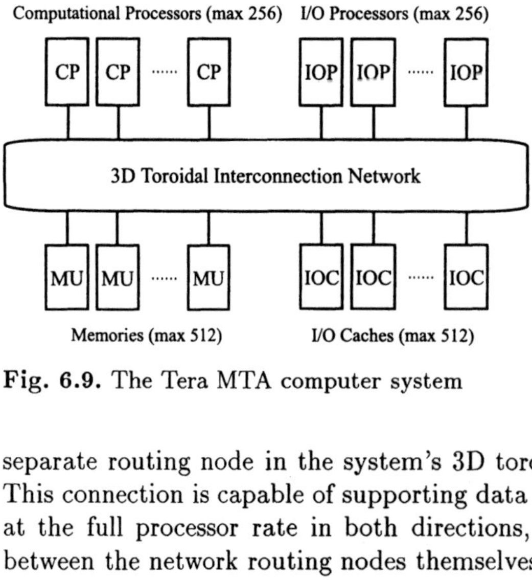

Fig. 6.9. The Tera MTA computer system

separate routing node in the system's 3D toroidal interconnection network. This connection is capable of supporting data transfers to and from memory at the full processor rate in both directions, as are all of the connections between the network routing nodes themselves. The three-dimensional torus topology used in MTA systems has 8 or 16 routing nodes per resource module with the resources sparsely distributed among the nodes. The network bandwidth scales excellently because there are several routing nodes per computational processor, rather than the several processors per routing node that many systems employ. Some MTA configurations are given in Table 6.1. For example, in the case of the MTA 256 model, there are 4 096 nodes, arranged in a $16 \times 16 \times 16$ mesh, consisting of 256 CPs, 256 IOPs, 256 IOC units, 512 MUs, and 2 816 routing nodes.

The Tera MTA custom chip CP (Fig. 6.10) is a multithreaded VLIW pipeline processor using the cycle-by-cycle interleaving technique. Each thread is associated with one 64-bit *stream status word* (SSW), thirty-two 64-

Table 6.1. Some MTA configurations

Model	Number of Processors	Memory (GB)	Performance (GFLOPS)	Bisection B/W (GB/s)	I/O B/W (GB/s)
MTA 16	16 CP + 16 IOP	16-32	16	179	6
MTA 32	32 CP + 32 IOP	32-64	32	179	12
MTA 64	64 CP + 64 IOP	64-128	64	358	25
MTA 128	128 CP + 128 IOP	128-256	128	717	51
MTA 256	256 CP + 256 IOP	256-512	256	1434	102

bit general registers, and eight 64-bit target registers. The processor switches context every cycle (3 ns cycle period) between as many as 128 distinct threads (called *streams* in the Tera), thereby hiding up to 128 cycles (384 ns) of memory latency. Since the context switching is so fast, the processor has no time to swap the processor state. Instead, it has 128 of everything, i.e., 128 SSWs, 4 096 general registers, and 1 024 target registers. In addition, each thread can issue as many as eight memory references without waiting for earlier ones to finish, further augmenting the memory latency tolerance of the processor. A thread implements a load/store architecture with three addressing modes and 32 general-purpose 64-bit registers. The 3-wide VLIW instructions are 64 bits. Three operations can be executed simultaneously per instruction: a memory reference operation (M-op), an arithmetic/logical operation (A-op), and a branch or simple arithmetic/logical operation (C-op). If more than one operation in an instruction specifies the same register or setting of condition codes, the priority is M>A>C. Dependences between instruction are explicitly encoded by the compiler using *explicit dependence lookahead*. Each instruction contains a 3-bit lookahead field that explicitly specifies how many instructions from this thread will be issued before encountering an instruction that depends on the current one. Since seven is the maximum possible lookahead value, at most eight instructions (i.e., 24 operations) from each thread can be concurrently executing in different stages of a processor's pipeline.

The clock speed is nominally 333 MHz, giving each processor a data path bandwidth of 10^9 64-bit results per second and a peak performance of 1 GFLOPS. The peak memory bandwidth is 2.67 GB/s, and it is claimed that the processor sustains well over 95 % of that rate.

The processor implements ANSI/IEEE 754-1985 arithmetic using the 64-bit double basic format. Hardware support for infinity arithmetic and denormalized operands is provided. Addition, subtraction, multiplication, division, and conversion to and from both signed and unsigned integer formats are supported directly. The type of integer rounding can be selected independent of the rounding mode. There are floating-point multiply-add, multiply-subtract and multiply-subtract-reverse operations. These operations round only once.

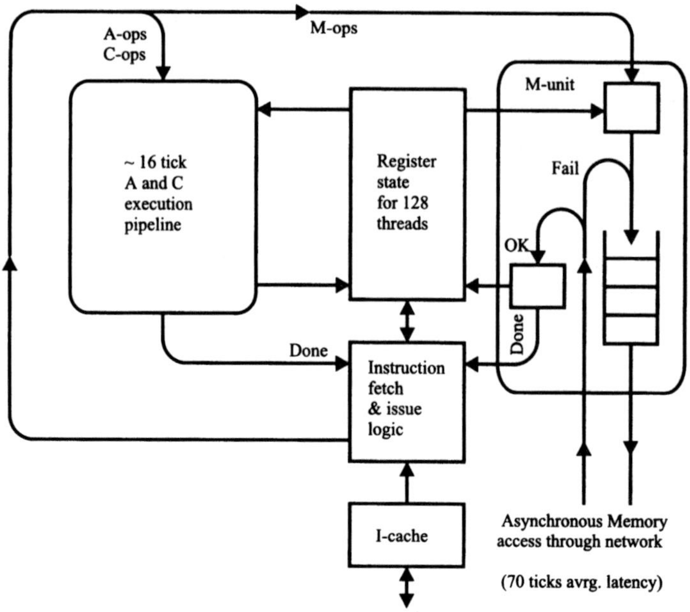

Fig. 6.10. The MTA processor

Division and square root are accomplished with the help of iterations of Newton's method. Floating-point maximum and minimum operations are also provided. Support for fast, 128-bit double-precision arithmetic is incorporated. In double precision two floating-point numbers are used to represent a single value, yielding approximately twice the precision of ordinary floating point. There are instructions that help to compute the double-precision sum, difference, product quotient, and square root in a few instructions.

Every processor has a clock register that is synchronized exactly with its counterparts in the other processors and counts up once per cycle. In addition, the processor counts the total number of unused instruction issue slots (measuring the degree of underutilization of the processor) and the time-integral of the number of instruction streams ready to issue (measuring the degree of overutilization of the processor). All three counters are user-readable in a single unprivileged operation. Eight counters are implemented in each of the protection domains of the processor. All are user-readable in a single unprivileged operation. Four of these counters accumulate the number of instructions issued, the number of memory references, the time-integral of the number of instruction streams and the time-integral of the number of messages in the network. These counters are also used for job accounting. The other four counters are configurable to accumulate events from any four of a large set of additional sources within the processor, including memory operations, jumps, traps and so on.

Thus, the Tera MTA exploits parallelism at all levels, from fine-grained ILP within a single processor to parallel programming across processors, to multiprogramming among several applications simultaneously. Consequently, processor scheduling occurs at many levels, and managing these levels poses unique and challenging scheduling concerns (*Alverson et al.* [8]).

After many delays the MTA reached the market in 1998. In April 1998 a two-processor MTA system was delivered to the San Diego Supercomputer Center. Currently the MTA processor runs at 145 MHz, but will scale to 333 MHz.

Other cycle-by-cycle interleaving processors

HEP (*Smith* [266], 1981): The Heterogeneous Element Processor (HEP) system was a MIMD shared-memory multiprocessor system developed by Denelcor Inc. (Denver, CO) between 1978 and 1985, and was a pioneering example of a multithreaded machine (see Fig. 6.11). Spatial switching occurred between two queues of processes; one of these controlled program memory, register memory, and the functional memory while the other controlled data memory. The main processor pipeline had eight stages, matching the number of processor cycles necessary to fetch a data item from memory in register. Consequently eight threads were in execution concurrently within a single HEP processor. In contrast to all other cycle-by-cycle interleaving

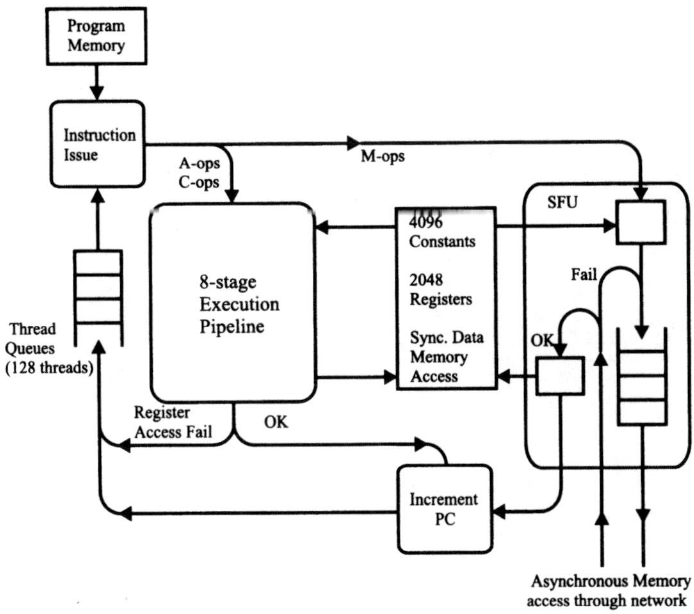

Fig. 6.11. The HEP processor

processors, all threads within a HEP processor shared the same register set. Multiple processors and data memories were interconnected via a pipelined switch and any register-memory or data-memory location could be used to synchronize two processes on a producer–consumer basis by a full/empty bit synchronization on a data memory word.

MASA (*Halstead and Fujita* [119], 1988): The Multilisp Architecture for Symbolic Applications (MASA) was a cycle-by-cycle interleaving multithreaded processor proposal for parallel symbolic computation with various features intended for effective Multilisp program execution. MASA featured a tagged architecture, multiple contexts, fast trap handling, and a synchronization bit in every memory word. Its principal novelty was the use of multiple contexts both to support interleaved execution from separate instruction streams and to speed up procedure calls and trap handling (in the same manner as register windows).

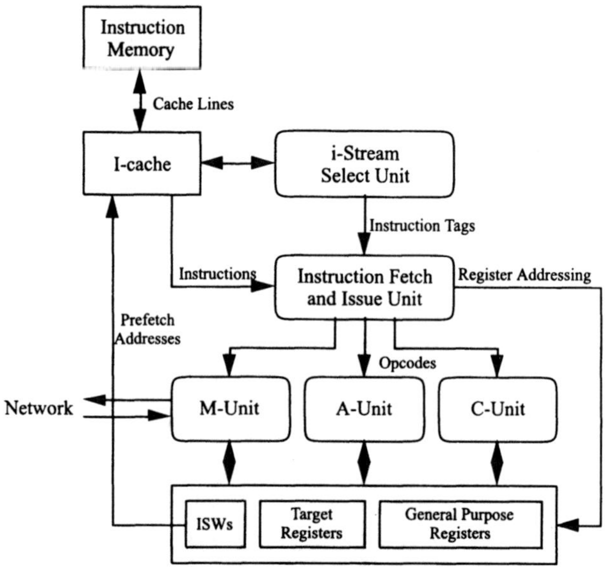

Fig. 6.12. The Horizon processor

Horizon (*Thistle and Smith* [294], 1988): This (paper) architecture was based on the HEP but extended to a massively parallel MIMD computer. The machine (see Fig. 6.12) was designed for up to 256 processing elements and up to 512 memory modules in a 16 × 16 × 6 node internal network. Like HEP it employed a global address space, and memory-based synchronization through the use of full/empty bits at each location. Each processor supported 128 independent instruction streams by 128 register sets with context switches occurring at every clock cycle. The Horizon is a preversion of the Tera MTA.

M-Machine (*Fillo et al.* [83], 1995): The MIT M-Machine supports both public and private registers for each thread, and uses cycle-by-cycle interleaving. Each processor supports four hardware resident user V-threads, and each V-thread supports four resident H-threads. All the H-threads in a given V-thread share the same address space, and each H-thread instruction is a 3-wide VLIW. Event and exception handling are each performed by a separate V-thread. Swapping V-threads between being resident on-chip and not resident in memory requires about 150 cycles (1.5 μs). M-Machine (like HEP, Horizon, and Tera MTA) employs full-empty bits for efficient, low-level synchronization. Moreover it supports message passing and guarded pointers with base and bounds for access control and memory protection.

SB-PRAM (*Bach et al.* [21]) and *HPP* (*Formella et al.* [90], 1996): The SB-PRAM (SB stands for Saarbrücken) or High Performance PRAM (HPP) is a MIMD parallel computer with shared address space and uniform memory access time due to its motivation: building a multiprocessor that is as close as possible to the theoretical machine model CRCW-PRAM. Processor and memory modules are connected by a butterfly network. Network latency is hidden by pipelining several so-called virtual processors on one physical processor node in cycle-by-cycle interleaving mode. A first prototype (SB-PRAM) is running with four processors, a second prototype (HPP) is under construction. Instructions of 32 so-called virtual processors are interleaved round-robin in a single SB-PRAM processor, which is therefore classified as 32-threaded cycle-by-cycle interleaving processor. The project is in progress at the University of Saarland (Saarbrücken, Germany).

6.3.4 Block Interleaving

The *block interleaving* approach (sometimes also called *coarse-grain multithreading*) executes a single thread until it reaches a situation that triggers a context switch. Usually such a situation arises when the instruction execution reaches a long-latency operation or a situation where a latency may arise. Compared to the cycle-by-cycle interleaving technique, a smaller number of threads is needed and a single thread can execute at full speed until the next context switch. Single-thread performance is similar to the performance of a comparable processor without multithreading.

In the following we classify block-interleaving processors by the event that triggers a context switch (see Fig. 6.13):

- *Static*: A context switch occurs each time the same instruction is executed in the instruction stream. The context switch is encoded by the compiler. The main advantage of this technique is that context switching can be

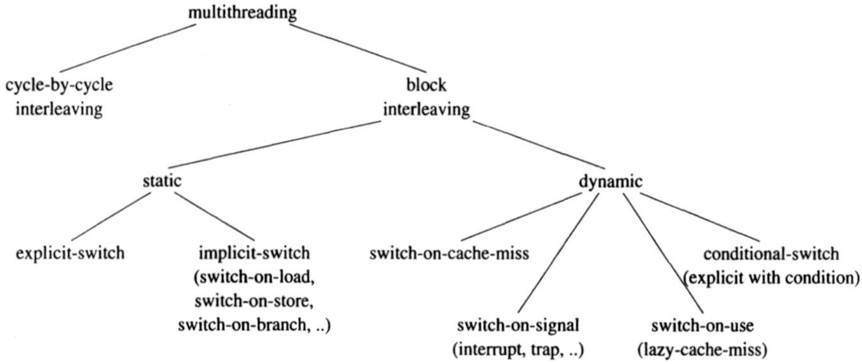

Fig. 6.13. Interleaving techniques

triggered already in the fetch stage of the pipeline. The context switching overhead is one (if the fetched instruction triggers the context switch and is discarded), zero (if the fetched instruction triggers a context switch but is still executed in the pipeline), and almost zero (if a context switch buffer is applied, see the Rhamma processor below in this section). There are two main variations of the static block-interleaving model:

- If an explicit context switch instruction exists, the model is called *explicit-switch*. This model is simple to implement and requires only one additional instruction.
- In the *implicit-switch* model, each instruction belongs to a specific instruction class and a context switch depends on the instruction class of the fetched instruction.

  The *switch-on-load* version switches after each load instruction to bridge memory access latency. However, assuming an on-chip D-cache, the thread switch occurs more often than necessary. This makes an extremely fast context switch necessary, preferably with zero-cycle context switch overhead.

  Switch-on-store switches after store instructions. This technique may be used to support the implementation of sequential consistency which means that the next memory access instruction can only be performed after the store has completed in memory.

  Switch-on-branch switches after branch instructions. The technique can be applied to simplify processor design by renouncing branch prediction and speculative execution. The branch misspeculation penalty is avoided, but single-thread performance is decreased. However, it may be effective for programs with a high percentage of branches that are hard to predict or even unpredictable.

- *Dynamic*: The context switch is triggered by a dynamic event. In general, all the instructions between the fetch stage and the stage that triggers the context switch are discarded, leading to a higher context switch overhead than static context switch strategies. Several dynamic models can be defined:

- The *switch-on-cache-miss* model switches the context if a load or store misses in the cache. The idea is that only those loads that miss in the cache have long latencies and cause context switches. Such a context switch is detected in a late stage of the pipeline. A large number of subsequent instructions have already entered the pipeline and must be discarded. Thus context switch overhead is considerably increased.
- The *switch-on-signal* model switches context on occurrence of a specific signal, for example, signaling an interrupt, trap, or message arrival.
- However, context switches sometimes also occur sooner than needed. If a compiler schedules instructions so that a load from shared memory is issued several cycles before the value is used, the context switch should not occur until the actual use of the value. This strategy is implemented in the *switch-on-use* model which switches when an instruction tries to use the (still missing) value from a load. This can be, for example, a load that missed in the cache. The switch-on-use model can also be seen as a *lazy* strategy that extends either the static switch-on-load strategy (*lazy-switch-on-load*) or the switch-on-cache-miss (*lazy-switch-on-cache-miss*).

  To implement the switch-on-use model, a *valid bit* is added to each register (by a simple form of scoreboard). The bit is cleared when the loading to the corresponding register is issued and set when the result returns from the network. A thread switches context if it needs a value from a register whose valid bit is still cleared.
- The *conditional-switch* model couples an explicit switch instruction with a condition. The context is switched only when the condition is fulfilled, otherwise the context switch is ignored. A conditional-switch instruction may be used, for example, after a group of load/store instructions. The context switch is ignored if all load instructions (in the preceding group) hit the cache; otherwise, the context switch is performed. Moreover, a conditional-switch instruction could also be added between a group of loads and their subsequent use to realize a lazy context switch (instead of implementing the switch-on-use model).

The explicit-switch, conditional-switch and switch-on-signal techniques enhance the ISA by additional instructions. The implicit switch technique may favor a specific ISA encoding to simplify instruction class detection. All other techniques are microarchitectural techniques without the necessity of ISA changes.

A previous classification by *Boothe and Ranade* [33, 34] concerns multithreading techniques only in a shared-memory multiprocessor environment and is restricted to only a few of the variations of multithreading techniques described above. In particular, the switch-on-load in [33] switches only on instructions that load data from remote memory, while storing data in remote memory does not cause context switching. Likewise, the *switch-on-miss* model is defined so that the context is only switched if a load from remote memory misses in the cache.

The MIT Sparcle (below in this section) and the MSparc (*Mikschl and Damm* [201]) processors switch context in case of remote memory accesses or failed synchronizations. They can be classified as block interleaving processors using *switch-on-cache-miss* and *switch-on-signal* techniques. Since both switching reasons are revealed in a late stage of the pipeline, the succeeding instructions that are already loaded in the pipeline cannot be used. Reloading of the pipeline and the software implementation of the context switch cause context switch cost of 14 processor cycles in the Sparcle. The MSparc processor is similar to the Sparcle except for hardware support for context switching. A context switch is performed one cycle after the event is recognized. In the case of the MIT Sparcle, context switches are used only to hide long memory latencies since small pipeline delays are assumed to be hidden by proper ordering of instructions by an optimizing compiler.

The multithreaded Rhamma processor (see below) decouples execution and load/store pipelines. Both pipelines execute instructions of different threads. In contrast to the Sparcle, the Rhamma processor is designed to bridge all kinds of latencies by a fast context switch applying various static and dynamic block-interleaving strategies.

Sparcle – Switch-on-Cache-Miss Processor

The MIT Sparcle processor (*Agarwal et al.* [2], 1993) is derived from a SPARC RISC processor. The eight overlapping register windows of a SPARC

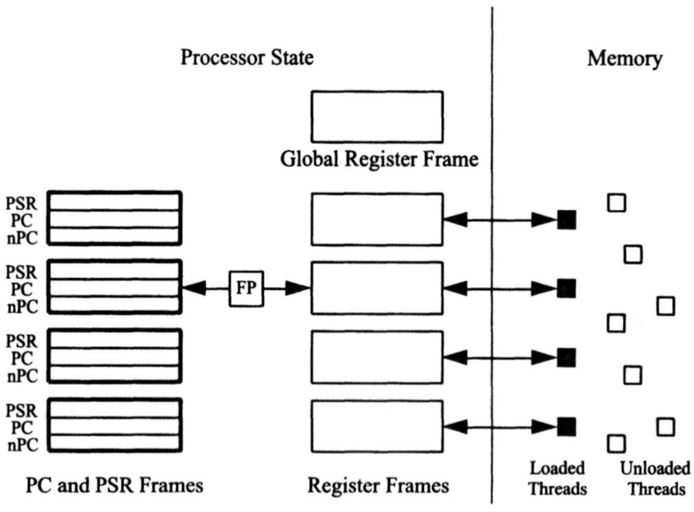

Fig. 6.14. Sparcle register usage

processor are organized in four independent non-overlapping thread contexts, each using two windows (one as a register set, the other as a context for trap and message handlers, see Fig. 6.14).

Context switches are used only to hide long memory latencies since small pipeline delays are assumed to be hidden by proper ordering of instructions by an optimizing compiler. The MIT Sparcle processor switches to another context in the case of a remote cache miss or a failed synchronization (switch-on-cache-miss and switch-on-signal strategies). Thread switching is triggered by external hardware, i.e., by the cache/directory controller (see Fig. 6.15). Emptying the pipeline of instructions of the thread that caused the context switch and organizational software overhead contribute a context switching penalty of 14 processor cycles.

The MIT Alewife DSM multiprocessor (*Agarwal et al.* [3]) is based on the multithreaded Sparcle processor. The multiprocessor has been operational since May 1994. A node in the Alewife multiprocessor comprises a Sparcle processor, an external floating-point unit, cache and a directory-based cache controller that manages cache-coherence, a network router chip, and a memory module (see Fig. 6.15).

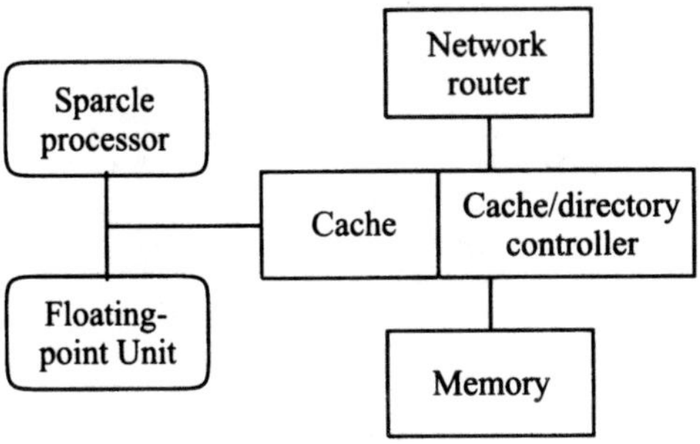

Fig. 6.15. An Alewife node

The Alewife multiprocessor uses a low-dimensional direct interconnection network. Despite its distributed-memory architecture, Alewife allows efficient shared-memory programming through a multilayered approach to locality management. Communication latency and network bandwidth requirements are reduced by a directory-based cache-coherence scheme referred to as LimitLESS directories. Latencies still occur although communication locality is enhanced by run-time and compile-time partitioning and placement of data and processes.

Rhamma Switch-on-Load Processor

The multithreaded Rhamma processor (*Grünewald and Ungerer* [108, 109]) was developed between 1993 and 1997 at the University of Karlsruhe

(Germany) as an experimental microprocessor that bridges all kinds of latencies by a very fast context switch. The Rhamma processor combines several static and dynamic block-interleaving techniques.

In the Rhamma processor (see Fig. 6.16), the *execution unit* (EU) and *load/store unit* (LSU) are decoupled and work concurrently on different threads. A number of register sets used by different threads are accessed by the LSU as well as the EU. Both units are connected by FIFO buffers for so-called continuations, each denoting the thread tag and the instruction pointer of a thread.

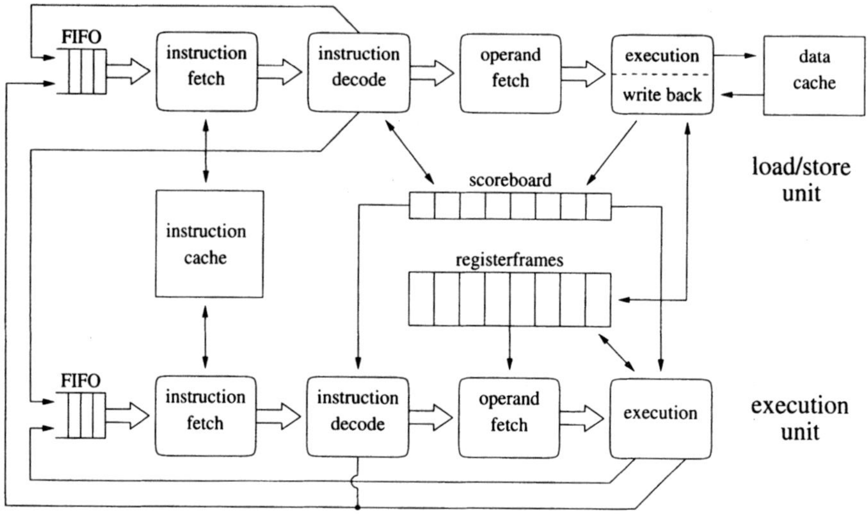

Fig. 6.16. Overall structure of the Rhamma processor

The EU and the LSU are modeled by a 4-stage instruction pipeline (instruction fetch IF, decode DE, operand fetch OF, and a combined execute and write-back EX/WB stage) with the full result-forwarding techniques. The EU is based on a conventional RISC processor (a variant of the DLX processor [134]). The original DLX instruction set is extended by thread management instructions. A scoreboard determines operand availability prior to the EX/WB stage. The result is written into the target register in the EX/WB stage.

In the implementation of the execution pipeline of Rhamma, each instruction that may cause a latency forces a context switch. Besides the load/store instructions the additional instructions concerned are the control instructions for jump, branch, thread management, and interrupt management. This design decision avoids complex hardware solutions which are necessary if, for example, a load/store instruction directly follows a branch instruction, thereby potentially causing a wrong context switch if the branch is taken. With this solution interlocking of the pipeline is avoided as well as speculative branch prediction and its rollback mechanism in the case of a wrong prediction. The single thread performance may be slowed down compared to an implementation with a proper branch prediction strategy. However, overall throughput is increased. The context switch is also used to handle instructions that may cause interrupts.

The EU of the Rhamma processor switches the thread whenever a load/store or control-flow instruction is fetched. Therefore, if the instruction format is chosen appropriately, a context switch can be recognized by *predecoding* the instructions already during the IF stage. Pipeline execution proceeds without bubbles when a control instruction that has to be executed by the EU causes a context switch. No cycle is lost.

In the case of a load/store, the continuation is stored in the FIFO buffer of the LSU, a new continuation is fetched from the EU's FIFO buffer, and the context is switched to the register set given by the new thread tag and the new instruction pointer. The LSU loads and stores data of a different thread in a register set concurrently to the work of the EU. Completion of a memory access may be signaled by a load/store acknowledgement.

There is a loss of one processor cycle in the EU for each sequence of load/store instructions. Only the first load/store instruction forces a context switch in the EU; succeeding load/store instructions are executed in the LSU. When context switching on a control instruction, the instruction is still fed into the pipeline (in contrast to a load/store instruction). In this case context switching is always performed without the loss of a cycle.

The loss of one processor cycle in the EU when fetching the first of a sequence of load/store instructions can be avoided by a so-called *context switch buffer* (CSB) whenever the sequence is executed the second time. The CSB is a hardware buffer, collecting the addresses of load/store instructions that have caused context switches. Before fetching an instruction from the I-cache, the IF stage checks whether the address can be found in the context switch buffer. In that case the thread is switched immediately and an instruction of another thread is fetched instead of the load/store instruction. No bubble occurs. If the instruction address is not in the CSB, the next instruction of the current thread is fetched. If a load/store instruction is recognized by predecoding (e.g., in case of the first execution of the load/store instruction), a loss of one cycle is caused due to the fetch of the load/store instruction that cannot be executed in the EU. The instruction address is written into the CSB.

The CSB is implemented similarly to a direct-mapped cache. The instruction address is divided into an index part and a tag part. The index part selects the buffer line where the tag part of the instruction address is stored. Since different instruction addresses may be mapped onto the same buffer line, the tag entry is necessary and must be compared with the tag part of the instruction address during buffer lookup. An additional bit per buffer line signals whether an entry is a valid tag or not. A lookup in the CSB yields one if the instruction address is a valid entry in the CSB. The direct-mapped buffer organization was chosen due to its low hardware complexity and fast lookup. All threads share a common CSB with the potential danger of thrashing buffer entries. Use of multiple CSBs, one for each thread, is possible, but multithreaded programs with a SPMD-structure may profit from a shared CSB.

The implementation of predecoding and a CSB only slightly enhances the hardware logic of the IF stage (see Fig. 6.17). The IF stage feeds the next instruction into the instruction pipeline. The instruction pointer used to access the I-cache is computed from the previous instruction of the current thread, i.e., the instruction pointer of the thread is increased. In the case of a context switch, the next instruction is denoted by a continuation fetched from the FIFO buffer. The instruction address is loaded into the instruction pointer register. The thread tag addresses the new current register set. In addition to the buses for the thread tag and the instruction pointer of the next thread, a signal line enables the other pipeline stages to force a context switch. When the context switch signal is high, a context switch is performed by the IF stage. Otherwise the current thread tag is stored in a latch and the new instruction pointer is computed from the old one.

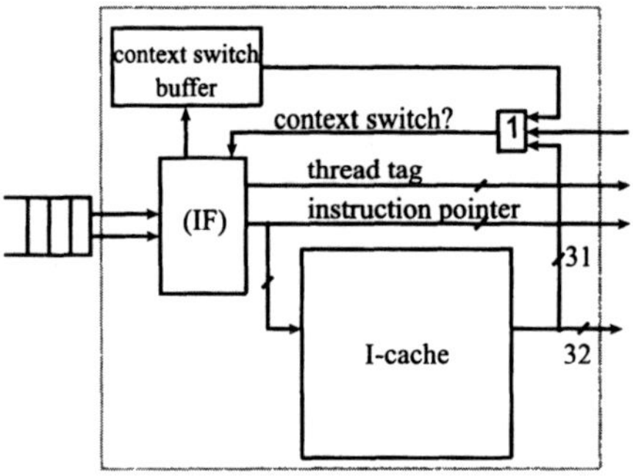

Fig. 6.17. The complete IF stage

To implement the predecoding, a 32-bit instruction format is used, and bit 31 – the highest order bit in the opcode – is set to one if the instruction forces a context switch in the pipeline of the EU. Otherwise the instruction fetch continues with the next instruction of the current thread. Bit 31 is used as a new control signal for the instruction fetch as shown in Fig. 6.17.

The CSB is integrated into the IF stage. As soon as the instruction pointer is incremented, it is made available for lookup in the CSB. The output signal is connected by an OR-gate with the context switch signal of the other pipeline stages and with the instruction predecoding. Thus the thread is switched if one of the three signals is activated. The CSB will be only updated by a context switch buffer miss, so the CSB influences execution only in the second and following runs of a basic block.

All solutions to optimize a fast context switch can be applied in an analogous way to the LSU.

The Rhamma processor combines several static and dynamic block interleaving techniques. Due to the decoupling of the execution and the load/store pipelines, a context switch is performed if an instruction is to be fetched but belongs to the other instruction class. The context switch is recognized by the predecoder in the instruction fetch stage by means of a tag in the instruction opcode (implicit-switch technique). A context switch is also performed in the EU after the fetch of a control flow (branch, jump, or context management) instruction (implicit-switch technique).

Additionally, two further context-switching techniques are applied in Rhamma. The availability of register values is tested with a scoreboard in the instruction decode stages of both pipelines. If an operand value is unavailable, pipeline interlocking is avoided by a context switch (switch-on-use technique). Due to full forwarding techniques implemented in the execution pipeline, the switch-on-use technique triggers a context switch only when a missing operand is not loaded fast enough from memory.

Second, a so-called `sync` instruction performs a context switch only when acknowledges of load/store instructions are pending (conditional-switch technique). The `sync` instruction is provided to ease the implementation of different consistency models.

In order to estimate the performance of the multithreading techniques in Rhamma, an execution-based software simulation compared the Rhamma processor with a single-threaded single-issue base processor that is similar to the DLX processor [134]. The base processor features the same execution pipeline and the same memory hierarchy (see Table 6.2) as assumed for the Rhamma processor. The base processor uses a static branch prediction technique. The Rhamma processor uses a CSB with 32 entries and eight register frames.

Table 6.2. Memory hierarchy assumed for Rhamma simulations

	access time	cycle time	size
I-cache	1	1	4 k
D-cache	2	1	4 k
Off-chip cache	9	9	1 M
DRAM	27	14	1 G

Multiple programs (partly from the Splash-2 benchmark suite) were executed by the simulator. The same program is executed by both processor configurations. Table 6.3 shows the speedup relative to the run time of the program on the base processor.

Table 6.3. Rhamma simulation results

Threads	1	2	4
Hilbert	0.55	0.84	1.54
Quicksort	0.64	1.14	1.67
Livermore	0.68	1.33	2.13
FFT	0.76	1.05	1.19
LU	0.77	1.20	1.49
Radix	0.78	1.27	1.79

The relative run times show clearly that the Rhamma processor with only one thread falls behind the base processor. This is because of the two clock cycles which are needed to exchange the threads between the pipelines. With two threads the Rhamma processor already outperforms the base processor (except for the **`Hilbert`** benchmark). With four threads the performance of the Rhamma processor is substantially better than the performance of the base processor.

As the next step in the design flow, a more detailed version of the Rhamma processor was implemented in VHDL code. The implementation limits according to ASIC or FPGA technology were determined with the help of VHDL simulation and synthesis by the Synopsys system. The VHDL-description of the Rhamma processor was evaluated by the execution of small workloads on the simulator. Because of long run times of the Synopsys simulator, larger workloads could not be tested. The automatic hardware synthesis performed by the Synopsys system generates from the VHDL-code a gate-level description of the circuit that is independent of specific chip technology. The ES2 1.0 μm library was used to estimate the chip area of the Rhamma implementation using ASIC technology. The synthesized circuit of the Rhamma processor with a CSB of 32 entries but without register sets, code, and D-caches needs an area of 38.35 mm^2 on an ASIC-chip with approximately 20 MHz cycle frequency. A CSB only lengthens the critical path of the instruction fetch by 1.8 ns, assuming a 50 ns processor cycle time. The pipeline cycle time can be reduced by an improved balancing of the pipelines. Moreover, an optimized hardware implementation, used for an industrial processor design as well as using modern processor chip technology, namely a full custom design, may increase cycle rate significantly.

The experiences with the Rhamma project showed that a consequent application of multithreading techniques to bridge all kinds of latencies avoids complex circuits for speculative execution (branch prediction as well as upcoming value speculation), for out-of-order issue and execution to bridge memory access latencies, as well as for traps and interrupts. This is made possible only due to the additional hardware solutions that reduce the cost of a context switch to one or even almost zero cycles. Fast context switching techniques may be applied in general-purpose microprocessors. However, also microcontrollers and signal processors may profit from these techniques.

Currently a multithreaded Java-microcontroller for real-time applications is under development at the University of Karlsruhe.

Other block interleaving processors

CHoPP (*Mankovich et al.* [191], 1987): The Columbia Homogeneous Parallel Processor (CHoPP) was a shared-memory MIMD with up to 16 powerful computing nodes. High sequential performance is due to issuing multiple instructions on each clock cycle, zero-delay branch instructions, and fast execution of individual instructions. Each node can support up to 64 threads.

MDP in J-Machine (*Dally et al.* [58], 1992): The MIT Jellybean Machine (J-Machine) is so-called because it is to be built entirely of a large number of "jellybean" components. The initial version uses an 8 × 8 × 16 cube network, with possibilities of expanding to 64K nodes. The "jellybeans" are message-driven processor (MDP) chips, each of which has a 36-bit processor, a 4 k word memory, and a router with communication ports for six directions. External memory of up to 1 M words can be added per processor. The MDP creates a task for each arriving message. In the prototype, each MDP chip has four external memory chips that provide 256 k memory words. However, access is through a 12-bit data bus, and with an error correcting cycle, the access time is four memory cycles per word. Each communication port has a 9-bit data channel. The routers provide support for automatic routing from source to destination. The latency of a transfer is 2 microseconds across the prototype machine, assuming no blocking. When a message arrives, a task is created automatically to handle it in 1 μs. Thus, it is possible to implement a shared memory model using message passing, in which a message provides a fetch address and an automatic task sends a reply with the desired data.

MSparc (*Mikschl and Damm* [201], 1996): An approach similar to the Sparcle processor was taken at the University of Oldenburg (Germany) with the MSparc processor. MSparc supports up to four contexts on chip and is compatible to standard SPARC processors. Switching is supported by hardware and can be achieved within one processor cycle. However, a four cycle overhead is introduced due to pipeline refill. The multithreading policy is block interleaving with the switch-on-cache-miss policy as in the Sparcle processor.

PL/PS-Machine (*Kavi et al.* [158], 1997): The PL/PS-Machine (Preload and Poststore) is most similar to the Rhamma processor. It also decouples memory accesses from thread execution by providing separate units. This decoupling eliminates thread stalls due to memory accesses and makes thread switches due to cache misses unnecessary. Threads are created when all data is preloaded into the register set holding the thread's context, and the results from an execution thread are poststored. Threads are non-blocking and each thread is enabled when the required inputs are available (i.e., data-driven at a coarse grain). The separate load/store/sync processor performs preloads and schedules ready threads on the pipeline. The pipeline processor executes the threads which will require no memory accesses. On completion the results from the thread are poststored by the load/store/sync processor.

6.3.5 Nanothreading and Microthreading

The *nanothreading* and the *microthreading* approaches use multithreading but spare the hardware complexity of providing multiple register sets.

Nanothreading (*Gwennap* [116], 1997) proposed for the DanSoft processor dismisses full multithreading for a nanothread that executes in the same register set as the main thread. The DanSoft nanothread requires only a 9-bit PC, some simple control logic, and it resides in the same page as the main thread. Whenever the processor stalls on the main thread, it automatically begins fetching instructions from the nanothread. Only one register set is available, so the two threads must share the register set. Typically the nanothread will focus on a simple task, such as prefetching data into a buffer, that can be done asynchronously to the main thread.

In the DanSoft processor nanothreading is used to implement a new branch strategy that fetches both sides of a branch. A static branch prediction scheme is used, where branch instructions include 3 bits to direct the instruction stream. The bits specify eight levels of branch direction. For the middle four cases, denoting low confidence on the branch prediction, the processor fetches from both the branch target and the fall-through path. If the branch is mispredicted in the main thread, the back-up path executed in the nanothread generates a misprediction penalty of only 1 to 2 cycles.

The DanSoft processor proposal is a dual-processor CMP, each processor featuring a VLIW instruction set and the nanothreading technique. Each processor is an integer processor, but the two processor cores share a floating-point unit as well as the system interface.

However, the nanothread technique might also be used to fill the instruction issue slots of a wide superscalar approach as in simultaneous multithreading.

The *microthreading* technique (*Bolychevsky et al.* [32], 1996) is similar to nanothreading. All threads share the same register set and the same runtime stack. However, the number of threads is not restricted to two. When a context switch arises, the program counter is stored in a continuation queue. The PC represents the minimum possible context information for a given thread. Microthreading is proposed for a modified RISC processor.

Both techniques – nanothreading as well as microthreading – are proposed in the context of a block-interleaving technique, but might also be used to fill the instruction issue slots of a wide superscalar approach as in simultaneous multithreading (see Sect. 6.4). The drawback to nanothreading and microthreading is that the compiler has to schedule registers for all threads that may be active simultaneously, because all threads execute in the same register set.

6.4 Simultaneous Multithreading

Cycle-by-cycle interleaving and block interleaving are multithreading techniques which are most efficient when applied to scalar RISC or VLIW processors. Combining multithreading with the superscalar technique naturally leads to a technique where all hardware contexts are active simultaneously, competing each cycle for all available resources. This technique, called *simultaneous multithreading*[4] (SMT), inherits from superscalars the ability to issue multiple instructions each cycle; and like multithreaded processors it contains hardware resources for multiple contexts. The result is a processor that can issue multiple instructions from multiple threads *each cycle*. Therefore, not only can unused cycles in the case of latencies be filled by instructions of alternative threads, but so can unused issue slots within one cycle.

Thread-level parallelism can come from either multithreaded, parallel programs or from individual, independent programs in a multiprogramming workload, while ILP is utilized from the individual threads. Because a SMT processor simultaneously exploits coarse- and fine-grain parallelism, it uses its resources more efficiently and thus achieves better throughput and speedup than single-threaded superscalar processors for multithreaded (or multiprogramming) workloads. The trade-off is a slightly more complex hardware organization.

The SMT approach combines a wide superscalar instruction issue with the multithreading approach by providing several register sets on the multiprocessor and issuing instructions from several instruction queues simultaneously. Therefore, the issue slots of a wide-issue processor can be filled by operations of several threads. Latencies occurring in the execution of single threads are bridged by issuing operations of the remaining threads loaded on the processor. In principle, the full issue bandwidth can be utilized. The SMT fetch unit can take advantage of the interthread competition for instruction bandwidth in two ways. First, it can partition this bandwidth among the threads and fetch from *several threads* each cycle. In this way, it increases the probability of fetching only nonspeculative instructions. Second, the fetch unit can be selective about *which threads* it fetches. For example, it may fetch those that will provide the most immediate performance benefit (see the ICOUNT feedback technique in SMT at the University of Washington below).

SMT processors can be organized in two ways:

- They may share an aggressive pipeline among multiple threads when there is insufficient ILP in any one thread to use the pipeline fully. Instructions of

[4] Simultaneous multithreading is also called multithreaded superscalar approach.

different threads share all resources like the fetch buffer, the physical registers for renaming registers of different register sets, the instruction window, and the reorder buffer. Thus SMT adds minimal hardware complexity to conventional superscalars; hardware designers can focus on building a fast single-threaded superscalar and add multithread capability on top. The complexity added to superscalars by multithreading include the thread tag for each internal instruction representation, multiple register sets, and the abilities of the fetch and the retire units to fetch/retire instructions of different threads.

- The second organizational form replicates all internal buffers of a superscalar such that each buffer is bound to a specific thread. Instruction fetch, decode, rename, and retire units may be multiplexed between the threads or be duplicated themselves. The issue unit is able to issue instructions of different instruction windows simultaneously to the FUs. This form of organization adds more changes to the organization of superscalar processors but leads to a natural partitioning of the instruction window and simplifies the issue and retire stages.

The main drawback to simultaneous multithreading may be that it complicates the issue stage, which is always central to the multiple threads. A functional partitioning as demanded for processors of the 10^9-transistor era cannot therefore be easily reached.

No simultaneous multithreaded processors exist to date. Projects simulating different configurations of simultaneous multithreading are discussed below.

6.4.1 SMT at the University of Washington

The SMT processor architecture, proposed in 1995 by *Tullsen et al.* [303] at the University of Washington (Seattle, WA), surveys enhancements of the Alpha 21164 processor. Simulations were conducted to evaluate processor configurations of an up to 8-threaded and 8-issue superscalar. This maximum configuration showed a throughput of 6.64 IPC due to multithreading using the SPEC92 benchmark suite and assuming a processor with 32 functional units (among them multiple load/store units).

The next approach was based on a hypothetical out-of-order issue superscalar microprocessor that resembles the MIPS R10000 and HP PA-8000 (*Tullsen et al.* [304], 1996; *Eggers et al.* [76], 1997). This approach evaluated more realistic processor configurations, and presented implementation issues and solutions to register file access and instruction scheduling for a minimal change to superscalar processor organization.

In the simulations of the latter architectural model (see Fig. 6.18) eight threads and an 8-issue superscalar organization are assumed. Eight instructions are decoded, renamed, and fed to either the integer or floating-point instruction window. Unified buffers are used in contrast to thread-specific queues in the Karlsruhe Multithreaded Superscalar approach (see below). When operands become available, up to eight instructions are issued out of order per cycle, executed, and retired. Each thread can address 32 architectural integer (and floating-point) registers. These registers are renamed to a large physical register file of 356 physical registers. The larger SMT register file requires a longer access time. To avoid increasing the processor cycle time, the SMT pipeline is extended by two stages to allow two-cycle register reads and two-cycle writes. Renamed instructions are placed into one of two instruction windows. The 32-entry integer instruction window handles integer and all load/store instructions, while the 32-entry floating-point instruction window handles floating-point instructions. Three floating-point and six integer units are assumed. All FUs are fully pipelined, and four of the integer units also execute load/store instructions. The I-cache and D-cache are multiported and multibanked, but common to all threads.

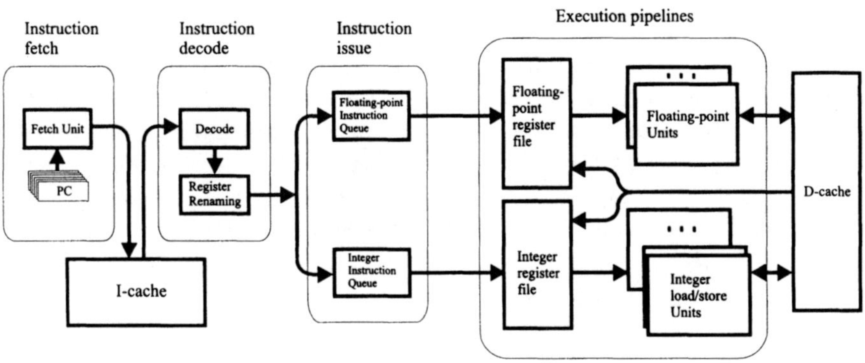

Fig. 6.18. SMT processor architecture

The multithreaded workload consists of a program mix of SPEC92 benchmark programs that are executed simultaneously as different threads. The simulations evaluated different fetch and instruction issue strategies.

An RR.2.8 fetching scheme to access multiported I-cache, i.e., in each cycle, two times 8 instructions are fetched in round-robin policy from two different threads, was superior to different other schemes like RR.1.8, RR.4.2, and RR.2.4 with less fetching capacity. As a fetch policy, the ICOUNT feedback technique, which gives highest fetch priority to the threads with the fewest instructions in the decode, renaming, and queue pipeline stages, proved superior to the BRCOUNT scheme which gives highest priority to those threads that are least likely to be on a wrong path, and the MISSCOUNT scheme which gives priority to the threads that have the fewest outstanding D-cache misses. The IQPOSN policy that gives lowest priority to the oldest instructions by penalizing those threads with instructions closest to the head of either the integer or the floating-point queue is nearly as good as ICOUNT and better than BRCOUNT and MISSCOUNT, which are all better than round-robin fetching. The ICOUNT.2.8 fetching strategy reached an IPC of about 5.4 (the RR.2.8 only reached about 4.2). Most interesting is that neither mispredicted branches nor blocking due to cache misses, but a mix of both and perhaps some other effects, proved to be the best fetching strategy.

In a single-threaded processor, choosing instructions for issue that are least likely to be on a wrong path is always achieved by selecting the oldest instructions, those deepest into the instruction window. For the SMT processor several different issue strategies have been evaluated, like oldest instructions first, speculative instructions last, and branches first. Issue bandwidth is not a bottleneck and all strategies seem to perform equally well, so the simplest mechanism still works. Also doubling the size of instruction windows (but not the number of searchable instructions for issue) has no significant effect on the IPC. Even an infinite number of FUs increases throughput by only 0.5 %.

Recently, simultaneous multithreading has been evaluated with database workloads (*Lo et al.* [186], 1998) and multimedia workloads (*Oehring et al.* [218], 1999), both achieving roughly a three-fold increase in instruction throughput with an eight-threaded SMT over a single-threaded superscalar with similar resources. *Wallace et al.* [318] presented the *threaded multipath execution* (TME) model, which exploits existing hardware on a SMT processor to execute simultaneously alternate paths of a conditional branch in a thread. Such a speculative execution increases single program performance by 14–23 %, depending on the misprediction penalty, for programs with a high branch misprediction rate.

6.4.2 Karlsruhe Multithreaded Superscalar

While the SMT processor by *Tullsen et al.* [303] in 1995 surveys enhancements of the Alpha 21164 processor, the *multithreaded superscalar* processor approach from the University of Karlsruhe (*Sigmund and Ungerer* [259, 260], 1996) is based on a simplified PowerPC 604 processor. Both approaches, however, are similar in their instruction-issuing policy.

The multithreaded superscalar processor uses various kinds of modern microarchitecture techniques such as separate I-cache and D-cache, BTAC, static branch prediction, in-order issue, independent FUs with reservation stations, rename registers, out-of-order execution, and in-order completion. The processor implements the 6-stage instruction pipeline of the PowerPC 604 processor (fetch, decode, issue, execute, complete, and write-back) (see Sect. 4.9.10) and extends it to employ multithreading. However, the instruction set is simplified, using an extended DLX instruction set (see *Hennessy and Patterson* [134]) instead. The processor uses static instead of dynamic branch prediction, and renounces the floating-point unit.

The processor is designed to be scalable: the number of parallel threads, the sizes of internal buffers, register sets and caches, and the number and type

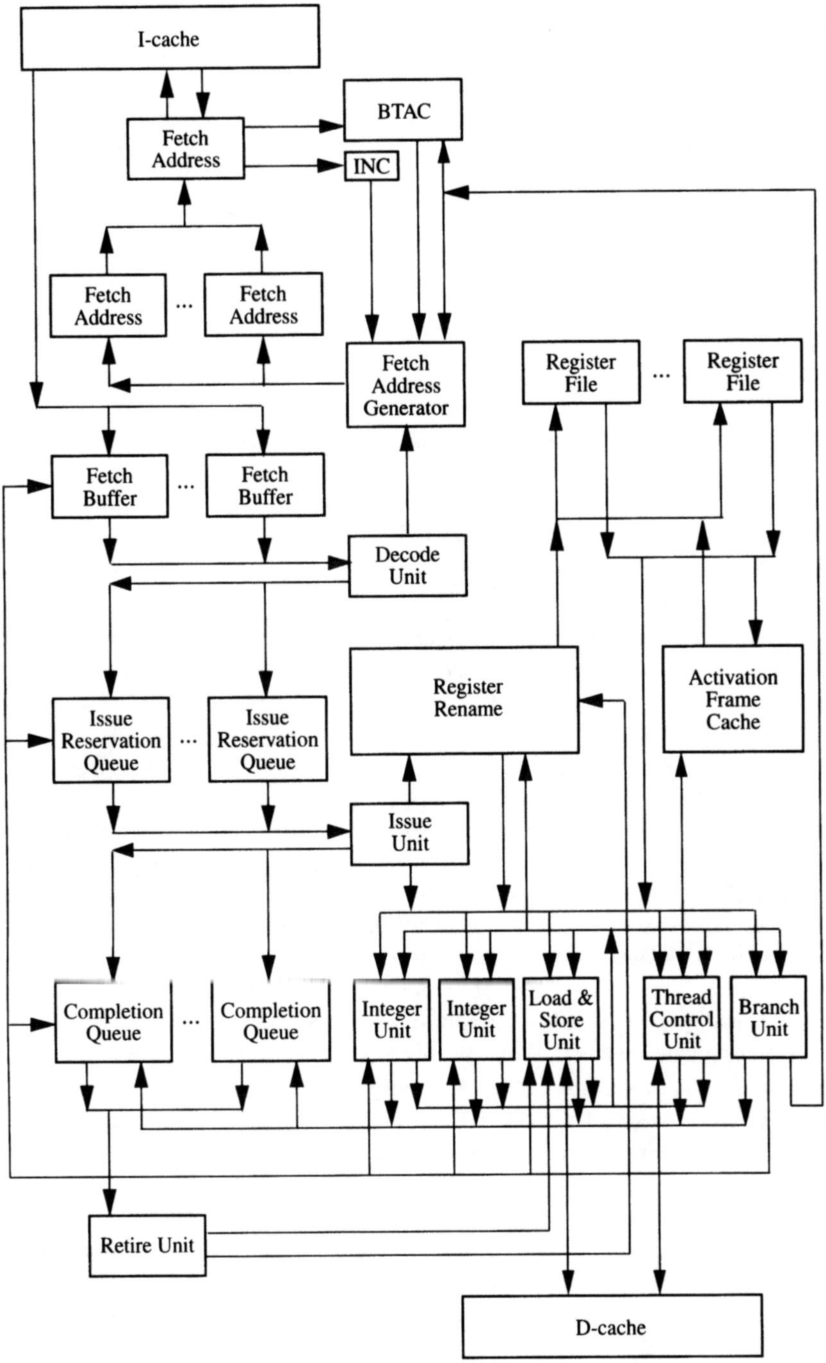

Fig. 6.19. The Karlsruhe Multithreaded Superscalar Processor

of FUs are not limited by any architectural specification. This allows experiments to find an optimized configuration for an actual processor depending on chip size and expected application workload.

The processor can be divided in four parts:

- the control pipeline consisting of the *fetch unit* (FEU), *decode unit* (DU), *issue unit* (IU), and *retire unit* (RU);
- several independent FUs (e.g., simple/complex integer, load/store, branch and thread control);
- caches, cache control, and memory access unit; and
- register files and activation frame caches.

Figure 6.19 shows a block diagram of an incarnation of the multithreaded superscalar processor, here with a single control pipeline equipped with an arbitrary number n of queue buffers that are able to host n threads, with two integer, one load/store, one thread control and one branch unit, and with n register files corresponding to the n queue buffers in the control pipeline.

Between FEU, DU, IU, and RU exist separate queue buffers for each thread. There may be several FEUs and DUs, so that several executions in these stages may be performed in parallel. The units can be allocated to the actual threads on demand. The I-cache is designed as non-blocking cache. The non-blocking-cache design decouples the FEU(s) from the actual instruction fetch from memory. When a fetch to the I-cache fails, fetch from memory is initiated and the FEU is free again to fetch instructions for another thread. The FEUs and DUs always work on a continuous block of instructions. The blocks actually fetched may not always be of maximum size, because of limitations imposed by the cache line size (instruction fetch can not overlap cache lines) and by branches that are predicted to be taken. The maximum block size of an instruction fetch is equal with the size of the fetch buffers and of the issue reservation queues, and is chosen corresponding to the instruction issue bandwidth. There are only a single IU and a single RU. The IU cannot be duplicated. The RU may simultaneously retire any number of instructions of any thread (up to a total maximum of retired instructions). The IU is not restricted with respect to the number of instruction issues according to each thread. The limitation is the maximum issue bandwidth – the maximum number of instructions that can be issued simultaneously. Instructions of a single thread are always issued in program order from the instruction window to the reservation stations of the FUs.

There is no fixed allocation between threads and FUs, otherwise the configuration would define a multiprocessor chip instead of a multithreaded superscalar processor. The instructions are executed out of order by the FUs. The use of separate reservation stations for the FUs simplifies the task of the IU, because only the lack of instructions, the lack of resources, and the maximum bandwidth can limit the simultaneous instruction issue rate. Data dependences are handled by rename registers and reservation stations. Instructions can be issued to the reservation stations without checking for control or data dependences. But an instruction will not be executed until all source operands are available.

The rename registers, the BTAC, and the I-cache and D-cache are shared by all active threads. There is no fixed allocation of any of these resources to specific threads. This allows for maximum performance with any number of active threads. Each thread executes in a separate register set. The contents of the registers of a register set describe the state of a thread and form a so-called activation frame. An activation frame is called active if the thread is currently being executed by the processor, i.e., the activation frame is represented in a register set. In addition to the active activation frames in the processor, an activation frame cache is provided, which holds activation frames that are not currently scheduled for execution. The activation frames in the activation frame cache and in the register sets can be interchanged without a significant penalty. The activation frame cache is also used to implement a derivation of a register window technique. A new activation frame is created for every subprogram activation, thereby significantly reducing the number of memory accesses to the D-cache.

Except for the load/store and the thread control unit, each kind of FU may be arbitrarily duplicated. The thread control unit is responsible for creation and deletion of threads, for synchronization and communication between threads. It would be quite difficult to duplicate this unit, and the gain would be small, as the percentage of thread control instructions is low. The reason for the single load/store unit is discussed later. All standard integer operations are executed in a single cycle by the simple integer units. Only the multiply and the divide instructions are executed in the complex integer unit. The execution of multiply instructions is fully pipelined, in contrast to the integer divide. Each branch unit may execute one branch instruction per cycle. Branch prediction starts in the FEU using a BTAC. It proceeds in the DU using a simple static branch prediction technique. Each forward branch is predicted as not taken, each backward branch as taken. A static branch prediction simplifies processor design, but reduces the accuracy and, consequently, increases the statistical penalty incurred by each conditional branch. Within a multithreaded processor latencies are almost all covered by other threads, so that the penalty created by static branch prediction should not affect average executions per cycle.

The memory interface is defined as a standard DRAM interface with configurable burst sizes and delays. All caches are highly configurable to test penalties due to cache thrashing caused by multiple threads.

Starting with the superscalar base processor a software simulation was conducted, evaluating various configurations of the multithreaded superscalar processor. All functional units were simulated with correct cycle behavior. The simulation workload was generated by a configurable workload generator, which created random high-level programs, and compiled them to the target machine language. The distribution of the machine instructions was similar to that generated from high-level programs, since typical program structures were observed by the workload generator. This approach allowed the creation of workloads for different types of programs without the need for a complete compiler. Optimized instruction scheduling by a compiler was not taken into account. For the performance results presented here, a multithreaded simulation workload was chosen, which represents general-purpose programs without floating-point instructions for a register window architecture. The instruction mix was derived from the SPECint92 benchmark suite, enhanced by thread control instructions. A lower percentage of load/store instructions was assumed because arguments of subprogram invocations were passed between activation frames located in register sets. Table 6.4 lists the percentage of instructions in the instruction mix applied for the simulation results presented below.

Table 6.4. Percentage of instructions for the multithreaded superscalar processor simulations

Instruction type	Average use (%)
integer	63.8
complex integer	1.1
load	13.2
store	7.0
branch	10.8
thread control	4.1

With a single load/store unit (used in this approach) the chosen workload has a theoretical maximum throughput of about 5 IPC (the frequency of load and store instructions sums to 20.2 %).

For the simulation results presented below the following was used:

- separate 8 kB 4-way set-associative I-cache and D-cache with 32-byte cache lines,
- a cache fill burst rate of 4-2-2-2,
- a fully associative 32-entry BTAC,
- 32 general-purpose registers per thread,
- 64 rename registers,
- a 12-entry completion queue,
- 4 simple integer units, single complex integer, load/store, branch, and thread control units, each FU with a 4-entry reservation station.

The number of simultaneously fetched instructions and the sizes of fetch and issue buffers were adjusted to the issue bandwidth. The issue bandwidth and the number of hosted threads were varied in each case from 1 up to 8, according to the total number of FUs.

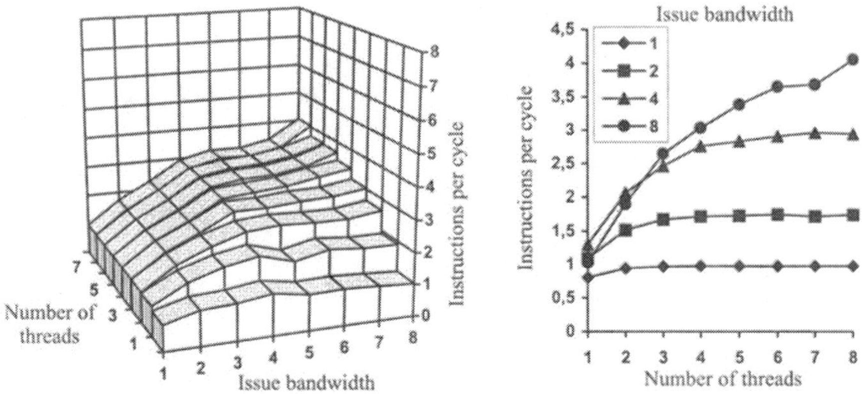

Fig. 6.20. Average instruction throughput per processor with one fetch and one decode unit

The simulation results in Fig. 6.20 show that the single-threaded 8-issue superscalar processor throughput (measured in average IPC) only reached a performance of 1.14 executed IPC. The 4-issue approach was slightly better with 1.28 IPC, because the need to fetch five or more instructions that were not necessarily aligned on a cache line boundary produces some penalty. Increasing the number of threads from which instructions were simultaneously issued to the eight issue slots also increased performance. The throughput reached a plateau at about 3.2 IPC, where neither increasing the number of threads nor the issue bandwidth significantly raised the number of executed instructions. When issue bandwidth was kept small (1 to 4 IPC), and four to eight threads were considered, a full exploitation of the issue bandwidth was expected. As seen in Fig. 6.20, however, the issue bandwidth was only utilized by about 75 %. Even a highly multithreaded processor seems unable to fully exploit the issue bandwidth. Further analysis revealed the single FEU and DU as bottlenecks, leading to starvation of the IU.

For further simulations two independent FEUs (identical to the RR.2.8 fetch policy of *Tullsen et al.* [304]) and two DUs were applied. The simulations results are shown in Fig. 6.21. The gradients of the graphs representing the multithreaded approaches are much steeper, indicating an increased throughput. The processor throughput in an 8-threaded 8-issue processor is about four times higher than in the single threaded 8-issue case. However, the range of linear gain, where the number of executed instructions equals the issue bandwidth, ends at an issue bandwidth of about 4 IPC. The throughput reaches a plateau at about 4.2 IPC, where neither increasing the number of threads nor the issue bandwidth significantly raises the number of executed instructions. Moreover, the diagram on the right of Fig. 6.21 shows a marginal gain in instruction throughput, when we advance from a 4-issue to an 8-issue bandwidth. This marginal gain is nearly independent of the number of threads in the multithreaded processor. Using the considered instruction

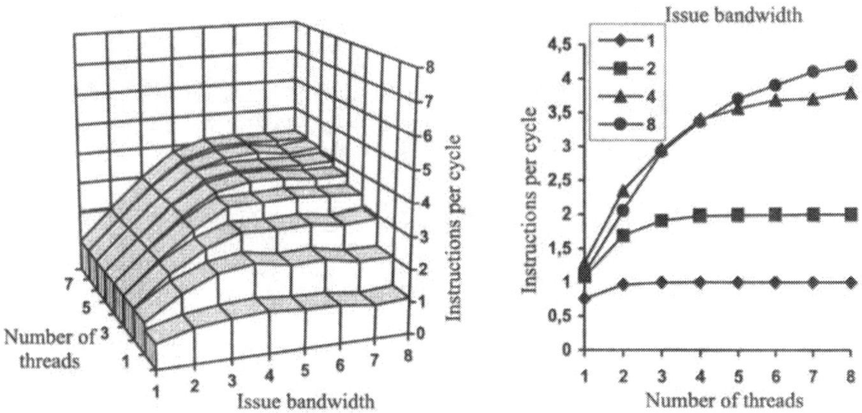

Fig. 6.21. Average instruction throughput per processor with two fetch and two decode units

mix with 20.2 % load and store instructions potentially allows a processor throughput of 5 instead of the measured 4.2 IPC with a single load/store unit. In order to locate and remove further bottlenecks, the reasons why the expected 5 IPC were not reached will be surveyed.

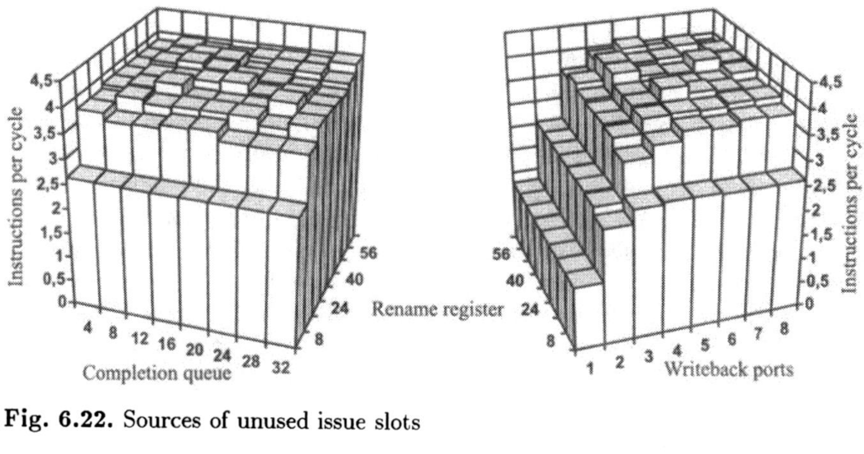

Fig. 6.22. Sources of unused issue slots

Further simulations (see Fig. 6.22) also showed that a performance increase was not found when the number of slots in the completion queues was increased over the 16 slots assumed in the simulations above. Instruction execution was limited by true data dependences and by control dependences that cannot be removed. Also, 4 write-back ports and 16 rename registers seem appropriate.

Further simulations varied the number of integer FUs from one to four. If only a single integer unit is used, this unit proves a bottleneck. Increasing the number of integer units up to four, the integer unit is joined as a bottleneck by the thread unit, the load/store unit, and the completion unit. The thread unit performs one thread control instruction per cycle. Duplicating the thread unit or increasing the performance of the thread unit is difficult to organize. The load/store unit and the memory subsystem remained as the main bottlenecks that may possibly be removed by a different configuration. The load/store unit was limited to the execution of a single IPC, assuming an ideal memory subsystem. Duplication of the load/store unit definitely increases performance. However, two or more load/store units accessing a single D-cache is difficult to implement because of consistency and thrashing problems.

To find out whether a different memory configuration might increase the throughput, different cache sizes, cache line sizes (8–128 bytes), cache schemes (direct-mapped, set-associative), workloads, and numbers of active threads were simulated. The simulations showed that all latencies caused by cache refills (4-2-2-2) are covered by the multithreaded execution model, and that the Karlsruhe Multithreaded Superscalar Processor was able to completely hide these latencies. It also reached the maximum throughput that was possible with a single load/store unit.

It is obvious that different processor configurations can only be compared if a measurement for their cost (e.g, in chip space) is used. Otherwise, unrealistic processors are compared with each other, simply stating that more units result in greater performance.

Table 6.5. Hardware cost formula

Unit type	Estimated cost
Integer unit	2
Load/Store unit	3
Fetch and decode unit	3
Branch unit	3
Caches	6
Registers	2 X Number of threads
Issue unit	2 X Issue bandwidth
Retire unit	1 X Number of threads X Issue bandwidth

To get a rough measurement of the cost for a processor configuration, a formula based on the Power PC 604 floor plan was proposed. The formula expresses hardware cost based on chip space usage per unit. Estimated cost per unit are given in Table 6.5.

The formula is only a rule of thumb. The formula for the IU is based on the required interconnections between the IU, the register sets, and the FUs. As each thread's register set and issue queue has to be connected with all

FUs, the required chip space is proportional to the product of the number of hosted threads and the issue bandwidth of the processor.

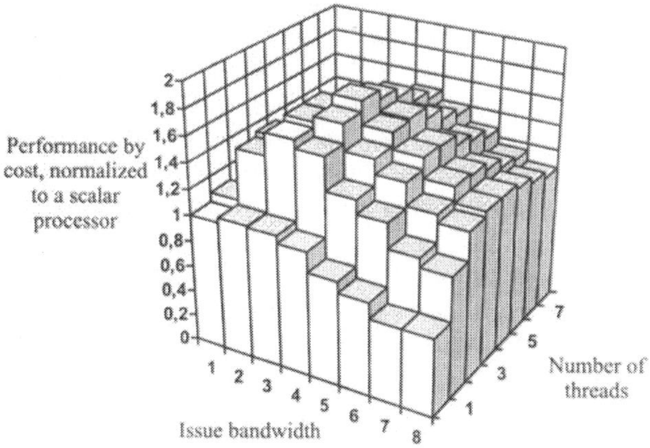

Fig. 6.23. Average instruction throughput in relation to chip cost

Figure 6.23 displays the IPC in relation to the hardware cost of a specific processor configuration. The solution with four threads and issue bandwidth four shows the best performance/cost relation. However, this observation is application- and design-specific. The advantage of multithreading is highly dependent on the ratio of load/store instructions to other instructions in the workload. Also the chip cost change with different architectural decisions.

Using an instruction mix with 20 % load and store instructions, the performance results show for an 8-issue processor with four to eight threads that two instruction FEUs, two DUs, four integer units, 16 rename registers, four register ports, and completion queues with 12 slots are sufficient. The single load/store unit proves the principal bottleneck because it cannot be easily duplicated. The multithreaded superscalar processor (8-threaded 8-issue) is able to hide completely latencies caused by 4-2-2-2 burst cache refills. It reaches the maximum throughput of 4.2 IPC that is possible with a single load/store unit.

Subsequent research explored microarchitecture models for a simultaneous multithreaded processor with multimedia enhancements. Simulations with a multithreaded MPEG-2 video decompression algorithm as workload showed that an 8-threaded 8-issue processor may yield up to a threefold performance increase over the single-threaded 8-issue model (see *Oehring et al.* [218], 1999).

6.4.3 Other Simultaneous Multithreading Processors

Media Research Laboratory Processor (*Hirata et al.* [138], 1992): The multithreaded processor of the Media Research Laboratory of Matsushita

Electric Ind. (Japan) was the first approach to simultaneous multithreading. Instructions of different threads are issued simultaneously to multiple functional units. Simulation results on a parallel ray-tracing application showed that using 8 threads a speedup of 3.22 in the case of one load/store unit, and of 5.79 in the case of two load/store units, can be achieved over a conventional RISC processor. However, caches or TLBs are not simulated, nor is a branch prediction mechanism.

Irvine Multithreaded Superscalar (*Gulati and Bagherzadeh* [111] and *Loikkanen and Bagherzadeh* [187], 1996): This multithreaded superscalar processor approach, developed at the University of California at Irvine, combines out-of-order execution within an instruction stream with the simultaneous execution of instructions of different instruction streams. A particular superscalar processor called the Superscalar Digital Signal Processor (SDSP) is enhanced to run multiple threads. The enhancements are directed by the aim of minimal modification to the superscalar base processor. Therefore, most resources on the chip are shared by the threads, as for instance the register file, reorder buffer, instruction window, store buffer, and renaming hardware. Based on simulations a performance gain of 20–55 % due to multithreading was achieved across a range of benchmarks.

SMV Processor (*Espasa and Valero* [79], 1997): The Simultaneous Multithreaded Vector (SMV) architecture, designed at the Polytechnic University of Catalunya (Barcelona, Spain) combines simultaneous multithreaded execution and out-of-order execution with an integrated vector unit and vector instructions. Figure 6.24 depicts SMV architecture. The fetch engine selects one of eight threads and fetches four instructions on its behalf. The decoder renames the instructions, using a per-thread rename table, and then sends all instructions to several common execution queues. Inside the queues, the instructions of different threads are indistinguishable, and no thread information is kept except in the reorder buffer and memory queue. Register names preserve all dependences. Independent threads use independent rename tables, which prevents false dependences and conflicts from occurring. The vector unit has 128 vector registers,[5] each holding 128 64-bit registers, and has four general-purpose independent FUs.

6.5 Simultaneous Multithreading versus Chip Multiprocessor

In this section, we compare the two main architectural principles capable of exploiting *multiple threads* of instructions, i.e., thread-level (or coarse-

[5] The number of registers is the product of the number of threads and the number of physical registers required to sustain good performance on each thread.

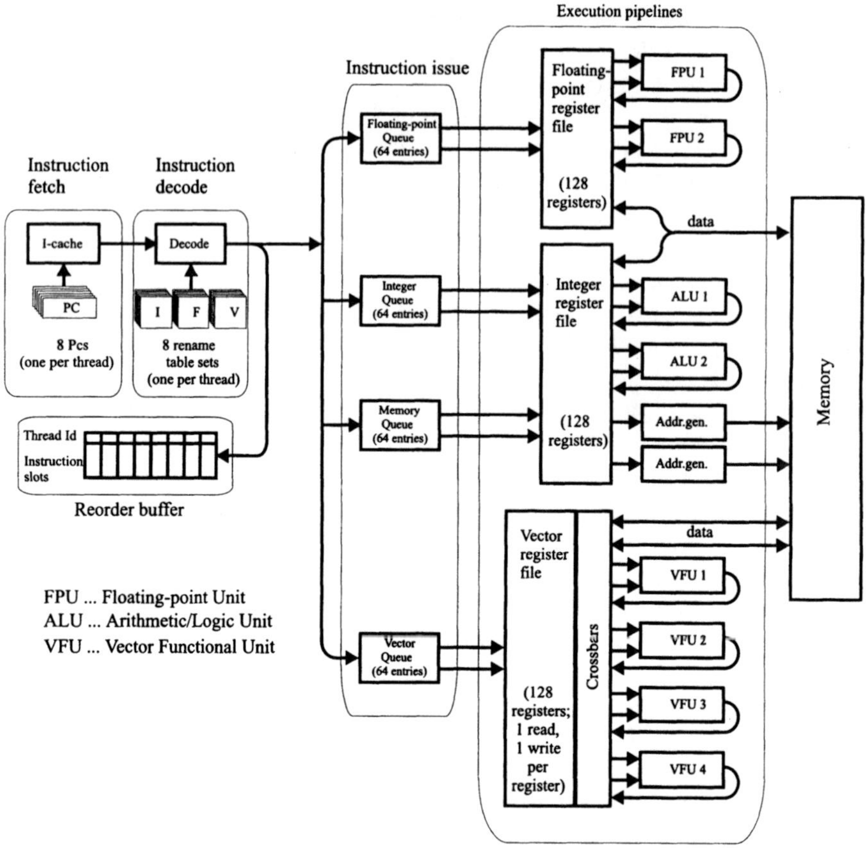

Fig. 6.24. Simultaneous Multithreaded Vector (SMV) architecture

grain) parallelism, namely the chip multiprocessor CMP and the simultaneous multithreading SMT approaches.

Simulations of *Sigmund and Ungerer* [259, 260] compare the simultaneous multithreading approach of Sect. 6.4.2 with a CMP approach. *Sigmund and Ungerer* simulated various configurations of the SMT model in combination with the multiprocessor chip approach and compared them to *Tullsen et al.*'s simulations of 1995 [303] (see Table 6.6). The simulations produced slightly different results to *Tullsen et al.*'s simulations, especially viewing the 8-threaded 8-issue superscalar approach $1\times(8,8)$ in Table 6.6 ($p\times(t,i)$ means p processors per chip, each processor equipped with t threads and i issue slots).

The reason for the difference in results follows from the high number of FUs in *Tullsen et al.*'s approach; for example, up to eight load/store units are used in *Tullsen et al.*'s simulation ignoring hardware cost and design problems, whereas the performance of *Sigmund and Ungerer*'s SMT model is restricted by the assumption of a single load/store unit. In *Tullsen et al.*'s

Table 6.6. IPC results of *Tullsen et al.* and of *Sigmund and Ungerer* simulations

Number X (Threads,Issue)	*Tullsen et al.*	*Sigmund and Ungerer*
1 X (8,8)	6.64	4.19
8 X (1,1)	5.13	6.07
2 X (4,4)	6.80	6.80
1 X (4,8)	4.15	3.37
4 X (1,2)	3.44	4.32
2 X (1,4)	1.94	2.56

simulations the SMT approach performs better than the CMP approach, whereas in *Sigmund and Ungerer*'s simulations the CMP reaches a higher throughput than the SMT approach, when using the same issue bandwidth and number of threads (comparing the SMT of 1×(8,8) with the CMP of 8×(1,1)). However, if chip costs are taken into consideration, a 4-threaded 4-issue superscalar processor shows the best performance/cost relation (see Sect. 6.4.2).

Further simulations of *Eggers et al.* [76] in 1997 compared SMT, wide-issue superscalar, cycle-by-cycle interleaving multithreaded superscalar, and two-CPU and four-CPU CMP. Comparison of the simulated processor architecture configurations is given in Table 6.7. The simulation results which are

Table 6.7. Processor architectures simulated by *Eggers et al.*

Features	superscalar	cycle-by-cycle superscalar	CMP2	CMP4	SMT
number of CPUs	1	1	2	4	1
CPU issue bandwidth	8	8	4	2	8
number of threads	1	8	1 per CPU	1 per CPU	8
number of architectural registers	32	32 per thread	32 per CPU	32 per CPU	32 per thread

given in Table 6.8 were obtained on a workload which consisted of a group of coarse-grained (parallel threads) and medium-grained (parallel loop iterations) parallel programs. The average instruction throughput of an 8-issue superscalar was an IPC of 3.3, which is already high compared to other measured superscalar IPCs (see Sect. 4.11), but rather low compared to the eight instructions possibly issued per cycle. The superscalar's inability to exploit more ILP or any thread-level parallelism contributed to its lower performance. By exploiting thread-level parallelism, a cycle-by-cycle interleaving multithreaded superscalar technique provided an average instruction throughput of 4.2 IPC. This IPC occurred with only four threads while performance fell with additional threads. One of the reasons is that a cycle-by-cycle interleav-

Table 6.8. Instructions per cycle (IPC) when executing a parallel workload

Threads	superscalar	CMP2	CMP4	cycle-by-cycle superscalar	SMT
1	3.3	2.4	1.5	3.3	3.3
2		4.3	2.6	4.1	4.7
4			4.2	4.2	5.6
8				3.5	6.1

ing multithreaded superscalar can issue instructions from only one thread each cycle and therefore cannot hide conflicts from interthread competition for shared resources. SMT obtained better speedups than CMP2 and CMP4, the latter being chip multiprocessors with respectively, two four-issue, and four two-issue CPUs. Speedups on the CMPs were hindered by the fixed partitioning of their hardware resources across the CPUs. Bridging of latencies is only possible in the multithreaded processor approaches, and not in CMP. CPUs in CMPs were idle when thread-level parallelism was insufficient. Exploiting large amounts of ILP in the unrolled loops of individual threads was not possible due to the CPU's smaller issue bandwidth in CMP. On the other hand, an SMT processor dynamically partitions its resources among threads, and therefore can respond well to variations in both types of parallelism, exploiting them interchangeably.

In contrast to *Eggers et al.* who compared architectures having constant total issue bandwidth (i.e., number of CPUs × CPU issue bandwidth), *Hammond et al.* [120] established a standard chip area and integration density, and determined the parameters for three architectures: superscalar, CMP, and SMT (Table 6.9). They argue that design complexity for a 16-issue CMP is similar to that of a 12-issue superscalar or a 12-issue SMT processor. In this

Table 6.9. Processor architectures simulated by *Hammond et al.*

Features	superscalar	CMP	SMT
number of CPUs	1	8	1
CPU issue bandwidth	12	2 per CPU	12
number of threads	1	1 per CPU	8
number of architectural registers	32	32 per CPU	32 per thread

case, CMP with eight 2-issue CPUs outperforms a 12-issue superscalar and a 12-issue, 8-threaded SMT processor on four SPEC95 benchmark programs. Figure 6.25 shows the performance of the superscalar, SMT, and CMP on the four benchmarks relative to a single 2-issue superscalar. The CMP achieved

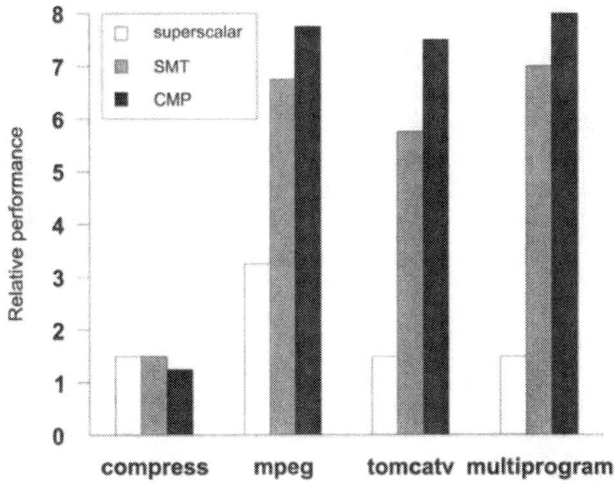

Fig. 6.25. Relative performance of superscalar, SMT, and CMP

higher performance than the SMT due to a total of 16 issue slots instead of 12 issue slots for the SMT.

6.6 Conclusions

The performance race between SMT and CMP has yet to be decided. Certainly, CMP will be easier to implement, but only SMT has the ability to hide latencies. A functional partitioning as required by the on-chip wire-delay of future microprocessors is not easily reached within a simultaneous multithreaded processor due to the centralized instruction issue. A separation of the thread queues as in the Karlsruhe Multithreaded Superscalar Processor is a possible solution, although it does not remove the central instruction issue.

A combination of simultaneous multithreading with the CMP is proposed by *Sigmund and Ungerer* [259, 260], and by *Krishnan and Torellas* [168]. Perhaps the simulations of *Sigmund and Ungerer* show the path towards a CMP consisting of moderately equipped (e.g., 4-threaded 4-issue superscalar) SMTs.

Usually, a CMP will feature separate L1 I-cache and D-cache per on-chip CPU and an optional unified L2 cache. If the CPUs always execute threads of the same process, the L2 cache organization will be simplified, because different processes do not have to be distinguished.

Moreover, the multiprocessor which is formed by the CMPs will be a symmetric multiprocessor (SMP) or even a distributed shared-memory multiprocessor (DSM).

Similarly, if all (hardware-supported) threads of a SMT processor always execute threads of the same process, preferably in SPMD fashion, a unified (primary) I-cache may prove useful, since the code can be shared between the threads. Primary D-cache may be unified or separated between the threads depending on the access mechanism used.

If CMP or SMT are the design choice of the future, the impact on multiprocessor development will favor shared-memory multiprocessors (either SMPs or DSMs) over message-passing machines. Since multithreading and message passing do not mix well even in state-of-the-art multiprocessor programs, there is an indication that message-passing programs using PVM or MPI will soon be outdated and produce a legacy problem.

7. Processor-in-Memory, Reconfigurable, and Asynchronous Processors

Looking further into the future, we envision a point at which off-chip communication is so expensive that all the system memory resides on the processor chip (or module). If a system designer wishes to provide more memory than is available on-chip, another of these homogeneous processor/memory modules is added ...

Doug Burger, James R. Goodman, and Alain Kägi
Memory Bandwidth Limitations of Future Microprocessors
(The 23rd Annual Int'l Symposium on Computer Architecture, May 1996)

Computing devices 10 years from now will include a strong mix of software-programmable hardware and hardware-configurable logic. ...

John Villasenor and William H. Mangione-Smith
Configurable Computing
(Scientific American, June 1997)

7.1 Processor-in-Memory

7.1.1 The Processor-in-Memory Principle

Processor-in-memory (PIM) or *intelligent RAM* (IRAM) chips integrate one or more processors with large, high-bandwidth, on-chip DRAM banks, which provide the processor(s) with sufficient bandwidth at a reasonable cost.

Technological trends have produced a large and growing gap between processor speed and DRAM access latency. Today, with processor performance increasing at a rate of about 60 % per year and memory latency improving by just 7 % per year, it takes dozens of cycles for data to travel between the CPU and main memory (*Kozyrakis et al.* [167]). The CPU-centric design philosophy has led to very complex superscalar processors with deep pipelines. Much of this complexity is devoted to hiding memory access latency (*Saulsbury et al.* [256]).

Extrapolating current trends suggests that soon a processor may be able to issue hundreds or even thousands of instructions while loading a single value into on-chip memory. Much research has focused on reducing or tolerating these large memory access latencies. Researchers have proposed many techniques for reducing the frequency and impact of cache misses, such as lockup-free caches, hardware and software prefetching, stream buffers, speculative loads and execution, and multithreading (*Burger et al.* [40]). The phenomenon that access times are increasingly limiting system performance is known as the *memory wall* (*Wulf and McKee* [326], *Wilkes* [324]).

Experiences with Sun's SPARCStation 5 workstation and its comparison with contemporary high-end workstations provide evidence for the possible benefits of tighter memory–processor integration (*Saulsbury et al.* [256], 1996). The SPARCStation 5 contains a single-scalar microSPARC processor (see Sect. 1.7.2) with 16 kB I-cache and 16 kB D-cache on-chip and no secondary cache. The memory controller is integrated onto the chip, so that DRAM devices are driven directly by logic on the processor chip. A separate I/O-bus connects the processor chip with peripheral devices, which can access memory only through the processor chip. A comparable high-end machine of the same era is the SPARCStation 10/61, containing a superscalar SuperSPARC processor with separate 20 kB I-cache and 16 kB D-cache, and a shared secondary cache of 1 MB. Compared to the SPARCStation 10/61, the SPARCStation 5 has an inferior SPEC92-rating, yet it outperforms it on a logic synthesis workload that has a working set of over 50 MB. The reason for this discrepancy is the lower main memory latency of the SPARCStation 5, which compensates for the slower processor. Codes that frequently miss the SPARCStation 10's large secondary cache have lower access times on the SPARCStation 5.

During program execution on a contemporary superscalar processor, cycles are lost due to pipeline hazards, wrong branch speculation, the lack of ILP, and in particular memory access latencies in the case of cache misses. Memory access latency is caused by *memory access latency stall times*, and *memory bandwidth stall times*. Memory access latency stall times can be roughly characterized by the cache access time plus the number of cycles the cache line travels from memory to the on-chip cache. Memory bandwidth stall times are caused by limited memory bandwidth between processor and DRAM memory, often due to insufficient pin bandwidth.

In conventional microprocessors the effective off-chip bandwidth can be increased by providing broader buses and more processor pins, but also by the use of compression for data, addresses, and code. All these schemes increase the effective bandwidth to memory at the expense of some extra hardware on the processor chip. Thus, memory bandwidth stall times are decreased, but the memory access latency stall times are not.

Most PIM approaches combine a processor with on-chip RAM and sometimes also with small on-chip caches. IRAM (*Kozyrakis et al.* [167], *Patterson et al.* [230], *Fromm et al.* [92]) is a PIM approach that implements dynamic RAM (DRAM) memory instead of SRAM-based cache memory on the processor chip. The approach to omit caches totally is based on the fact that DRAM can accommodate 30–50 times more data than the same chip area devoted to caches. Moreover, cache memory is just a redundant copy of information that would not be necessary if main memory could be accessed at processor speed.

PIM systems have several potential advantages:

- The main advantage is that on-chip memory can support high bandwidth and low latency by using a wide interface and eliminating the delay of pads and buses that arises with off-chip memory access. On-chip memory could reduce processor–memory latency by factors of 5–10 and increase memory bandwidth by factors of 50–200 [230].
- The processor–DRAM gap in access speed increases in future. PIM provides higher bandwidth and lower latency for memory accesses, thus simultaneously decreasing memory access latency stall times and memory bandwidth stall times.
- In many cases the entire application will fit in the on-chip storage. Having the entire memory on the chip, coupled to the processor through a high bandwidth and low-latency interface, allows for processor designs that demand fast memory systems [230].
- Due to memory integration, PIM needs less off-chip traffic than conventional microprocessors. While the majority of pins in conventional microprocessors are devoted to wide memory interfaces, a PIM chip can have a much less powerful memory interface requiring fewer pins. Serial interfaces may be directly attached to the PIM chip and provide enough I/O bandwidth without being limited by conventional memory buses.
- Energy consumption is a problem for portable computers. PIM decreases energy consumption in the memory system due to the reduction of off-chip accesses [92].
- Finally, the processing unit with its wide interface to memory can operate as a parallel built-in self-test engine for the memory array, significantly reducing the DRAM testing time and the associated cost. A processor test requires a memory subsystem, and a memory is tested with processor-like accesses. All that may be required in a PIM system is to download a self test program.

One challenge to the PIM approach is scaling a system beyond a single PIM. The amount of DRAM that can fit on a single PIM chip is bounded. It may be sufficient for portable computers, but not for the high-end workstations in the future. A potential solution is to back up a single PIM chip with commodity-external DRAM, using the off-chip memory as secondary storage with pages swapped between on-chip and off-chip memory. Alternatively, multiple PIMs could be interconnected with a high-speed network to form a parallel computer. Fortunately, historical trends indicate that the end-user demand for memory will scale at a lower rate than the available capacity per chip. Thus, over time a single PIM chip will be sufficient for increasingly larger systems, from portable and low-end PCs to workstations and even servers (*Kozyrakis et al.* [167]).

The DRAM technology today does not allow on-chip coupling of high performance processors with DRAM memory since the clock rate of DRAM memory is too low. Logic and DRAM manufacturing processes are fundamentally different. DRAM processes typically support multiple layers of polysilicon, few metal layers, and 3D structures for maximizing capacitor area. Logic processes tend to have more metal levels and are optimized for transistor switching speed. DRAM cells in a logic process would not be particularly dense: estimates vary from 4 to 20 times less dense than in an optimized DRAM process. Gates in a DRAM process are much slower than their logic process counterparts. The benefits of processor memory integration may drive the development of hybrid processes that improve system performance even though they do not match their optimized counterparts for density or speed (*Burger et al.* [41], 1997). The upcoming 0.25 μm DRAM processes, with two or three metal layers, are already capable of supporting a simple 200 MHz CPU core (*Saulsbury et al.* [256]).

The PIM approach can be combined with most processor organizations. The processor(s) itself may be a simple or moderately superscalar standard processor; it may also include a vector unit as in the vector IRAM type which is described below. It may also follow any of the execution modes as described in Sects. 5.2–5.5. To harness coarse-grained parallelism, several closely coupled processors could be integrated with memory banks to form a shared-memory chip multiprocessor. The difference to the CMP approach (see Sect. 6.2) lies in the closer coupling of the processors to the memory system, potentially rendering caches unnecessary.

If PIM technology proves successful, even more dramatic processor and memory integrations may be devised, distributing portions of processors closer to the individual memory banks. A more radical technique than PIM is to build computational ability into the memory system. The processor would be able to issue primitives more powerful than simple loads or stores to the memory system. Such a *smart memory system*, exemplified here by the Active Page approach (see Sect. 7.1.4), performs computations locally and returns the results only. The challenge for such innovative systems is software migration.

On the other hand, all attempts to add more capabilities to DRAMs, such as video-buffers (VDRAM), integrated caches (CDRAM), graphic support (3D-RAM), and smart, higher-performance interfaces (RamBus, SDRAM) are penalized by the extra cost for the non-memory areas [256].

If processing becomes cheap enough and/or communication becomes too expensive, it is likely that designers will put functional units wherever there is storage. These are called *memory-centric architectures* (*Burger et al.* [41]). *Burger et al.* [42] have developed a memory-centric architecture – the DataScalar model (see Sect. 5.6) – which targets scalar, hard-to-parallelize codes. Processors are coupled with regions of the main memory, and each processor runs the program redundantly, broadcasting operands from its local memory to the other processors. If a processor needs an operand from a remote memory bank, it does not need to request that operand. It simply waits for the operand to arrive, since it will be sent by the processor at the remote memory.

Looking further into future, *Burger et al.* [40, 41] envision a point at which off-chip communication is so expensive that all of the system memory resides on the processor chip. If a system designer wishes to provide more memory than is available on-chip, another of these homogeneous, processor/memory modules is added. Off-chip accesses thus simply become communication with another processor, and accesses to remote data have more in common with a page fault than with a cache miss. Whether this point is reached by migrating computational ability into the DRAM systems, or by migrating DRAM onto the processor (or both), the end result is the same.

7.1.2 Processor-in-Memory approaches

Processor and memory integration can start with a scalar or superscalar microprocessor chip that is enhanced by RAM memory rather than cache memory. Two proposals for coupling a RISC processor with DRAM memory were made by Sun Microsystems and Mitsubishi Electric Corporation.

Sun PIM Processor

Saulsbury et al. [256] of Sun Microsystems propose a design that reserves about 10 % of a DRAM die size for a processor and additional logic. Utilizing 10 % of a current 256 Mbit DRAM chip is slightly more than the size of a MIPS R4300i processor shrunk to a 0.25 μm CMOS process. In addition, the number of pads and interface circuitry are reduced by using high-speed serial-link-based communication.

A block diagram of the proposed integrated processor and memory device is shown in Fig. 7.1. The processor core uses a standard 5-stage pipeline similar to the R4300i or the microSPARC-II processors. The chip is dominated by the DRAM section, which is organized into multiple banks to improve speed. Sixteen independent bank controllers are assumed in a 256 Mbit device. Memory access time is assumed to be 6 cycles of the 200 MHz clock. Each bank is capable of transferring 4 kbits between the sense amplifier array of its DRAM cell and three 512-byte column buffers. The column buffers form the processor I-cache and D-cache. Two column buffers per bank are used for a two-way set-associative D-cache making a total of thirty-two 512-byte lines spread across the 16 banks. The remaining 16 column buffers make up a direct-mapped I-cache with 512-byte lines. The high transfer rate between DRAM memory and the column buffers combined with much shorter DRAM access latency can dramatically improve the cache performance.

The performance of the 16 kB D-cache is enhanced with a fully associative victim cache of sixteen 32-byte lines with a LRU replacement policy. A *victim*

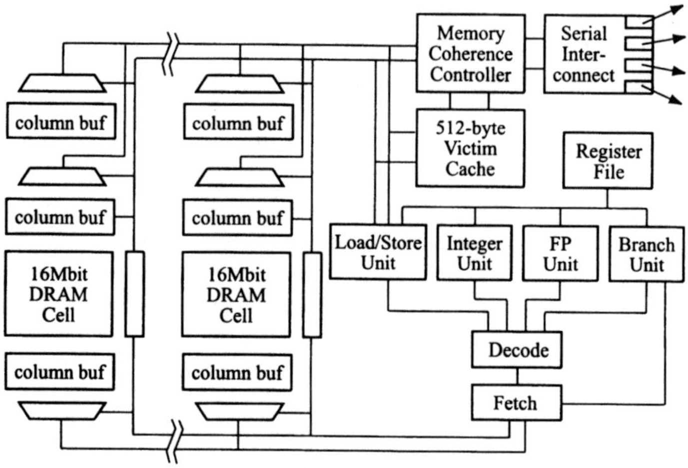

Fig. 7.1. The Sun PIM processor

cache (*Jouppi* [151], 1990), implemented as a small, fully associative buffer, is used to hold cache lines most recently evicted from the D-cache. The victim cache receives a copy of the most recently accessed 32-byte block of a column buffer whenever a column buffer is reloaded.

Two independent 64-bit data paths connect the column buffers with the processor core, one for data and one for instruction access. These buses operate synchronously with the 200 MHz processor clock, and each provides 1.6 GB/s of memory access bandwidth. All off-chip communication is handled via a scalable serial link interconnection system, which can operate at 2.5 Gbits/s. Four links provide a peak I/O bandwidth of 1.6 GB/s, which matches the internal memory bandwidth. All I/O transfer and communication with other processing elements are controlled by two specialized protocol engines. These engines execute a downloadable microcode and can provide a message-passing or cache-coherent shared-memory functionality (*Saulsbury et al.* [256]).

Mitsubishi PIM Processor M32R/D

It is not unusual for microcontrollers to implement a system-on-a-chip approach with a processor kernel, a small DRAM memory, and I/O controllers. The special approach of the M32R/D of Mitsubishi Electric Corporation (*Nunomura et al.* [212], 1997) is to put a relatively large DRAM memory instead of peripheral controllers on the processor chip. The M32R/D integrates a simple RISC CPU core with 1 or 2 MB of DRAM memory on a single chip. The RISC CPU is connected via an internal 128-bit-wide bus with the memory. The chip additionally comprises a 4 kB cache, an instruction queue of two 128-bit entries, a 128/32 bit selector that adjusts the data transfer between the 128-bit bus and the 32-bit CPU, and a peripheral bus interface. The chip design aims at embedded systems applications. The internal memory bus results in low power dissipation in contrast to an off-chip memory solution.

7.1.3 The Vector IRAM approach

The Vector IRAM approach was proposed by *Kozyrakis et al.* [167] in 1997 as a basic architectural technique for future 10^9-transistor microprocessors. In a Vector IRAM a scalar processor is combined with a vector unit and the memory system on a single die. Large, high-bandwidth, on-chip DRAM banks provide the vector unit with sufficient bandwidth. The vector unit contains vector registers and multiple parallel pipelines operating concurrently.

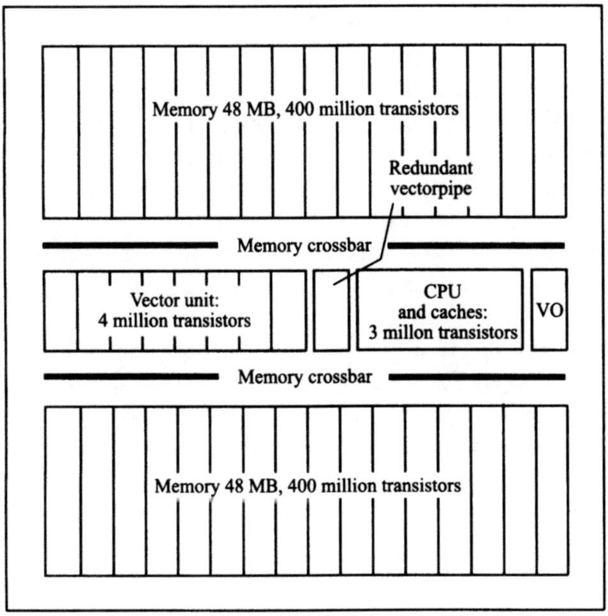

Fig. 7.2. The Vector IRAM processor

Figure 7.2 shows the Berkeley Vector IRAM design which combines a fast in-order processor for scalar operations with a vector unit to create a general-purpose processor that is able to deliver high performance. A potential configuration for a 0.13 μm, 400 mm^2 Vector IRAM chip may include:

- a vector unit with two load, one store and two arithmetic units,
- a dual-issue processor with L1 I-cache and D-cache,
- 96 MB memory organized in 32 sections each comprising sixteen 1.5 Mbit banks, and

- a crossbar switch connecting memory banks with the vector unit and the scalar processor.

Assuming a pipelined synchronous-DRAM interface with 20 ns latency and 4 ns cycle time provides a 192 GB/s bandwidth to the vector unit, a Vector IRAM processor may reach 16 GFLOPS peak performance with a 1 GHz clock rate.

The vector unit itself may contain multiple parallel pipelines operating concurrently. Increasing the number of pipelines provides a straightforward way to scale performance, as the capacity of integrated circuits increases, without requiring changes to the instruction issue logic or recompilation. Because of the simplicity of their circuits, vector processors can operate at higher clock speeds than other architectural alternatives. The floor plan in Fig. 7.2 indicates another advantage of Vector IRAM: the design is highly regular with a few unique pieces used repeatedly across the die. Thus the development cost of Vector IRAM could be much lower than it would be for conventional designs with 10^9 transistors.

Vector IRAM processors are not restricted to scientific applications, but can also be modified for multimedia, database accesses, data mining, and many other applications. Emerging applications like multimedia (video, image, and audio processing) are inherently vectorizable: a vector instruction set can be used to express concurrent operations on arrays of data, like pixels or audio samples. Many database operations, like sort, search, and hash-join, can be vectorized, and memory-intensive database applications like decision support and data mining could benefit from IRAM systems with a vector processor. Even integer applications can often achieve significant speedup through vectorization of their inner loops. For example, the SPECint95 benchmark `m88ksim` and data decompression achieve speedups respectively of 42 % and 36 % through vectorization. In pretty good privacy (PGP) encryption, a vector microprocessor has been shown to outperform significantly an aggressive superscalar processor while occupying less than one-tenth of the die area (*Kozyrakis et al.* [167]).

7.1.4 The Active Page model

Recently, *Oskin et al.* ([220], 1998) proposed the Active Page model, a smart memory approach that does not integrate a processor with the RAM chip. Instead, it shifts data-intensive computing to the memory system while the processor stays off-chip. An *active page* consists of a data page and a set of associative functions that can operate on the data. The Active Page model is implemented on *Reconfigurable Architecture RAM* (RADram), a memory system based upon the integration of DRAM and reconfigurable logic (see the next section).

To use active pages, computation must be partitioned between processor and memory. Active page functions can be used, for example, to gather operands for a sparse-matrix operation and then pass those operands onto the processor for execution. To perform such a gather function, the matrix data and the gathering function must first be loaded into the RADram-based memory system. The processor then starts the gather functions in the memory system. As the operands are gathered within the Active Page memory system, the processor reads them from user-defined output areas in each active page, performs the ALU operations, and writes the results back to the array data structure in memory.

Active pages are intended for simple, application-specific operations, leaving the more complex computations to the microprocessor. Examples of operations that are suitable for active page functions are the multimedia instruction primitives. Implementing these within the Active Page memory system potentially leads to very wide instruction operands. While, for example, a MMX instruction is restricted to 64-bit registers, a RADram MMX instruction could produce up to 256 kB of data per instruction.

The RADram implementation with its reconfigurable function registers naturally builds a bridge from processor and memory integration to reconfigurable computing which is covered in the next section.

7.2 Reconfigurable Computing

7.2.1 Concepts of Reconfigurable Computing

Instead of integrating processor and memory within a single chip, a processor can also be combined with reconfigurable units to perform application-dependent tasks that occasionally change due to environment demands with high performance.

Computer designers face a constant struggle to find the right balance between speed and generality. The reconfigurable computing approach exploits the fact that most of the processing time for computing-intensive tasks is spent in a relatively small portion of the code, and hardware acceleration can significantly improve performance over a general-purpose microprocessor for many applications.

Application-specific integrated circuits (ASICs) are one alternative to implementing hardware accelerators. However, an ASIC contains fixed circuits specialized to a particular task. ASICs provide precisely the functionality needed for a specific task, but nothing else.

Another alternative are *field programmable gate arrays* (FPGA). FPGAs consist of arrays of configurable ("programmable") logic cells that implement the logical functions. In FPGAs both the logic functions performed within the logic cells and the connections between the cells can be altered by sending signals to the FPGA. FPGAs are the most common devices used for reconfigurable computing today (*Villasenor and Mangione-Smith* [313]).

The usual FPGA technology only permits FPGAs to be configured once (using fusable links to yield a read-only FPGA) or to be reconfigured before program start, but not during run-time. Today, configurable FPGAs can be reconfigured application-dependent within milliseconds. Newer FPGA technology can be reconfigured much faster. These FPGAs – exemplified by the XC6200 FPGA family of Xilinx [327] – allow the dynamic reconfiguration of the FPGA or parts of the FPGA during run-time. The XC6264 and the XC6216 reconfigurable processing units (RPUs) of Xilinx have been optimized for reconfigurable logic applications such as real-time adaptive filtering. While traditional FPGAs are usually applied to implement individual logic designs, RPUs are intended to support multiple designs in a single FPGA by dynamically changing the hardware logic. The XC6200 features fast partial reconfiguration, a built-in microprocessor interface, and an open bit stream format.

Although to date the XC6264 FPGA of the Xilinx 6200 series offers the high integration of 64 000 gates, it is just viewed by the reconfigurable research community as a starting point to envision future reconfigurable FPGA chips with much higher integration density. A broader application of reconfigurable computing devices as today may be reached by the design of new FPGA chips that can be reconfigured extremely fast. *Villasenor and Mangione-Smith* [313] expect to see devices with configuration times as low as 100 μs within two years. Ultimately, computing devices may be able to adapt their hardware almost continuously in response to changes in the input data or processing environment.

Computing devices can make use of reconfigurable elements in many different ways. One way to characterize the differences among reconfigurable computing devices is to consider the frequency with which a system may be reconfigured for executing different applications. Reconfiguration is either *static* (execution is interrupted), *semi-static* (also called *time-shared*) or *dynamic* (in parallel with execution):

- *Static* configuration involves hardware changes at the slow rate of hours, days, or weeks, a capability of FPGAs typically used by hardware engineers to evaluate prototype chip implementations.
- *Time-sharing* is a more interesting approach. If an application can be pipelined, it might be possible to implement each phase in sequence on the reconfigurable hardware. The switch between the phases is on command: a single FPGA performs a series of tasks in rapid succession, reconfiguring itself between each one. Such designs operate the chip in a time-sharing mode and swap between successive configurations rapidly.
- The *dynamic* reconfiguration of FPGAs is the most powerful form of reconfigurable computing. It involves the hardware reconfiguring itself on-the-fly as it executes a task, refining its own programming for improved performance (*Villasenor and Mangione-Smith* [313]).

Reconfigurable computing systems (often also called configurable computing systems) combine software programmable general-purpose computing with reconfigurable hardware. Several different levels of integration of processor and reconfigurable hardware can be devised, starting from the coupling of a standard microprocessor with FPGA chips available today, then its coupling with projected future fast and large reconfigurable FPGA chips, to on-chip integration of fixed and reprogrammable circuits. A reconfigurable system is called *remote* if the system's host processor is not on the same chip as the programmable hardware, or it is called *local* if the host processor and programmable logic reside within the same chip (*Singh et al.* [264], 1998).

Commodity (re)configurable computers combine a standard microprocessor with (dynamically reconfigurable) FPGAs. The microprocessor executes its program and delegates the computing-intensive tasks to specially configured FPGAs. Using dynamically reconfigurable FPGAs allows the processor to execute software commands that alter the FPGA circuits as needed to perform a variety of tasks.

Existing systems typically use the I/O bus to provide a coprocessor-like structure. Figure 7.3 illustrates the basic architectural components of such reconfigurable computers (*Mangione-Smith et al.* [190], 1997).

The state-of-the-art FPGA operates at a much lower cycle rate than microprocessors and are difficult to interface closely with standard microprocessors.

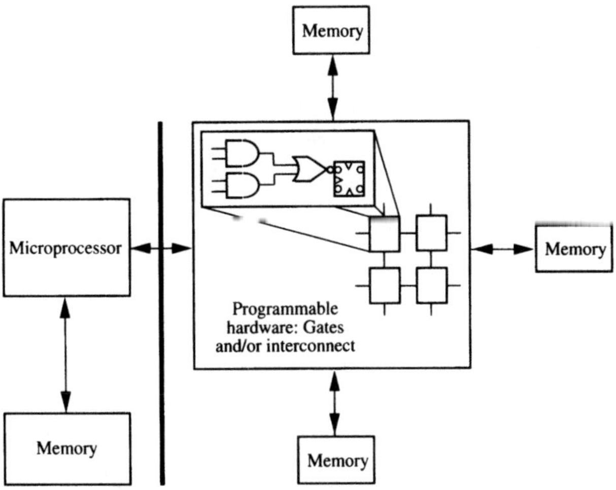

Fig. 7.3. Coupling of a microprocessor with programmable hardware

Programmable hardware could also be coupled much more tightly to the processor, yielding a truly reconfigurable processor. The reconfigurable hardware could be integrated onto the processor chip as one or more hardware cells, like, for example, the functional units of a processor. The reconfigurable hardware cells may be fed by the internal data paths of the processor, operate directly on the processor's registers and communicate directly with the processor's functional units. However, as in the PIM approach, two different styles of hardware logic must be mixed within a single chip leading to fabrication problems. Integrating FPGA technology with fast processor logic shows similar technological problems such as processor and DRAM integration.

A last resort would be to build whole computers from reconfigurable hardware. This is not feasible today, but may lead to innovative parallel operating principles in future. Today's FPGAs are slow circuits compared to the high clock rate of today's microprocessors. Also the gate capacity of an FPGA is very limited. Moreover, FPGAs are not general-purpose, but well capable of performing small sets of highly parallel and highly regular tasks demanding relatively slow reconfiguration speed to switch between different tasks. FPGA-based computations achieve their speeds by exploiting fine-grained parallelism and fast static communication. An FPGA compiler accesses the low-level FPGA hardware structures, allowing the software to optimize mapping of the user application. However, compilation for FPGAs is slow because compilers must target their low-level hardware structures.

The *depth of programmability* (single vs multiple) is defined as the number of *configuration planes* resident in a reconfigurable system. The number of configuration planes determines the number of different configurations which the reconfigurable hardware is able to adopt. Some systems may have only a single resident configuration plane. This means that the system is statically reconfigurable; its functionality is limited to a single context. On the other hand, a system may have multiple configuration planes. In this case, different tasks may be performed by choosing varying planes for execution. Multiple context systems (with multiple configuration planes) favor dynamic reconfiguration (*Singh et al.* [264], 1998).

Reconfigurable computers can be roughly partitioned into two classes according to the level of abstraction, which is expressed by the *granularity* of operations, i.e., bit-level vs word-level. Bit-level operations correspond to *fine granularity*, whereas word-level operations imply *coarse granularity* (see *Singh et al.* [264]).

Fine-grained devices are also called *netlist computers*, according to *Mangione-Smith et al.* [190]. A typical netlist computing device today is an FPGA containing thousands of FPGA cells that are composed of single flip-flops and logic gates. FPGAs have a programmable interconnect that is manipulated as individual wires. Because of their fine granularity, netlist computers are the most flexible reconfigurable computers; their elements can be used to implement state machines, data paths, and almost any digital circuit. This flexibility is bought with additional silicon, and it results in lower performance on certain classes of problems, compared to more coarse grained-devices (see below).

Examples of netlist computers are the DECPeRLE-1 (*Bertin et al.* [26], 1989), Splash (*Gokhale et al.* [103], 1991), and Wildfire (*McHenry* [198], 1995) which is commercially available.

The netlist computer presents a number of serious challenges to application development. The developers must be concerned with the size and usage of the FPGA devices, the size and usage of the memories, and finally the overall interconnection of all devices on the platform during all phases of the design process. The design process is therefore more difficult and time-consuming. Furthermore, modifications can require a significant amount of CAD compilation time (*Mangione-Smith et al.* [190]).

To raise the level of abstraction, the programmable hardware can be limited to the interconnect and, instead of gates and flip-flops, higher-level components such as arithmetic logic units (ALUs) or multipliers can be provided. In the second class of reconfigurable computers are those which are based on more coarse-grained function units such as complete ALUs and multipliers. Due to the more coarse-grained units, configuration times are shorter than with the fine-grained approach and less configuration data is needed. These units are also called *chunky function unit architectures* [190] and often limit the programmable hardware to the interconnect between the function units. Chunky function unit architectures address the problem of the poor hardware efficiency of fine-grained FPGAs by providing highly optimized function units with a programmable interconnect. Advanced FPGAs may contain memory, arithmetic processing units and other special-purpose blocks of circuitry. Here the building blocks for new highly parallel architectural principles are provided. The *computation model* of most coarse-grained reconfigurable systems may be described as either SIMD or MIMD. Some systems may follow the VLIW model.

Such new architectural solutions are being pursued by a number of research projects, including the MorphoSys System (see Sect. 7.2.2) at the University of California at Irvine, the Raw machine containing multiple, simple reconfigurable processors (see Sect. 7.2.3), the Xputer project which defines a non-von Neumann paradigm implemented on a reconfigurable Datapath Architecture (also called KressArray) at the University of Kaiserslautern (see Sect. 7.2.4), RaPiD at the University of Washington (*Ebeling et al.* [72], 1996), and MATRIX at MIT (*Mirsky and DeHon* [205], 1996; *DeHon et al.* [62], 1997).

Each of these architectures presents an abstraction that is much higher than logic gates and flip-flops, and more towards simple but highly interconnected processing elements. Consequently, highly regular applications map well to such implementations and are likely to achieve high performance. Several regular applications can be mapped onto the same implementation, for example, by breaking wide-word operations into a composition of the narrower, native hardware function units, or by simply wasting the upper bits of the fixed data path to handle narrow-word operations. However, highly irregular computations will strain gate capacity limits and cause performance problems.

Most reconfigurable computing prototypes today are based on (fine-grained) FPGAs. Limitations of such FPGAs that may lead to limitations of reconfigurable computing systems themselves are seen by *Mangione-Smith et al.* [190] as follows:

- *Insufficient gate capacity*: Today's FPGAs provide the equivalent of 10 k to 500 k gates which is often large enough to experiment with the basic strategies for configuration, but with a limited scope of design. FPGAs have become a common component in a wide range of commercial products, but the vast majority of FPGAs are used in place of more efficient, but more costly, ASICs to reduce development cost and time. The capacity limits are due to the use of conservative semiconductor processes, particularly with limited metal layers. Recent FPGA devices have moved to leading-edge process technology, resulting in FPGAs with up to 500 k gates (exemplified by the Xilinx XC4000XV family). The problem of gate density is likely to decrease in the future as on-chip delays drive designers toward partitioned designs that can be implemented efficiently on multiple chips.
- *Low reconfiguration speed*: Most existing FPGAs use relatively slow serial paths for device configuration, even when a parallel interface is presented through the I/O pins. The reconfiguration time is important for many models of computation. In particular, those using fast design swapping might eventually be limited by reconfiguration time. The industry is moving toward FPGAs that reconfigure faster.
- *Lack of on-chip memory*: Furthermore, most FPGAs currently provide very little on-chip memory for storage of intermediate results in computations; thus, many reconfigurable computing applications require large external memories.
- *Lack of memory interfaces*: Moreover, most FPGAs have no external memory interface that can be accessed from the active circuit, which forces system designers to sacrifice some FPGA space to build an application-specific memory interface. A more efficient approach would be to implement standard memory interfaces on FPGAs in dedicated hardware. The transfer of data to and from the FPGA increases power consumption and may slow down the computations.

Reconfigurable computing is an area of active research. Reconfigurable computing systems have demonstrated the potential for achieving high performance for a range of applications, including image filtering, convolution, morphology, feature extraction, and object tracking. FPGAs are well suited to algorithms composed of bit-level operations, such as pattern matching and integer arithmetic, but they are ill-suited to numeric operations, such as high-precision multiplication or floating-point calculations. The earliest commercial successes are likely to involve signal processing, particularly image processing.

FPGAs will never replace microprocessors for general-purpose computing tasks, but the concept of reconfigurable computing is likely to play a growing role in the development of high-performance computing systems. The computing power of reconfigurable hardware will make them the devices of choice for applications involving algorithms in which rapid adaptation to the input is required.

Current FPGAs still do not come close to exploiting the full possibilities of the reconfigurable computing technique. Future FPGAs will be much larger; as with many other integrated circuits, the number of gates on a single FPGA has doubled roughly every 18 months. Before the decade is out, *Villasenor and Mangione-Smith* [313] expect to see FPGAs that have a million logic elements. Such chips will have much broader application, including highly complex communications and signal-processing algorithms.

In addition, the line between programmable processors and FPGAs may become less distinct. There are next-generation microprocessors under development whose hardware supports limited amounts of FPGA-like reconfiguration. One of these is the *Complexity-adaptive Processor* (CAP) approach (*Albonesi* [5], 1998) that implements configurability deep within the pipeline of an otherwise conventional processor (see end of Sect. 7.2.5).

Future machines might download new hardware configurations as they are needed. Computing devices ten years from now may include a strong mix of software-programmable hardware and hardware-configurable logic.

7.2.2 The MorphoSys system

The MorphoSys project at the University of California at Irvine (*Singh et al.* [264], 1998) is focused on developing dynamically reconfigurable computing systems. Its objective is to design and build a processor with an accompanying reconfigurable circuit chip which tolerates much slower operation than the processor. The MorphoSys computing system is targeted at image processing applications.

The MorphoSys reconfigurable computing system M1 (see Fig. 7.4) is composed of a control processor with I-cache/D-cache, a reconfigurable array with an associated control memory, a data buffer (usually acting as a frame buffer), and a DMA controller. The main thread of control is managed by the on-chip control processor. The programmable part is an 8×8 array of reconfigurable cells with multiple context words and operating in SPMD or SIMD fashion. MorphoSys is coarse-grain (chunky function unit approach) by its reconfigurable cells which resemble small processors with 16-bit data paths.

The main component of MorphoSys is the *reconfigurable cell* (RC) array. It has 64 reconfigurable cells, arranged as an 8×8 array. Each cell has an ALU/multiplier and a 16-bit wide register file. The RC array functionality and interconnection network are configured through 32-bit context words. The context words are stored in a context memory in two blocks, one for

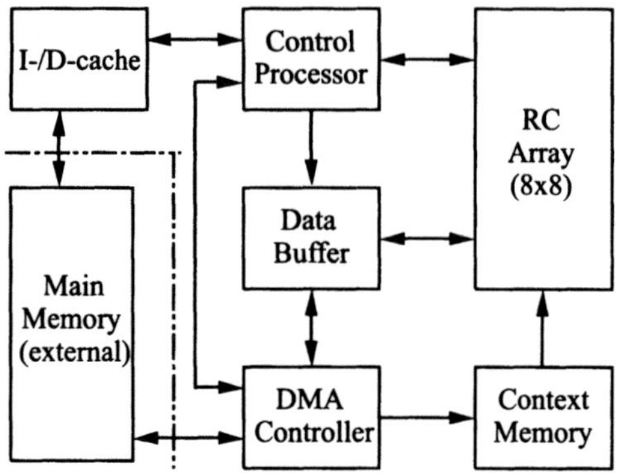

Fig. 7.4. The MorphoSys system M1

rows and the other for columns. The context memory can store up to 16 contexts corresponding to an individual row and 16 contexts corresponding to an individual column. Thus, each block has eight sets of 16 contexts. The MorphoSys design provides the option of broadcasting contexts across rows or columns.

The *control processor* of MorphoSys is a 32-bit RISC processor, called Tiny RISC. Tiny RISC controls the RC array operations, data transfer to and from the RC array, as well as to and from the data buffer. The control processor and the RC array are on the same chip. This prevents I/O limitations from affecting performance. In addition, the memory interface is through an on-chip DMA controller, for faster data transfers between external memory and the data buffer. It also helps in decreasing the configuration loading time.

The MorphoSys system operates as follows. The Tiny RISC processor loads the configuration data from the external main memory into the on-chip context memory via the DMA controller. Next, it enables the data buffer to be loaded with image data from the main memory. This data transfer is also performed by the DMA unit. At this point, both configuration and data are ready. Now, the Tiny RISC processor issues instructions to the RC array for execution. These instructions specify the particular context (among the multiple contexts in the context memory) to be executed. The Tiny RISC can also enable selective functioning of individual rows or columns, and can access data from selected RC outputs.

The RC array follows the SPMD model of computation. The mode of operation can be SIMD where all RC cells perform the same operation on different data, but it is not necessarily SIMD, because different control words specifying different operations can be assigned to the individual row (or column) control memory and be activated by the control processor. Each row or column is configured by one context, which serves as an instruction word. However, each of these cells operates on different data. This model serves the target applications (i.e., applications with a large number of data-parallel operations) for MorphoSys very well.

Each cell of the RC array function is configured by the context word. The context word specifies one of several instruction opcodes for the RC array, and provides control bits for input multiplexers. It also specifies constant values that are needed for computations.

Dynamic reconfiguration capability is achieved by changing some portion of the context memory while the RC array is executing contexts from a different portion. For example, while the RC array is operating on the 16 contexts in row broadcast mode, the other 16 contexts for column broadcast mode can be reloaded. Context loads and reloads are done through Tiny RISC instructions.

7.2.3 Raw Machine

The Raw architecture approach, which was proposed by *Waingold et al.* [316] in 1997 (see also *Lee et al.* [178], 1998) for the envisioned 10^9-transistor era, defines a coarse-grained or chunky function unit architecture. A *Raw machine* consists of hundreds of very simple processors, called *tiles*, each with some reconfigurable logic on a single chip (see Fig. 7.5). The principal idea is to eliminate the traditional instruction set interface and instead expose the details of a simple replicated architecture directly to the compiler. This allows the compiler to customize the hardware to each application. Execution and communication is controlled almost entirely in software. *Waingold et al.* call systems based on this approach *Raw* architectures, because they implement only a minimal set of mechanisms in hardware.

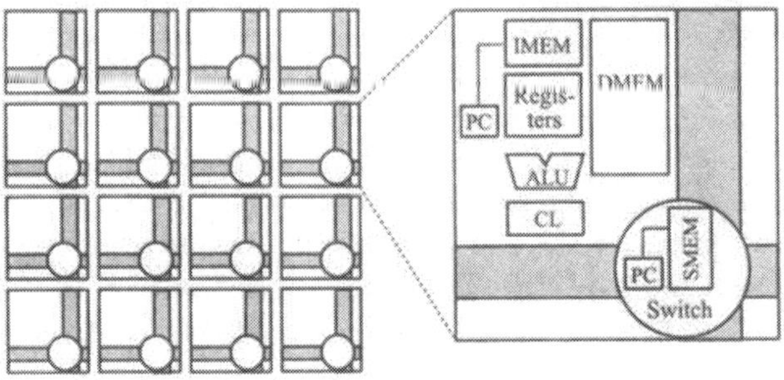

Fig. 7.5. The RAW processor

Each tile in a Raw machine contains *instruction memory* (IMEM), *data memories* (DMEM), an *arithmetic logic unit* (ALU), *registers*, *(re)configurable logic* (CL), and a *programmable switch* with its associated instruction memory (SMEM). The tile works internally as a simple, RISC-like pipeline. The tiles are connected with programmable, tightly integrated interconnects over a pipelined, point-to-point network. The programmable switches support both dynamic and static routing. The synchronous network interface of a Raw machine allows for inter-tile communication with short latencies similar to those of register accesses. Static scheduling guarantees that operands are available when needed, eliminating the need for explicit synchronization.

Static RAM (SRAM) portions are distributed across the tiles to eliminate the memory bandwidth bottleneck and to provide significantly shorter latency to each memory module. The focus is on keeping each tile small to maximize the number of tiles that can fit on a chip.

A potential 10^9-transistor configuration could consist of 128 tiles. Each tile uses 5 million transistors for memory: 16 kB IMEM, 16 kB SMEM, and 32 kB first-level DMEM. Each tile uses 2 million transistors for CPU (MIPS R2000 equivalent) and reconfigurable logic. Switched interconnects are applied between tiles instead of buses. The switches are integrated directly into the processor pipelines to support single-cycle message injection and receive operations. The processor communicates with the switch using distinct opcodes to distinguish between accesses to static and dynamic network ports. No signal in a Raw machine travels more than a single tile width within a clock cycle. Because the interconnect allows inter-tile communication to occur at nearly the same speed as a register read, compilers can schedule single-word data transfers and exploit ILP, in addition to coarser forms of parallelism. The switch multiplexes two logically distinct networks, one static and one dynamic, over the same set of physical wires. The dynamic wormhole router makes routing decisions based on each message's header, which includes additional lines for flow control.

Each tile includes two sets of control logic and instruction memories. The first set controls the processors, the second is dedicated to sequencing routing instructions for the static switch. Separate controls for the processor and the switch let the processor take arbitrary, data-dependent branches without disturbing the routing of independent messages passing through the switch. Loading a different program into the switch instruction memory changes the switch schedule. Programming network switches with compile-time schedules lets the compiler statically schedule the computations in each tile, thus providing the reconfiguration capability of the Raw machine. A drawback of the Raw approach is that it must revert to a software-simulated broadcast when the compiler cannot statically establish a schedule.

Each tile supports bit-level, byte-level, and word-level operations and programmers can use the reconfigurable logic in each tile to construct operations uniquely suited to a particular application. The byte-granular reconfigurable logic permits the ALUs to be used for either a few wide-word operations or many narrow-word computations similar to multimedia operations. The small amount of reconfigurable logic in each Raw tile – for which software can select the data path width – permits a Raw processor to support multigranular operation. Raw machines combine bit-level or byte-level parallelism with special communication paths among the individual bits or bytes. This permits significantly more powerful multigranular operations than those supplied by MMX-like instruction sets.

A compiler for a Raw machine starts with a sequential or multithreaded high-level language program and has full access to the underlying Raw hardware mechanisms. The Raw compiler views the set of tiles in a Raw machine as a collection of functional units for exploiting ILP. The principal compiler steps are partitioning, placement, routing, global scheduling, and configuration selection for the reconfigurable logic.

Partitioning generates parallel code for an idealized Raw machine with the same number of tiles as the physical architecture. This phase, however, assumes an idealized, fully connected switch, an unbounded number of virtual registers per tile, and symbolic data references. Placement selects a one-to-one mapping from threads to physical tiles. The placement algorithm minimizes latency and bandwidth costs and is a variant of a VLSI cell-placement algorithm. Routing and global scheduling allocate physical network resources to minimize the overall program execution time. This phase produces a program for each tile and switch.

The configuration selection phase selects an application-specific configuration for loading into the reconfigurable logic. For each custom operation, the configuration phase must both output a specification for the reconfigurable logic and rewrite the intermediate code. This replaces each compound operation by a call to the appropriate custom instruction. The compiler will invoke a logic synthesis tool to translate a custom operation specification into the appropriate bit sequence for the reconfigurable logic. Thus the compiler resembles more a hardware synthesis tool than a high-level language compiler. The burden on the compiler is extreme. It is unclear how this complexity could be handled; programs will probably need very long compile-times, in the order of several hours or days.

A RawLogic prototype and an associated compilation system was implemented by *Waingold et al.* [316] in 1997, using commercial FPGA-based logic emulation technology. The RawLogic prototype consists of a logic emulator coupled with a Sun SPARCStation 10/51. The emulator also has a SCSI interface for downloading configurations and controlling clock speed and consists of five boards, each with 64 directly connected Xilinx 4013 FPGAs.

The RawLogic prototype does not support all Raw architecture features. It has simple replicated tiles and supports statically scheduled, tightly integrated communication, multigranularity, and configurability. But it does not support the instruction processing of a more general Raw machine. Each static control sequence is converted into an individual state machine and hardwired into the RawLogic hardware. As a result, the prototype has the problems associated with FPGA-based systems – it lacks flexibility and has long compilation times. To compile applications for RawLogic, a framework was developed that specifies the dependence structure of a program's loops (in C) and the computation it performs (in behavioral Verilog). The program specification is used to generate automatically a behavioral Verilog netlist for the program. A commercial behavioral compiler automatically synthesizes a gate-level netlist, which is then processed by a VirtualWires compiler. This compiler partitions, places, and schedules the logic to produce binary code for the FPGA hardware. The RawLogic prototype achieved a 10–1 000-fold speedup over an all-software Raw machine simulator. The FPGA compilation step took several hours per board using ten workstations for the larger benchmarks.

In the near future, the Raw architecture approach will be best suited for stream-based signal-processing computations. *Waingold et al.* [316] believe that, in 10 to 15 years, 10^9-transistor chip densities, faster switching speeds, and growing compiler sophistication will allow a Raw machine's performance-to-cost ratio to surpass that of traditional architectures for future, general-purpose workloads. Because Raw architectures are field-programmable, they may also become a cost-effective alternative to custom hardware, replacing ASICs. Thus they may offer a universal solution for both general-purpose and special-purpose applications.

7.2.4 Xputers and KressArrays

The Xputer principle developed by *Hartenstein et al.* [124, 125, 126, 127, 128] at the University of Kaiserslautern (Germany) was specially designed to reduce the address and data manipulation overhead in von Neumann computing. This overhead contributes a significant amount (up to 90 % in image processing, 58 % in digital signal processing) to the run time of many types of algorithms in digital signal processing, image processing, electronic design automation, and mathematical computation-intensive problems.

The Xputer paradigm and its data-sequencing operation principle reverses the control-driven von Neumann paradigm. The Xputer replaces the program counter – the instruction sequencer of the von Neumann architecture – by a *data sequencer*, and the hardwired ALUs in a von Neumann computer by reconfigurable ALUs. The data sequencer operates in a data-driven way, but in a strictly procedural way that uses a deterministic selection of executables. In contrast, dataflow computers also operate in a data-driven way, but instructions with all input operands available are arbitrarily selected for execution, which leads to a nondeterministic operation (see Chap. 2). The Xputer principle supports the use of reconfigurable data paths, such as in a reconfigurable ALU [126], whereas the tight coupling between the instruction sequencer and the ALU in von Neumann machines hinders the use of reconfigurable data paths.

The basic structure of an Xputer module consists of three major parts:

- a *reconfigurable arithmetic logic unit* (rALU),

- a reconfigurable data sequencer comprising several *generic address generators* (GAG), and
- a 2D organized data memory.

The Xputer principle was first published by *Hartenstein et al.* [124] in 1987 as a reconfigurable procedurally data-driven machine architecture called PIxel-oriented System for image Analysis (PISA) and later renamed to Map-oriented Machine (MoM). The hardware provided a reconfigurable *problem-oriented logic unit* (POLU) to update the so-called scan cache, later called the Scan Window. The POLU was activated each time a new set of input data was available in the Scan Window. The POLU looked for matches with scan pattern in the Scan Window. A scan pattern is the counterpart to control-flow in von Neumann languages. The address generating *Move Control Unit* of the PISA machine only supported a single form of scan pattern, a video scan with an absolute addressing of pixels.

Several Xputer architectures, MoM-1 through MoM-3, have been developed at the University of Kaiserslautern supporting 2D memory organization and multiple data sequencer parallel usage. The second generation prototype was called MoM-2. The scope of applications was enhanced beyond pattern matching-based applications to parallel arithmetic operations and an even far wider range of applications. The MoM-2 still provided only a single address generator to update the Scan Window, but the repertory of scan patterns now included variations of video-scans, shuffle-scans, linear scans, and relative jumps of the Scan Window. The Scan Window had a variable size and shape, bounded by a rectangle, in order to adapt to the requirements of the application's data distribution.

The POLU supported pattern matching and hardware-controlled modification of the input data. A custom-designed NMOS circuit supported SRAM-based pattern matching which allowed the exchange of patterns during runtime. Although the term FPGA was coined later on, this circuit can be considered as one of the early SRAM-based FPGAs, which turned out to be an important technology platform for Xputers [128]. In the MoM-2, the rALU had to be a combinational hardware without any registers. The timing of the rALU directly influenced the clock speed of the address generator. A task sequencer was introduced in the architecture to combine the address generator's scan patterns to arbitrary address sequences. The new machine paradigm was called Xputer from then on.

On the software side, a graphical CAD tool supported an easy way of programming the reference and result patterns for the POLU, as well as arithmetic operations (although these were not included in the prototype hardware due to the lack of suitable FPGAs at that time). This CAD tool generated the structural code for the MoM-2 and the sequential code was compiled from MoPL-1 (MoM Programming Language 1), the first data procedural language for Xputers.

The latest Xputer prototype built at Kaiserslautern University is the MoM-3 [128]. It introduces multiple Scan Windows, whose registers are integrated into the rALU for optimized packaging. The shape of the Scan Windows can be completely arbitrary and their size is increased to 64 memory words of 32 bits at most. Each Scan Window is controlled by a GAG, featuring an even richer set of scan patterns than in the MoM-2 version. The GAGs are integrated in a coordinating microprogrammable data sequencer. A standard interface between the rALU and the data sequencer allows both to run at their full speed, so that virtually any FPGA board for custom computing machines can be adapted as a rALU for the MoM-3. The rALU no longer has to be a combinational net, which allows the use of pipelining to speed up large operators.

The MoM-3 system architecture allows both direct memory access to the host computer's main memory, and multiple parallel accessible local memory modules. Pipelining and 32-bit wide arithmetic are supported by their own custom FPGA circuit, called *reconfigurable Datapath Architecture* (rDPA) or also KressArray-I. The KressArray consists of a regular array of identical *data path units*. Each data path unit consists of an ALU, a microprogrammable control and four registers. There are two levels of interconnection: local nearest neighbor connections (a mesh network of short wires) and a global bus system. The KressArray is dynamically reconfigurable.

The main difference between the KressArray and conventional FPGAs is its more coarse-grained structure, implementing 32-bit wide data paths and all operators of the C programming language. The KressArray aims for better performance of word-level operations through data paths wider than typical FPGA data paths. Moreover, the KressArray features (partial) in-system reconfiguration at run-time, fully transparent expansion across chip-boundaries, fully parallel internal communications and a global bus for quick distribution of input (and output) data into the array.

A compiler for Xputers generates code from an extension of C called X-C. The compiler generates two kinds of machine code: sequential code to program the data sequencer (a kind of software code), and reconfiguration code to set up application-specific data paths (configured hardware) within the KressArray platform. Therefore, a partitioning problem has to be solved that corresponds to the partitioning problem in hardware/software co-design. From this point of view the data sequencer architecture comprises a particular hardware/software interface platform.

Xputers can be used for acceleration. In particular, for algorithms with regular or semi-regular data dependences the Xputer paradigm is much more efficient than the von Neumann paradigm.

There exists a strong relationship between Xputers and systolic arrays. The KressArray can be viewed as a generalization of systolic arrays. The layout concept of the KressArray is the same as for systolic arrays, but for synthesis simulated annealing is used instead of the linear projection methods used for systolic arrays. A data path synthesis tool (DPSS) compiles application data paths into structural code for the KressArray hardware. The KressArray is much more area-efficient than an FPGA due to the avoidance of long-range connections.

Currently a new Xputer prototype called MoM-PDA (Map-oriented Machine with Parallel Data Access) is under development at the University of Kaiserslautern. The MoM-PDA features a new data sequencer and a new memory architecture with banked memory access. Also the KressArray has developed into KressArray-III. Besides the underlying concepts, the hardware implementation of a *field-programmable ALU array* (FPAA), the KrAA-III, is explained by *Hartenstein et al.* ([123], 1998).

7.2.5 Other Projects

Configurable computing was first proposed in the late 1960s, but it is still a young and active field of research. In this section we give some further examples of reconfigurable computing projects from the wide field of research activities.

The DECPeRLE-1 and Splash computers were among the first research efforts in configurable computing. Both can be classified as netlist computers and are constructed as attached accelerators alongside workstations.

DECPeRLE-1 (*Bertin et al.* [26], 1989): DECPeRLE-1 is organized as a 2D mesh and consists of a 4 × 4 array of FPGAs. Each FPGA has connections to its nearest neighbors as well as to a column bus and a row bus.

Splash (*Gokhale et al.* [103], 1991): The Splash conceptually consists of a linear array of processing elements. This topology makes Splash a good candidate for linear-systolic applications, which stress neighbor-to-neighbor communications. Because of limited routing resources, Splash has not proved as effective at implementing multichip applications that are not linear systolic.

Neither DECPeRLE-1 nor Splash provide general-purpose routing networks between FPGAs. Instead they require the designer to partition the circuit manually during the design phase, ensuring that the available interconnect is used as efficiently as possible (*Mangione-Smith et al.* [190], 1997). Both systems are fine-grained, with remote interface, single configuration, and static reconfigurability. Other research prototypes with fine-grain granularity include DPGA and Garp.

DPGA (*Tau et al.* [289], 1995): The Dynamically Programmable Gate Array (DPGA) is a fine-grain prototype system that uses traditional 4-input lookup tables as the basic array element. Each cell can store 4 context words. DPGA supports rapid run-time reconfiguration. Small collections of array elements are grouped as subarrays that are tiled to form the entire array. A subarray has complete row and column connectivity. Configurable crossbars are used for communication between subarrays.

Garp (*Hauser and Wawrzynek* [129], 1997): Garp is a fine-grained approach that has been designed to fit into an ordinary processing environment, where a host processor manages the main thread of control while only certain loops and subroutines use the reconfigurable array for speedup in performance. The host processor is responsible for loading and execution of configurations on the reconfigurable array. The instruction set of the host processor has been expanded to accommodate instructions for this purpose. The array is composed of rows of blocks that resemble CLBs of the Xilinx 4000 series. There are at least 24 columns of blocks, while the number of rows is implementation specific. The blocks operate on 2-bit data. There are vertical and horizontal block-to-block wires for data movement within the array. Separate memory buses move information (data as well as configuration) in and out of the array.

Systems with coarse-grain granularity include the KressArray, Raw, and Morphosys described already, as well as PADDI, MATRIX, and RaPiD.

PADDI (*Chen and Rabaey* [47], 1992): PADDI has a set of concurrently executing 16-bit functional units (EXUs). Each of these has an 8-word instruction memory. The communication network between EXUs uses crossbar switches for flexibility. Each EXU has dedicated hardware for fast arithmetic operations. Memory resources are distributed among the EXUs.

MATRIX (*Mirsky and DeHon* [205], 1996; *DeHon et al.* [62], 1997): The MATRIX project of MIT falls into the class of the chunky function architectures. MATRIX aims to unify resources for instruction storage and computation. The *basic functional unit* (BFU) can serve either as a memory or a computation unit. The 8-bit BFUs are organized in an array, where each BFU has a 256-word memory, ALU-multiply unit, and reduction control logic. The interconnection network has a hierarchy of three levels (nearest neighbor, length four bypass connection, and global lines).

RaPiD (*Ebeling et al.* [73], 1997): RaPiD was developed at the University of Washington. It is a linear array of functional units, which is configured mostly to form a linear computation pipeline. The identical array cells each have an integer multiplier, three ALUs, six registers, and three small local memories. A typical array has 8 to 32 of these cells. It uses segmented buses for efficient utilization of interconnection resources.

FPGA-Based ATR System (*Villasenor et al.* [314], 1996): At the University of California (Los Angeles) *Villasenor et al.* have built an FPGA-based system for automatic target recognition (ATR), i.e., a single-chip video transmission system that reconfigures itself four times per video frame. It falls in the class of the time-share mode, and requires only a quarter of the hardware that would be needed for a fixed ASIC. The FPGA first stores an incoming video signal in memory, then applies two different image-processing transformations and finally transforms itself into a modem to send the signal onward.

DISC (*Wirthlin and Hutchings* [325], 1995): Another approach to the time-shared mode is the Dynamic Instruction Set Computer (DISC) project at Brigham Young University which effectively marries a microprocessor to an FPGA and demonstrates the potential of automatic reconfiguration using stored configurations. DISC involves a custom FPGA-based processor as well as configuration components for an FPGA accelerator. The FPGA caches configuration components and executes a demand-driven fetch and reload in response to a fault. Time sharing also has been used to improve performance for a number of applications, including a video communications system and image recognition. As a program runs, the FPGA requests reconfiguration if the designated circuit is not resident. DISC allows a designer to create and store a large number of circuit configurations and activate them much as a programmer would initiate a call to a software subroutine in a microprocessor.

Wormhole (*Bittner and Athanas* [28], 1997): The Colt Group of Virginia Polytechnic Institute and State University is investigating a run-time reconfiguration technique called Wormhole that lends itself to distributed computing. The unit of computing is a stream of data that creates custom logic as it moves through the reconfigurable hardware.

CAP (*Albonesi* [5], 1998): In contrast to the Active Page approach already described at the end of Sect. 7.1.4, the Complexity-adaptive Processors (CAP) implements configurability deeply in the processor. CAPs employ reconfigurable hardware for the core superscalar control and cache hierarchy structures of the processor, potentially selecting the best configuration at critical run-time points. To implement this approach the processor, caches, and external interface consist of conventional fixed hardware structures intermixed with complexity-adaptive structures under the control of a configuration manager.

7.3 Asynchronous Processors

Superscalar processors are also being developed in asynchronous design (in addition to asynchronous scalar processors). Therefore, let us briefly describe the research in asynchronous processor design.

Conventional synchronous architectures are based on global clocking whereby global synchronization signals control the rate at which different elements operate. For example, all functional units operate in lockstep under the control of a central clock.

As the clocks get faster, the chips get bigger and the wires get finer. As a result, it becomes increasingly difficult to ensure that all parts of the processor are ticking along in step with each other. Even though the electrical clock pulses are travelling at a substantial fraction of the speed of light, the delays in getting from one side of a small piece of silicon to the other can be enough to throw the chip's operation out.[1] For example, the 1997 National Technology Roadmap for Semiconductors [336] forecasts that CMOS technology will reach a point where the switching delay for a single gate will be close to 10 ps while a single chip area will be nearly 7.5 cm^2. It will take 30 clock cycles for the electric signal to cross such a chip. Moreover, the interchip clock skew already represents a major problem.

The clock-related timing problems have been recently attacked by *asynchronous* (or *self-timed*) design techniques. These asynchronous processors do away with the idea of having a single central clock keeping the chip's functional units in step. Instead, each part of the processor – for example, the arithmetic units, the branch units, etc. – all work at their own pace, negotiating with each other whenever data needs to be passed between them.

Without a global clock, asynchronous systems enjoy (*Takamura et al.* [288], 1997):

- Data-dependent cycle time rather than worst case cycle time: The conventionally clocked chip has to be slowed down so that the most sluggish function does not get left behind. To deal with this problem one can either use some extra circuitry to try to speed up these slow special cases, or alternatively just accept it and slow everything down to take account of the lowest common denominator. Either way the result is that resources are wasted or the chip's speed is determined by an instruction that may hardly ever be executed. In the asynchronous approach the chip only becomes more sluggish when a tricky operation is encountered.
- Potential for low power consumption: The conventional processors are becoming increasingly power consuming. For example, DEC's Alpha and the PowerPC 620 emit around 20 W to 30 W in normal operation. If we were to continue to use 5 V supplies, we could expect by the end of 1999 a 0.1 µm processor dissipating 2 kW. Reducing the supply to 3 V (or 2 V) would only reduce the power dissipation to 660 W (or 330 W). One of the reasons is that many of the logic gates switch their states simply because they are being driven by the clock, and not because they are doing any useful work. Removing the clock in asynchronous processors also removes the unnecessary power consumption as CMOS gates only dissipate energy when they are switching.

[1] Even if the clock were injected optically to avoid the wire delays, the signals issued as a result of the clock would still have to propagate along wires in time for the next clock pulse, and a similar problem would remain.

- Ease of modular composition, i.e., circuits can be assembled as plug-and-play.
- Optimization of frequent operations while rare operations can spend more time.
- No need for clock alignment at the interfaces.
- Timing fault-tolerance.

There are also several shortcomings to the asynchronous approach:

- clock-based computers are easier to build than asynchronous;
- it is easier to verify a synchronous design due to its deterministic operation (by comparison, verifying an asynchronous design, with each part working at its own pace, is difficult).

7.3.1 Asynchronous Logic

Virtually all digital design today is based on a synchronous approach whereby each subsystem is a clocked finite state machine that changes its states on the edges of a regular global clock. Such a system behaves in a discrete and deterministic way, provided the delays are managed so that the flip-flop set-up and hold times are met under all conditions.

As a contrast, in asynchronous design there is no clock to govern the timing of state changes. Subsystems exchange information at mutually negotiated times with no external timing regulation.

Figure 7.6a shows the structure of a synchronous pipeline with latches and combinational logic blocks. All latches are controlled by a single global clock signal and operate simultaneously.

An asynchronous implementation of the pipeline is shown in Fig. 7.6b. The latches and the combinational logic block are the same as in the synchronous pipeline. The timing, however, is controlled differently. Each latch has an associated *latch control circuit* (LCC) which opens and closes the latch in response to *request* signals from the previous stage and *acknowledge* signals from the following stage. There are a few key features which describe most current approaches:

- *Delay-insensitive* vs *speed-independent* design: Delay-insensitive designs make no assumptions about delays within the system. That is, any gate or interconnection may take an arbitrary time to propagate a signal. Speed-independent systems are tolerable to variations in gate speeds but assume instantaneous transmissions along wires.
- *Dual-rail encoding* vs *data bundling* communication protocol: In dual-rail encoded data, each Boolean is implemented as two wires. This allows the value and the timing information to be communicated for each data bit. Bundled data, on the other hand, has one wire for each data bit and a separate wire to indicate the timing.

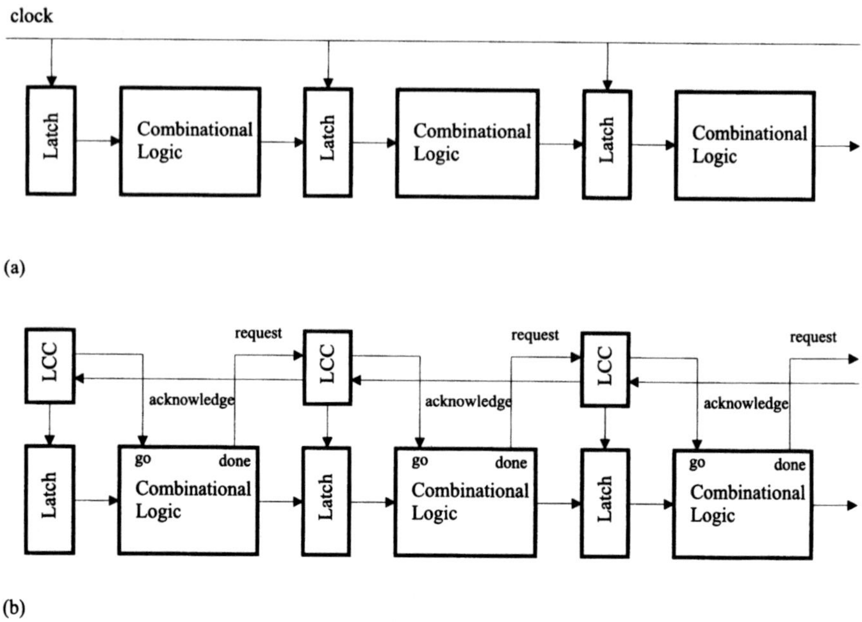

Fig. 7.6. A simple pipeline **(a)** synchronous **(b)** asynchronous

- *Level* vs *transition* encoding: Level-sensitive circuits typically represent a logic one by a high voltage and a logic zero by a low voltage. In transition signaling, only changes in the level of signals are taken into account.

Delay-insensitive circuits with dual-rail communication and encoding with transition signaling proved to be ideal for automatic transformation into a silicon layout, as the delays introduced by the layout compiler cannot affect the functionality. The most popular form in recent years has been dual-rail encoding with level-sensitive signaling. Delay insensitivity is achieved at the cost of more power dissipation than with transition signaling. The advantage of this approach over transition signaling is that the logic processing elements can be much simpler. A well-known form of delay-insensitive circuit with bundled data communication and encoding with transition signaling is the *micropipelined* approach, which was proposed by *Sutherland* [285], 1989) and adopted in the AMULET project (see below).

It has been predicted that asynchronous techniques will find their way into certain niches, in particular, embedded applications where the work required is extremely burst-intensive or where power-saving requirements make the approach attractive. Clocked chips with some asynchronous parts may also be expected.

Table 7.1. Recent asynchronous microprocessors

Processor	Design Style	ISA	Organization
CAP	4-phase, dual-rail, delay-insensitive	own 16-bit RISC-like	fetch-execute pipeline
FAM	4-phase, dual-rail, delay-insensitive	own RISC-like	pipelined
STRiP	variable clock, synchronous	MIPS-X	pipelined, forwarding
ST-RISC	dual-rail, delay-insensitive	own	fetch-execute pipeline
NSR	2-phase, bundled data	own 16-bit RISC-like	pipelined, no forwarding, decoupled branch and load/store
CFPP	2-phase, bundled data	SPARC	pipelined, multiple execution stages, single issue, result pipeline, forwarding using counter-flow
AMULET1	2-phase, bundled data	ARM	pipelined, no forwarding
TITAC-1	2-phase, dual-rail, quasi delay-insensitive	own 8-bit	nonpipelined
Fred	2-phase, bundled data	based on 88100	pipelined, multiple functional units, single issue, no forwarding, decupled branch and load/store
Hades	unspecified	own	pipelined, multiple functional units, multiple issue, forwarding
ECSTAC	fundamental mode	own variable length	pipelined, no forwarding
AMULET2e	4-phase, bundled data	ARM	pipelined, forwarding
SCALP	4-phase, bundled data	own	pipelined, multiple functional units, multiple issue, explicit forwarding
TITAC-2	2-phase, dual-rail, scalable delay-insensitive	own 32-bit	pipelined, multiple functional units
AMULET3i	4-phase, bundled data	ARM	pipelined, branch prediction, out-of-order completion, unrestricted forwarding

7.3.2 Projects

A number of asynchronous microprocessors have been proposed or built recently. The processors described can be divided broadly into two categories:

- Those that were built using a conservative timing model, suitable for formal synthesis or verification, but with a simple architecture. Among these are CAP, TITAC, and ST-RISC.
- Those that were built with a less cautious timing model using an informal design approach, but with a more ambitious architecture. These include the AMULET processors, NSR, Fred, CPP, Hades, ECSTAC, STRiP, and SCALP.

Table 7.1 summarizes these characteristics.

Let us describe the architecture and asynchronous design of the asynchronous superscalar processor SCALP, and only briefly, some other projects. We follow the presentation of *Endecott* [78].

Superscalar Asynchronous Low-Power Processor

The first asynchronous superscalar processor was designed in 1996 by *Endecott*[78] from the University of Manchester. The processor was named SCALP, for Superscalar Asynchronous Low-Power Processor. SCALP's main architectural innovation is its lack of a global register file and its result forwarding network. Most SCALP instructions do not specify the source of their operands and the destination of their results by means of register numbers. Instead, *Endecott* introduced the idea of *explicit forwarding* whereby each instruction specifies the destination of its result. That destination is the input to another FU consuming the value. Instructions do not specify the source of their operands at all; they implicitly use the values provided for them by the preceding instructions.

Figure 7.7 shows the organization of the SCALP processor. SCALP does have a register file; it constitutes one of the FUs. It is accessed only by read and write instructions which transfer data to and from other FUs by means of the explicit forwarding mechanism. Several instructions are fetched from memory at a time. Each instruction has a *functional unit identifier*, which is a small number of easily decoded bits that indicate which FU will execute the instruction. The instructions are statically allocated to FUs. If there is more than one FU capable of executing a particular instruction, one must be chosen by the compiler. This simplifies the instruction issuer and is essential to the explicit forwarding mechanism. The instruction issuer is responsible for distributing the instructions to the various FUs on the basis of the functional unit identifier. Each FU has a number of input queues: one for instructions and one for each of its possible operands. An instruction begins execution once it and all of its necessary operands have arrived at the FU. The FU sends the result, along with the destination address, to the result-routing

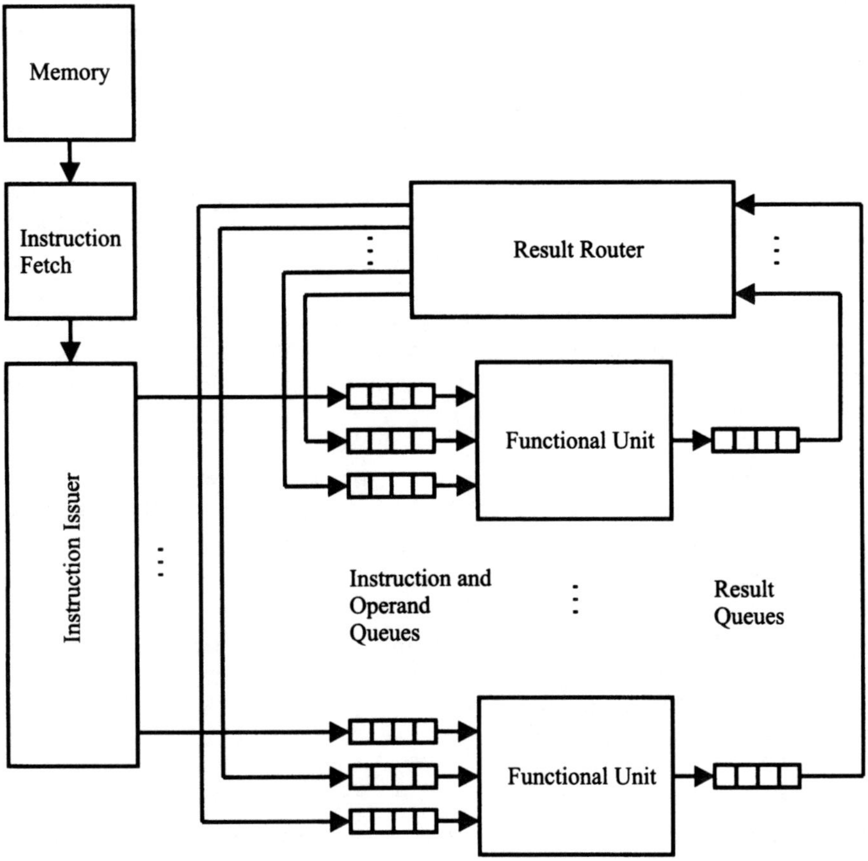

Fig. 7.7. SCALP overall organization

network. This places the result into the appropriate input queue of another FU.

There are some similarities between the SCALP approach and dataflow computing (see Chap. 2). In particular, it is possible to describe SCALP programs by means of dataflow graphs.[2] Nevertheless, the flow of control in SCALP is determined by a conventional control-flow mechanism, not a dataflow mechanism.

Other Projects

CAP (*Martin et al.* [193], 1989): The Caltech Asynchronous Processor (CAP) was built at California Institute of Technology (Pasadena, CA). The circuit design was delay-insensitive with dual-rail encoded communication.

[2] Note the similarity between the asynchronous control-driven SCALP and the synchronous data-driven machine MADAME (see p. 94).

The processor featured a RISC-like load/store instruction set with 16 registers. A number of concurrent processes were responsible for instruction fetch, operand read, ALU operate, etc. The processor was implemented in a 1.6 μm CMOS process, and operated at 18 MIPS at room temperate and 5 V. The circuit continued to function at very low supply voltages, with optimum energy per operation at around 2 V. It was also tested in liquid nitrogen at 77 K when its performance reached 30 MIPS. More recently, a GaAs version of CAP has been implemented (*Tierno et al.* [297], 1994).

FAM (*Cho et al.* [48], 1992): FAM was a dual-rail asynchronous processor with a RISC-like load/store instruction set. It had a 4-stage pipeline but register read, ALU, and register write occurred in a single stage eliminating the need for any forwarding.

STRiP (*Dean* [61], 1992): STRiP was built at Stanford University. Its instruction set was that of the MIPS-X processor. STRiP is included here even though it has a global clock signal and could be considered synchronous. It is unusual in that the speed of the global clock is dynamically variable in response to the instructions being executed, giving much of the advantage of an asynchronous system. The performance of STRiP was typically twice that of an equivalent synchronous processor. By maintaining global synchrony STRiP was able to implement forwarding in the same simple way as synchronous processors.

ST-RISC (*David et al.* [59], 1993): ST-RISC was an architecture from the Israel Institute of Technology. It was delay-insensitive with a dual-rail data path. ST-RISC had a 2-stage pipeline (fetch and execute) and a 3-address-register-based instruction set.

AMULET (*Furber et al.* [93, 94, 95], 1993–1998): At the University of Manchester (UK) several asynchronous processors called AMULET1, AMULET2, and AMULET3 were implemented. AMULET1 was the first asynchronous implementation of a commercially important ISA (ARM's ISA version 3 used in the ARM6 processor) [93], and appeared in early 1993. It was designed using a 2-phase bundled data design style, with a 5-stage pipeline and no result forwarding. An interlocking was used to stall instructions at the register read stage until their operands had been written by previous instructions. After being fetched, a branch instruction had to pass through ten pipeline or FIFO stages before the target address was sent to memory. This resulted in large numbers of prefetched instructions that were discarded and a significant number of bubbles. As a result, the pipeline throughput was low. AMULET1 permitted out-of-order completion of load instructions relative to other instructions. AMULET1 was about 70 % the speed of a 20 MHz ARM6 processor, but with faster simple operations (e.g., with three times faster multiplication). In October 1996, the AMULET2 processor was designed [94], based on the ARM ISA version 4. The processor used a 4-phase bundled data design style because this was believed to have benefits in terms of speed, size, and power relative to AMULET1. The processor had a slightly shorter pipeline than AMULET1 and employed both forwarding and branch prediction. It also incorporated limited forwarding by employing a last-result register at the output of the ALU, and forwarding mechanisms to use the result of a load instruction in a following instruction. A more sophisticated register-interlocking mechanism was used to remove WAW hazards. The AMULET2e chip consisted of 454 000 transistors including a 4 kB fully associative cache. The synchronous equivalent ARM810 used almost twice as many transistors, but also an 8 kB cache. With its 40 MIPS, the AMULET2 was 3.2 times faster than AMULET1, and with half the performance of a 75 MHz ARM810. The next in the AMULET line, AMULET3, is expected to be a commercial product in 1999 [95]. It is expected that the key feature of this microprocessor will be a reorder buffer capable of solving the problems of result forwarding and exception handling in an asynchronous pipeline. This will allow a high degree of flexibility in operation, such as out-of-order completion, whilst avoiding RAW hazards (by stalling until the relevant value appears) and WAW hazards (averted by the reorder buffer).

NSR (*Brunvand* [38], 1993): The Non-synchronous RISC (NSR) processor was built using FPGA technology at the University of Utah. It was implemented using a 2-phase bundled data protocol. NSR was a pipelined processor with pipeline stages separated by FIFO queues. The idea of the FIFO queues is that they decouple the pipeline stages so that an instruction that spends a long time in one stage need not hold up any following instructions. The disadvantage of this approach is that the latency of the queues themselves is significant and, because of the dependencies within a processor pipeline, the increase in overall latency is detrimental. NSR used a locking mechanism to stall instructions that need operands produced by previous instructions. There was no forwarding mechanism. NSR had a 16-bit data path and 16-bit instructions. The instructions included three 4-bit register specifiers and a 4-bit opcode. Some aspects of its instruction set were specialized for the asynchronous implementation: branch, load, and store instructions made the FIFOs that interconnected the functional units visible to the programmer. Conditional branch instructions were decoupled from the ALU that executed comparison instructions by a Boolean FIFO. Computed branch instructions also used a FIFO to store computed branch addresses. Load instructions had two parts. One instruction specifies a load address. A subsequent instruction used the load result by reading from a special register `r1`. There could be an arbitrary separation between the two instructions, and it was possible to have several load operations outstanding at one time. Store instructions worked similarly by writing the data to `r1`.

CFPP (*Sproull et al.* [277], 1994): The Counterflow Pipeline Processor (CFPP) was based on extensions of the techniques proposed in [285]. The CFPP executed SPARC instructions. Its novel contribution was the result forwarding in an asynchronous pipeline. CFPP had two pipelines. In one pipeline instructions flowed upwards; in the other results flowed downwards. As instructions flowed upwards they watched out for results of previous instructions that they had to use as operands flowing downwards. If they spotted any such operands they captured them; otherwise, they eventually received a value that flowed down from the register file which was at the top of the pipelines. When an instruction obtained all of its operands it continued to flow upwards until it found a pipeline stage where it could compute the result. Once computed, the result was injected into the result pipeline for use by any following dependent instructions and was also carried forward in the instruction pipeline to be written into the register file. The counterflow pipeline neatly solved the problem of result forwarding.

TITAC (*Nanya et al.*[209], 1994; *Takamura et al.* [288], 1997): TITAC-1 was a simple 8-bit asynchronous processor built at the Tokyo Institute of Technology (Japan). It was based on the quasi delay-insensitive (QDI) timing model (where additional assumptions are introduced into the delay-insensitive model) and so had to use dual-rail encoding communication. This resulted in about twice as many gates in the data path compared to an equivalent synchronous data path. Architecturally, TITAC-1 was very straightforward with no pipelining and a simple accumulator-based instruction set. In 1997, a 32-bit asynchronous microprocessor TITAC-2 was built whose ISA was based on the MIPS R2000. It uses a scalable delay-insensitive (SDI) model, which unlike the QDI model, assumes that the relative delay ratio between any two components is bounded. SDI circuits can run faster than equivalent QDI circuits. The measured performance of TITAC-2 was 52.3 MIPS using the Dhrystone benchmark.

Fred (*Richardson and Brunvand* [241], 1995): Fred is a development of NSR and also built at the University of Utah. Like NSR, Fred is implemented using 2-phase data bundling. It is modeled using VHDL. Fred extends the NSR to have a 32-bit data path and 32-bit instructions, based on the Motorola 88100 instruction set. Fred has multiple functional units. Instructions from the functional units can complete out of order. However, the instruction issuer can only issue one instruction at a time, and the register file is only able to provide operands for one instruction at a time. This allows for a relatively straightforward instruction issue and a precise exception mechanism, but limits the attainable level of parallelism. There is no forwarding mechanism; instructions are stalled at the instruction issuer until their operands have been written to the register file. There is no out-of-order issue. Like the NSR, Fred uses programmer-visible FIFO queues to implement decoupled load/store and branch instructions. This arrangement has the possibility of deadlock if the program tries to read from an empty queue or write to a full one. Fred chooses to detect this condition at the instruction issuer and generate an exception.

Hades (*Elston et al.* [77], 1995): Hades is a proposed superscalar asynchronous processor from the University of Hertfordshire (UK). It is in many ways similar to a conventional synchronous superscalar processor; it has a global register file, forwarding, and a complex (though in-order) instruction issue. Its forwarding mechanism uses a scoreboard to keep track of which result is available and from where.

ECSTAC (*Morton et al.* [206], 1995): ECSTAC is an asynchronous microprocessor designed at the University of Adelaide (Australia). ECSTAC is implemented using fundamental mode control circuits. It is deeply pipelined with a complex variable-length instruction format. It has 8-bit registers and ALU. The variable-length instructions and the mismatch between the address size and the data path width made the design more complex and slower. There is no forwarding mechanism within the data path, and a register interlocking scheme is used to stall instructions until their operands are available.

7.4 Conclusions

Processor-in-memory, reconfigurable computing, and asynchronous processors are the most exotic techniques described in this book. All three approaches are destined for future chip generations envisioning chip technologies that must still be developed – integration of processor chip technology with DRAM technology in the case of the processor-in-memory approach, and with reconfigurable logic in the case of reconfigurable computing.

Viewed from an architectural point of view, most of the PIM or reconfigurable computing approaches are still rather conventional. However, both research directions have the potential to overcome the von Neumann architectural limitations and lead to highly parallel architectural principles in future.

The asynchronous processor paradigm has the potential to solve the clocking problems in large processor chips.

Acronyms

A

AGP	Accelerated graphics port
ALU	Arithmetic logic unit
AMD	Advanced Micro Device
ANSI	American National Standards Institute
ARB	1. Address resolution buffer 2. Address reorder buffer
ARM	Advanced RISC Machines
AS	Activity store
ASCII	American Standard Code for Information Interchange
ASIC	Application specific integrated chip
ATR	Automatic target recognition

B

BCD	Binary coded decimal
BERT	Branch effect reduction technique
BHR	Branch history register
BHT	Branch history table
BTAC	Branch target address cache
BTB	Branch target buffer
BTC	Branch target cache

C

CAD	Computer-aided design
CAM	Computer-aided manufacturing
CAP	Complexity-adaptive processor
CC-NUMA	Cache-coherent non-uniform memory access
CCR	Condition code register
CDB	Common data bus
CDC	Control Data Corporation
CDRAM	Cache dynamic random access memory

CFG	Control flow graph
CISC	Complex instruction set computer
CMOS	Complementary metal oxide semiconductor
CMP	Chip multiprocessor
CN	Communication network
CP	Computational processor
CPI	Clock cycles per instruction
CPU	Central processing unit
CQ	Continuation queue
CRCW-PRAM	Concurrent read concurrent write parallel random access machine
CSB	Context switch buffer
CWP	Current window pointer

D

D-cache	Data cache
DCM	Data control memory
DEA	Direct external access
DEC	Digital Equipment Corporation
DISC	Dynamic instruction set computer
DMA	Direct memory access
DMEM	Data memory
DRAM	Dynamic random-access memory
DSM	Distributed shared-memory multiprocessor
DSS	Decision support system
DSP	Digital signal processing
DU	Decode unit
DVD	Digital versatile disc

E

ECU	External cache unit
EDO	Extended data output
EDVAC	Electronic discrete variable automatic computer
EMIF	External memory interface
EPIC	Explicitly parallel instruction computing
ETS	Explicit token store
EX	Execution (pipeline stage)
EU	Execution unit

F

FEU, FU	Fetch unit

FIFO	First-in, first-out
FLOPS	Floating-point operations per second
FM	Frame memory
FMAC	Floating-point multiply accumulate
FMU	Fetch/matching unit
FPAA	Field programmable ALU array
FPGA	Field programmable gate array
FPU	Floating-point unit
FTU	Form token unit
FU	1. Functional unit 2. Fetch unit

G

G	Giga (times 10^9)
GAG	Generic address generator
GB	Gigabyte
GM	Global memory
GRU	Graphics unit

H

HDTV	High-definition television
HLL	High-level language
HP	Hewlett-Packard
HPI	Host-port interface

I

IA-32	Intel 32-bit architecture
IA-64	Intel 64-bit architecture
IBM	International Business Machines
IBU	Input buffer unit
I-cache	Instruction cache
ICM	Instruction control memory
ID	Instruction decode (pipeline stage)
IDT	Integrated Device Technology
IEEE	Institute of Electrical and Electronics Engineers
IEU	Integer execution unit
IF	Instruction fetch (pipeline stage)
IFU	Instruction fetch unit
ILP	Instruction-level parallelism
IM, IMEM	Instruction memory
I/O	Input/output

IOC	Input/output cache
IOP	Input/output processor
IPC	Instructions per clock cycle
IQ	Instruction queue
IRAM	Intelligent random access memory
IRB	Instruction reorder buffer
IS	Issue (pipeline stage)
ISA	Instruction set architecture
ISSE	Internet streaming SIMD extension
ISU	Instruction store unit
IU	1. Integer unit 2. Issue unit

J

JVM	Java virtual machine

K

k	Kilo (times 10^3)
K	Kelvin
kB	Kilobyte
KNI	Katmai new instructions

L

L1	Level-one (primary) cache
L2	Level-two (secondary) cache
L3	Level-three (tertiary) cache
LCC	Latch control circuit
LIFO	Last-in, first-out
LIW	Long instruction word
LRU	Least recently used
LSU	Load and store unit

M

M	Mega (times 10^6)
MAX	Multimedia acceleration extension
MB	Megabyte
McBSP	Multi-channel buffered serial port
MCU	Memory control unit
MDC	Miss distance counter
MDCE	Multidimensional directed cycles ensemble

MDMX	MIPS digital media extensions
MDP	Message driven processor
MEM	Memory access (pipeline stage)
MHz	Megahertz
MIMD	Multiple-instructions, multiple-data
MIPS	1. Microprocessor without interlocking pipeline stages
	2. Millions of instructions per second
MIT	Massachusetts Institute of Technology
MMU	Memory management unit
MMX	Matrix manipulation extensions
MoM	Map-oriented machine
MP	Master processor
MPEG	Motion Picture Experts Group
MPI	Message-passing interface
MU	1. Matching unit
	2. Memory unit
MUX	Multiplexer
MVI	Motion video instructions
MVP	Multimedia video processor

N

n	Nano (times 10^{-9})
NCC-NUMA	Non-cache-coherent non-uniform memory access
nm	Nanometer
NOW	Network of workstations
NS	National Semiconductors
NUMA	Non-uniform memory access

O

OFU	Operand fetch unit
OLTP	Online transaction processing
OSU	Operand store unit
OU	Operation unit

P

PA	Precision architecture
PA-RISC	Precision architecture RISC
PBHT	Per-address branch history table
PC	Program counter
PCI	Peripheral component interconnect
PDP	Programmed data processor

PDU	Prefetch and dispatch unit
PE	Processing element
PGP	Pretty good privacy (encryption)
PHT	Pattern history table
PIM	Processor-in-memory
PISA	Pixel-oriented system for image analysis
PM	Program memory
POLU	Problem-oriented logic unit
POWER	Performance optimization with enhanced RISC
PP	Parallel processor
P-RISC	Parallel reduced instruction set computer
PS	Program store
PSR	Processor state register
PTREQ	Packet transfer request
PU	Processing unit
PVM	Parallel virtual machine

Q

QDI	Quasi delay-insensitive

R

rALU	Reconfigurable arithmetic logic unit
RAS	Return address stack
RAM	Random-access memory
RAW	Read after write
RC	Reconfigurable cell
RCA	Radio Corporation of America
rDPA	Reconfigurable data path architecture
RGB	Red-green-blue
RICA	Reduced interprocessor-communication architecture
RISC	Reduced instruction set computer
RO	Read operand (pipeline stage)
ROM	Read-only memory
ROMP	Research Office Products Division microprocessor
RS	Reservation station
RU	1. Receive unit 2. Retire unit

S

SB-PRAM	Saarbrücken Parallel Random Access Machine
SCSI	Small computer system interface

SDI	Scalable delay-insensitive
SDRAM	Synchronized dynamic random access memory
SGI	Silicon Graphics International
SFU	Special-function unit
SIA	Semiconductor Industry Association
SIMD	Single-instruction, multiple-data
SLDRAM	Synchronous-line dynamic random access memory
SLRAM	Synchronous-line random access memory
SMEM	Switch with instruction memory
SMP	Symmetric Multiprocessor
SMT	Simultaneous multithreading
SPARC	Scalable processor architecture
SPEC	System performance evaluation cooperative
SPMD	Single-program, multiple-data
SPSD	Single-program, single-data
SSW	Stream status word
SQL	Structured Query Language
SRAM	Static random-access memory
SU	1. Send unit 2. Switching unit 3. Synchronization unit

T

T	Tera (times 10^{12})
TAP	Test access port
TB	Terabyte
TC	Transfer controller
TI	Texas Instruments
TLB	Translation lookaside buffer
TME	Threaded multipath execution
TQ	Token queue

U

UMA	Uniform memory access
UU	Update unit

V

VC	Video controller
VDRAM	Video random-access memory
VHDL	Very high-speed integrated circuit hardware description language

VIS	Visual instruction set
VLIW	Very large instruction word
VLSI	Very large-scale integration

W

W	1. Word 2. Watt
WAR	Write after read
WAW	Write after write
WB	Write-back (pipeline stage)
WMU	Wait-match unit

μ

μ	Micro (times 10^{-6})
μm	Micron

Glossary

A

address generation interlock – a mechanism to stall the pipeline for one cycle when an address used in one machine cycle is being calculated or loaded in the previous cycle. Address generation interlocks cause the CPU to be delayed for a cycle.

addressing – a mechanism to refer to a device or storage location by an identifying number, character, or group of characters, which may contain a piece of data or a program step.

addressing mode – defines how a processor determines the destination address for an operation. The different addressing modes of a processor determine the variety of ways that an operand or its address can be referenced by an instruction.

addressing range – defines the number of memory locations addressable by the CPU. For a processor with one address space, the range is determined by the number of signal lines on the address bus of the CPU.

alignment – placement of operand values in a memory, at addresses relative to their sizes or length. By naturally aligned data (or aligned, for short) we understand that a data item's lowest-addressed byte must reside in the memory at an address that is a multiple of the size of the data item (in bytes). Thus, a properly aligned value is positioned at an address equal to an integral multiple of its size. For example, the address of a naturally aligned long word is a multiple of four.

antidependence – a potential conflict between two instructions when the second instruction alters an operand which is read by the first instruction. For correct results, the first instruction must read the operand before the second alters it. Also called a write-after-read hazard.

architectural state – the value of registers, flags, and memory as viewed by the programmer.

architecture – an image of a computing system as seen by a programmer capable of programming in machine language. Includes all registers accessible by any instruction, including the privileged instructions, the complete instruction set, all instruction and data formats, addressing modes, and other details that are necessary in order to write a machine language program.

arithmetic instruction – a machine instruction that performs computation, such as addition or multiplication.

arithmetic logic unit (ALU) – the logic circuitry that performs arithmetic calculations on binary numbers and makes logical decisions based on Boolean operations.

associative memory – a memory in which each storage location is selected by its contents and then an associated data location can be accessed. Requires a comparator with each storage location and hence is more complex that random-access memory. Used in fully associative cache memory and in some translation lookaside buffers. Also called content addressable memory.

associativity, in a cache – the number of lines in a set. An n-way set-associative cache has n lines in each set. The term block is also used for line.

asynchronous – a system (e.g. computer, circuit, device) in which events are not executed in a regular time relationship. They are timing-independent. Each event or operation is performed upon receipt of a signal generated by the completion of a previous event or operation, or upon availability of the system resources required by the event or operation.

B

barrel shifter – a shifter, which contains $\log_2$(*max number of bits shifted)* stages, where each stage shifts the input by a different power of 2 number of positions. It lends itself well to being pipelined.

benchmark – standard tests that are used to compare the performance of computers, processors, circuits, or algorithms.

big-endian – a storage scheme in which the most significant unit of data or an address is stored at the lowest memory address.

binary operator – any mathematical operator that requires two data elements with which to perform the operation.

block, in a cache – a group of sequential locations held as one unit in a cache and selected as whole. Also called a line.

branch history table (BHT) – a buffer that is used to hold the history of previous branch paths taken during the execution of individual branch instructions. The BHT is used to improve prediction of the correct branch path whenever a branch instruction is encountered.

branch penalty – the delay in a pipeline after a branch instruction when instructions in the pipeline must be cleared from the pipeline and other instructions fetched. Occurs because instructions are fetched into the pipeline one after the other and before the outcome of branch instructions is known.

branch prediction – a mechanism used to predict the outcome of branch instructions prior to their execution. Pipelined machines must fetch the next instruction before they have completely executed the previous instruction. If the previous instruction was a branch, the next instruction fetch could have been from the wrong place. Branch prediction is a technique that attempts to infer the proper next instruction address, knowing only the current one, typically using an associative memory called a branch target buffer.

branch recovery – when a branch is mispredicted, the speculative state of the machine must be flushed and fetching restarted from the correct target address.

branch target address – the address of the instruction to be executed after a branch instruction if the conditions of the branch are satisfied.

branch target buffer (BTB) – a hardware component that holds the branch target addresses of previously executed branch instructions. Used to predict the outcome of branch instructions when these instructions are next encountered.

C

cache coherence – state of a multiprocessor computer system having multiple caches ensuring (by a cache coherence protocol) that a read-data access of a processor will always deliver the last memory word written by any other processor to that memory address.

cache, direct-mapped – a cache using random-access memory in which each cache line and the most significant bits of its main memory address (the tag) are held together in the cache at a location given by the least significant bits of the memory address (the index). After the cache line is selected by its index, the tag is compared with the most significant bits of the required memory address to find whether the line is the required line and to access the line.

cache, fully associative – a cache using associative memory in which the addresses of the lines are stored with the lines. All the addresses stored in the cache are compared with the incoming address simultaneously to find whether the line is in the cache and to access the line.

cache hit – occurs when the processor requests data from memory and the data requested is already in the cache memory.

cache line – a block of data associated with a cache tag.

cache memory – a small, fast, redundant memory used to store the most frequently accessed parts of the main memory.

cache miss – occurs when the processor requests data from memory and the data requested is not in the cache memory. When this occurs it is necessary to access the next level in the memory hierarchy (potentially the main memory) to retrieve the data.

cache, set-associative – a cache which is divided into a number of sets, each set consisting of groups of lines and each line has its own stored tag (the most significant bits of the address). A set is accessed first using the index (the least significant bits of the address). Then all the tags in the set are compared with that of the required line to find whether the line is in the cache and to access the line.

cache, unified – a cache which can hold both instructions and data.

central processing unit (CPU) – a part of a computer which performs the actual data processing operations and controls the whole computing system.

CISC processor – a processor with a large quantity of instructions, some of which may be quite complicated, as well as a large quantity of different addressing modes, instruction and data formats, and other attributes. A CISC processor usually has a relatively complicated control unit. Most CISC processors are microprogrammed.

clock cycle – one complete event of a synchronous system's timer, including both the high and low periods.

context switching – an operation that switches the CPU from one process to another, by saving all of the CPU registers for the first and replacing them with the CPU registers for the second.

coprocessor – a processor that is connected to a main processor and operates concurrently with the main processor, although under the control of the main processor. Coprocessors are usually special-purpose processing units, such as floating-point, array, DSP, or graphics data processors.

D

data dependence – the situation between two sequential instructions in a program when the first instruction produces a result that is used as an input operand by the second instruction. To obtain the desired result, the second instruction must not read the location that will hold the result until the first has written its result to the location. Also called a read-after-write dependence or a flow dependence.

dataflow architecture – an architecture that operates by having source operands trigger the issue and execution of each operation, without relying on the traditional, sequential von Neumann style of fetching and issuing instructions.

dataflow computer – a computer in which instructions are executed when the operands that the instructions require become available rather than being selected in sequence by a program counter as in a traditional von Neumann computer. Usually, more than one processor is present to execute the instructions simultaneously when possible.

dataflow graph – a directed graph consisting of named nodes, which represent instructions, and arcs, which represent data dependences between instructions. During the execution of the program, data propagate along the arcs in data packets, called tokens.

D-cache – a cache that only holds the data of a program (not instructions).

delayed branch instruction – a form of conditional branch instruction in which one or more instructions immediately following the branch instruction are executed irrespective of the outcome of the branch. The branch then takes effect. Used to reduce the branch penalty.

delay slot – in a pipelined processor, a time slot following a branch instruction. An instruction issued within this slot is executed regardless of whether the branch condition is met, so it may appear that the program is executing instructions out of order. Delay slots can be filled (by compilers) by rearranging the program steps, but when this is not possible, they are filled with no-op instructions.

demand-driven – execution where an instruction is executed if there is a demand for the result.

dependence – a logical constraint between two operations based on information flowing between their source and/or destination operands; the constraint imposes an ordering on the order of execution of (at least) portions of the operations.

digital signal processor (DSP) – a microprocessor specifically designed for processing digital signals.

dynamic scheduling – issuing instructions to functional units out of program order. The processor can dynamically issue an instruction as soon as all its operands are available and the required functional unit is not busy. Thus, an instruction is not delayed by a stalled previous instruction unless it needs the results of that previous instruction.

E

exception – an event that causes suspension of normal program execution. Types include addressing exception, data exception, operation exception, overflow exception, protection exception, and underflow exception.

explicit token store – the concept of allocating a separate frame for each active loop iteration or subprogram invocation in the token memory of a dataflow processor.

F

fetch cycle – the period of time during which an instruction is retrieved from memory.

field programmable gate array (FPGA) – a programmable logic device which consists of a matrix of programmable cells embedded in a programmable routing mesh. The combined programming of the cell functions and the routing network define the function of the device.

firing rule – a computational rule of the dataflow model that specifies when an instruction can actually be executed.

floating-point unit (FPU) – a circuit that performs floating-point computations, which are generally addition, subtraction, multiplication, or division.

flush (pipeline) – the act of clearing out all actions being processed in a pipeline structure. This may be achieved by aborting all of those actions, or by refusing to issue new actions to the pipeline until those present in the pipeline have left the pipeline because their processing has been completed.

forwarding – to provide the result of the previous instruction immediately to the current instruction, before the result is written to the register file. Also called bypass.

functional unit (FU) – a module in which actual instruction execution takes place. There may be a number of functional units of different types within a single CPU, including integer units, floating-point units, load/store units, and branch units.

G

general-purpose register – a digital storage element inside the CPU which is used to hold values temporarily for later transfer to the ALU or memory. General-purpose registers are typically not equipped with any dedicated logic to operate on the data stored in the register.

H

Harvard architecture – a computer design feature where there are two separate memory units: one for instructions and the other for data.

I

I-cache – a cache that only holds the instructions of a program (not data). I-caches generally do not need a write policy.

in-order issue – the situation in which instructions are sent to be executed in the same order as they appear in the program.

instruction decoder unit – the module that receives an instruction from the instruction fetch unit, identifies the type of instruction from the opcode, assembles the complete instruction with its operands, and sends the instruction to the appropriate functional unit, or to an instruction pool to await execution.

instruction fetch unit – the module that fetches instructions from memory, usually in conjunction with a bus interface unit, and prepares them for subsequent decoding and execution by one or more functional units. If an I-cache is existent, the instructions are fetched from the I-cache.

instruction format – the specification of the number and size of all possible instruction fields in an instruction set architecture.

instruction issue – the act of initiating the performance of an instruction (not its fetch). Issue policies are important design decisions in systems that use parallelism and execution out of program order to achieve more speed.

instruction-level parallelism (ILP) – the concept of executing two or more instructions in parallel (generally instructions taken from a sequential, not parallel, stream of instructions).

instruction pipeline – a structure that separates the execution of instructions into multiple phases, and executes separate instructions in each phase simultaneously.

instruction reordering – a technique in which the CPU executes instructions in an order different from that specified by the program, with the purpose of increasing the overall execution speed of the CPU.

instruction scheduling – the relocation of independent instructions in order to maximize instruction-level parallelism (and/or minimize instruction stalls).

instruction set – the collection of all the machine-language instructions available to the programmer.

instruction window – for an out-of-order issue mechanism, a buffer holding a group of instructions being considered for issue to functional units. Instructions are issued from the instruction window when dependences have been resolved.

integer unit – a type of functional unit designed specifically for the execution of integer-type instructions.

interleaving, block – instructions of a thread are executed successively until an event occurs that may cause latency. This event induces a context switch. Also called coarse-grained multithreading.

interleaving, cycle-by-cycle – an instruction of another thread is fetched and fed into the execution pipeline at each processor cycle. Also called fine-grained multithreading.

internal forwarding – a mechanism in a pipeline which allows results from one pipeline stage to be sent directly back to one or more waiting pipeline stages. The technique can reduce stalls in the pipeline.

I-structure – may be viewed as a data repository obeying the single-assignment rule.

J

Java Virtual Machine – the (abstract) engine that actually executes a Java program compiled to Java bytecode.

L

L1 cache – in systems with two separate sets of cache memory between the CPU and standard memory, the set nearest the CPU. L1 cache is often provided within the same integrated circuit that contains the CPU. In operation, the CPU accesses L1 cache memory; if L1 cache memory does not contain the required reference, it accesses L2 cache memory, which in turn accesses standard memory, if necessary.

L2 cache – in systems with two separate sets of cache memory between the CPU and standard memory, the set between L1 cache and standard memory.

line, bus – one wire of a bus, which may be used for transmitting a datum, a bit of an address, or a control signal.

line, cache – a group of words from successive locations in memory stored in cache memory together with an associated tag, which contains the starting memory reference address for the group.

little-endian – a storage scheme in which the least significant unit of data or an address is stored at the lowest memory address.

load instruction – an instruction that requests a datum from a memory address to be placed in a specified register.

load/store architecture – a system design in which the only processor operations that access memory are simple register loads and stores.

load/store unit – a functional unit used to process instructions that load data from memory or store data to memory.

local bus – the set of wires that connects a processor to its local memory module.

logical operation – the machine-level instruction that performs Boolean operations.

lookahead – the number of instructions that can be accessed for issue by the scheduler of the instruction window (usually corresponding to the length of the instruction window).

M

machine code – the machine format of a compiled executable, in which individual instructions are represented in binary notation.

main memory – the level of memory hierarchy farthest from the processor.

memory data register – the processor register that holds data being written to or read from memory.

memory hierarchy – the separation of memory systems into groups based on cost and access times. The "top" of the hierarchy is usually the most expensive, fastest and smallest, from where information "percolates" as it is used.

memory latency – the time between the initiation of a memory request and its completion.

memory management unit (MMU) – a part of a processor, or a separate component, that implements virtual memory functions. A MMU translates virtual addresses from the processor into real addresses for the memory.

memory-reference instruction – an instruction that communicates with memory, writing to it (store) or reading from it (load).

memory word – the total number of bits that may be stored in each addressable memory location.

M.E.S.I. protocol – a cache coherence protocol for a single-bus multiprocessor. Each cache line exists in one of four states, modified (M), exclusive (E), shared (S), or invalid (I).

microarchitecture, processor – refers to the internal organization of the processor. Several specific processors with different microarchitectures may share the same architecture.

microcode – a collection of low-level operations that are executed as a result of a single instruction being issued.

MIMD architecture – a parallel processing system architecture where there is more than one processor and where each processor performs different instructions on different data values simultaneously.

multiprocessor – a computer system that has more than one internal processor capable of operating collectively on a computation. Normally associated with those systems where the processors can access a common main memory.

multithreaded architecture – supports execution whereby several enabled instructions from different threads all become candidates for execution.

N

no fetch on write, in a cache – in a write-through cache policy, a line is not fetched from the main memory into the cache on a cache miss, if the reference is a write reference. Also called non-allocate on write as space is not allocated in the cache on write misses.

no-op – a computer instruction that performs no operation. It can be used to put a delay between the execution of other instructions.

O

opcode – a part of an assembly language instruction that represents an operation to be performed by the processor.

operand – specification of a storage location that provides data to or receives data from an operation.

operand address – the location of an element of data that will be processed by the computer.

operand address register – the internal CPU register that points to the memory location that contains the data element that will be processed by the computer.

out-of-order issue – the situation in which instructions are sent to be executed not necessarily in the order that they appear in the program. An instruction is issued as soon as any data dependences with other instructions are resolved.

output dependence – the situation when two sequential instructions in a program write to the same location. To obtain the desired result, the second instruction must write to the location after the first instruction. Also known as write-after-write hazard.

P

parallel architecture – a computer system architecture made up of multiple CPUs.

parallel computing – performed on computers that have more then one CPU operating simultaneously.

PC-relative addressing – an addressing mechanism for machine instructions in which the address of the target location is given by the contents of the program counter and an offset held as a constant in the instruction, added together. Allows the target location to be specified as a number of locations from the current (or next) instruction. Generally only used for control transfer instructions (e.g. jumps and branches).

physical address – the actual address of a value in the physical memory.

physical register set – an additional register set to hold the results of speculative instruction execution until the instruction retires. Physical registers (also called rename registers) are used to prevent conflicts between instructions that would normally use the same registers. See also: speculative execution.

pipeline hazard, control – arises from branch, jump, and other control-flow change instructions. For example, if a branch is to be taken, the flow of instructions into the pipeline has to be interrupted, and the branch target must be fetched before the pipeline can resume execution.

pipeline hazard, data – arises because of the unavailability of an operand.

pipeline hazard, structural – arises from some combinations of instructions that cannot be accommodated because of resource conflicts.

pipeline interlock – a hardware mechanism to prevent instructions from proceeding through a pipeline when a data dependence or other conflict exists.

pipeline latency – the number of cycles between the time an instruction is issued and the time a dependent instruction (which uses its result as an operand) can be issued.

pipeline machine cycle – the time required to move an instruction one step down the pipeline.

pipeline processor – a processor that executes more than one instruction at a time, in pipelined fashion. The execution of each instruction is divided into a sequence of simpler suboperations. Each suboperation is performed by a separate hardware section called a stage, and each stage passes its result to a succeeding stage. Normally, each instruction only remains at each stage for a single cycle, and each stage begins executing a new instruction as previous instructions are being completed in later stages. Thus, a new instruction can often begin during every cycle. Pipelines greatly improve the rate at which instructions can be executed, as long as there are no dependences. The efficient use of a pipeline requires that several instructions be executed in parallel, however the result of any instruction is not available for several cycles after that instruction has entered the pipeline. Thus, new instructions must not depend on the results of instructions which are still in the pipeline.

pipeline repeat rate – the number of cycles that occur between the issuance of one instruction and the issuance of the next instruction to the same functional unit.

pipeline throughput – the number of instructions that can leave a pipeline per cycle.

pipelining – splitting the CPU into a number of stages, which allows multiple instructions to be executed concurrently.

pop instruction – an instruction that retrieves contents from the top of the stack and places the contents in a specified register.

postincrementation – an addressing mode in which the address is incremented after accessing the memory value. Used to access elements of arrays in memory.

precise interrupts – an implementation of the interrupt mechanism such that the processor can restart after the interrupt at exactly where it was interrupted. All instructions that have started prior to the interrupt should appear to have completed before the interrupt takes place and all instructions after the interrupt should not appear to start until after the interrupt routine has finished.

predecrementation – an addressing mode using an index or address register in which the contents of the address are reduced by the size of the operand before the access is attempted.

prediction (of branches) – the act of guessing the likely outcome of a conditional branch decision. Prediction is an important technique for speeding up execution in overlapped processor designs. Increasing the depth of the prediction (the number of branch predictions that can be unresolved at any time) increases both the complexity and speed.

prefetching – the act of fetching instructions prior to being needed by the CPU.

prefetch queue – a queue of instructions which have been prefetched.

preincrementation – an assembly language addressing mode in which the address is incremented prior to accessing the memory value. Used to access elements of arrays in memory.

Princeton architecture – a computer architecture in which the same memory holds both data and instructions.

processor-in-memory (PIM) – integrates one or more processors with large on-chip memory, which provides the processor(s) with sufficient bandwidth at a reasonable cost.

program counter (PC) – a CPU register that contains the address of the next instruction in sequence to be executed.

push instruction – an instruction that stores the contents of a specified register(s) on the stack.

Q

quadword – a data unit formed from four words.

queue – a data structure maintaining a first-in, first-out discipline of insertion and removal.

R

random-access memory (RAM) – a memory that allows access to any element in the same period of time.

read-only memory (ROM) – a form of random access memory in which storage locations can only be accessed for reading, not for writing. Normally also has non-volatile characteristics.

reconfigurable computing system – combines programmable general-purpose computing with reconfigurable hardware.

register – a circuit formed from identical flip-flops or latches and capable of storing several bits of data.

register direct addressing – an instruction addressing method in which the memory address of the data to be accessed or stored is found in a general-purpose register.

register file – a collection of CPU registers addressable by number.

register indirect addressing – an instruction addressing method in which the register field contains a pointer to a memory location that contains the memory address of the data to be accessed or stored.

register renaming – dynamically allocating a location in a special register file for an instance of a destination register appearing in an instruction prior to its execution. Used to remove antidependences and output dependences.

register window – a set, or window, of registers selected out of a larger group.

relative addressing – an addressing mechanism in which the address of the target location is given by the contents of a specific register and an offset held as a constant in the instruction, added together.

reorder buffer – a set of storage locations holding instructions (and sometimes also the result values) in program order.

reservation station – a storage location placed in front of the functional units and provided to hold instruction and associated operands until the functional units become available.

resource conflict – the situation when a component such as a register or functional unit is required by more than one instruction simultaneously.

retire unit – the unit used to assure that instructions are completed in program order, even though they may have been executed out of order.

RISC processor – a processor implementing the computer design philosophy of a relatively simple control unit design with a reduced number of instructions (selected to be simple), data and instructions formats, and addressing modes. The processor is pipelined. One of the particular features of a RISC processor is the restriction that all memory accesses should be by load and store instructions only (the so-called load/store architecture). All arithmetic logic operations in a RISC are register-to-register, meaning that both the sources and destinations of all operations are CPU registers. All this tends to reduce CPU-to-memory data traffic significantly, thus improving performance. In addition, RISCs usually have the following properties: most instructions execute within a single cycle, all instructions have the same size, the control unit is hardwired (to increase the speed of operations), and there is a CPU register file of considerable size.

S

scalar processor – a CPU that issues at most one instruction at a time.

scoreboard – a centralized control unit which enables out-of-order execution of instructions. It holds various information to detect dependences.

shared-memory architecture – an organization of a computer system having more than one processor in which each processor can access a common main memory.

SIMD architecture – a parallel processing architecture where more than one processor performs the same instruction on different data simultaneously.

simultaneous multithreading (SMT) – when instructions are simultaneously issued from multiple threads to the functional units of a superscalar processor.

single-address instruction – an instruction defining an operation and exactly one address of an operand or another instruction.

single-assignment rule – means that a variable may appear on the left-hand side of an assignment only once within the area of the program in which it is active.

(single)-chip multiprocessor or multiprocessor chip (CMP) – integrates two or more complete processors on a single chip.

source operand – in ALU operations, one of the input values.

spatial locality, cache – when items whose addresses are near one another tend to be referenced close together in time.

SPEC benchmarks – suites of test programs created by the System Performance and Evaluation Cooperative. The cooperative was formed by four companies, Apollo, Hewlett-Packard, MIPS, and Sun Microsystems, to evaluate smaller computers. The programs are actual scientific and engineering applications.

speculative execution – a technique in which instructions are executed speculatively and are discarded if speculation was wrong.

stack – a hardware or software data structure in which items are stored in a last-in, first-out manner.

stack architecture – an architecture that accesses data as though it were in a pile and only the top-most elements are directly accessible.

stall, in a pipeline – a pause in processing instructions in a pipeline, usually caused by an instruction dependence or resource conflict.

static prediction – a method of branch prediction which uses machine-fixed prediction (e.g., predict always taken/not taken) or which relies on the compiler selecting one of the two alternative instructions for execution after the branch instruction (either the next instruction or that at the target location specified in the branch instruction). A bit is provided in the branch instruction which is set to a 0 for one alternative and 1 for the other. The processor then follows this advice when it executes the branch instruction.

store instruction – a machine instruction which copies the contents of a register into a memory location.

superpipelining – a pipeline design technique in which for every external clock cycle two or more pipeline stages are processored within the processor; because this is standard in contemporary processors, it often means just a long pipeline.

superscalar processor – a processor able to issue multiple instructions dynamically each clock cycle from a conventional linear instruction stream.

symmetric multiprocessor (SMP) – a multiprocessor system where all processors are connected by a global memory system (in contrast to a distributed shared-memory multiprocessor where all memory modules are physically distributed to the processors).

synchronous – an operation or operations that are controlled or synchronized by a clocking signal.

system bus – in digital systems, the main bus over which information flows.

T

tag, in caches – a part of a memory address held in a direct mapped or set-associative cache next to the corresponding line, generally the most significant bits of the address.

temporal locality, cache – when recently accessed items are likely to be accessed in the near future.

threaded dataflow – a technique where the dataflow principle is modified so that instructions of certain instruction streams are processed in succeeding machine cycles.

Tomasulo's scheme – a hardware-dependent resolution scheme that allows out-of-order execution of instructions in the presence of hazards.

trace cache – a new paradigm for caching instructions.

translation lookaside buffer (TLB) – for a paging system, a high-speed hardware lookup table for the conversion of virtual addresses generated by the processor into real addresses. The table is of limited size and only holds recently used page addresses.

two-address instruction – a class of instruction in which two operands addresses are specified and the third one is implicit. One of the two addresses is also used to store the result of the ALU operation.

two-port memory – a memory system that has two access paths, one path is usually used by the CPU and the other by I/O devices. This is also called dual-port memory.

two-way interleaved – in memory technology, a technique that provides faster access to memory values by interleaving memory values in two separate modules.

U

U-interpreter – a method in dataflow computing used for assigning tags to each execution of an instruction.

unary operation – an operation a computer performs that involves only one data element.

unconditional branch – an instruction that causes a transfer of control to another address without regard to the state of any condition flags.

user-visible register – an alternative name for general-purpose registers, emphasizing the fact that these registers are accessible to the instructions in user programs. The counterpart to user-visible registers are registers that are reserved for use by privileged instructions, particularly within the operating system.

V

virtual address – the address generated by the processor in a paging (virtual memory) system.

virtual memory – a system to handle the memory hierarchy, providing an automatic method of transferring the contents of blocks of memory (pages) into the main memory when needed. Relies on using two addresses for each stored word, a virtual address which is generated by the processor and the corresponding real address for accessing the memory.

VLIW processor – a computer architecture that performs no dynamic analysis on the instruction stream of long instruction words and executes operations precisely as packed by the compiler into a long machine word.

W

write-back cache – locations in cache memory are grouped together in blocks and when it is necessary to update main memory to reflect changes in the cache, the entire block of main memory is updated rather than just individual locations.

Z

zero-address instruction – an instruction in which the operands are kept on a first-in, first out stack in the CPU, and thus require no explicit addresses.

References

1. Ackerman W.B., Dennis J.B. (1979) VAL - A Value-Oriented Algorithmic Language, Preliminary Reference Manual. Tech. Report TR 218, Laboratory for Computer Science, MIT, Cambridge, MA
2. Agarwal A., Babb J., Chaiken D., D'Souza G., Johnson K.L., Kranz D., Kubiatowicz J., Lim B.-H., Maa G., Mackenzie K. (1993) Sparcle: A Multithreaded VLSI Processor for Parallel Computing. Lecture Notes in Computer Science 748:359, Springer-Verlag, Berlin
3. Agarwal A., Bianchini R., Chaiken D., Johnson K.L., Kranz D., Kubiatowicz J., Lim B.-H., Mackenzie K., Yeung D. (1995) The MIT Alewife Machine: Architecture and Performance. In: Proceedings of the 22nd Annual International Symposium on Computer Architecture (June), Santa Margherita Ligure, 2–13
4. Agerwala, T., Cocke J. (1987) High Performance of Reduced Instruction Set Processors. IBM Technical Report (March)
5. Albonesi D. (1998) Dynamic IPC/Clock Rate Optimization. In: Proceedings of the 25th Annual International Symposium on Computer Architecture (June-July), Barcelona, 282–292
6. Alpert D., Avnon D. (1993) Architecture of the Pentium Microprocessor. IEEE Micro 13:11–21 (June)
7. Alsup M. (1990) Motorola's 88000 Family Architecture. IEEE Micro 10:48–66 (June)
8. Alverson G., Kahan S., Korry R., McCann C., Smith J.B. (1995) Scheduling on the Tera MTA. Lecture Notes in Computer Science 949:19–44, Springer-Verlag, Berlin
9. Alverson R., Callahan D., Cummings D., Koblenz B., Porterfield A., Smith J.B. (1990) The Tera Computer System. In: Proceedings of the 1990 International Conference on Supercomputing (June), Amsterdam, 1–6
10. Amdahl G.M., Blaauw G.A., Brooks Jr. F.P. (1964) Architecture of the IBM System/360. IBM J Res Dev 8:87–101 (April)
11. Anderson D.W., Sparacio F.J., Tomasulo R.M. (1967) The IBM System/360 Model 91: Machine Philosophy and Instruction Handling. IBM J Res Dev 11(1):8–24
12. Ang B.S., Arvind, Chiou D. (1995) StarT the Next Generation: Integrating Global Caches and Dataflow Architecture. In: Gao G.R., Bic L., Gaudiot J.-L. (eds.) Advanced Topics in Dataflow Computing and Multithreading. IEEE Computer Society Press, Los Alamitos, CA, 19–54
13. Ang B.S., Chiou D., Rudolph L., Arvind (1996) Message Passing Support on StarT-Voyager. Technical Report MIT/CSG Memo 387, Laboratory for Computer Science, MIT, Cambridge, MA
14. Arvind, Gostelow K.P., Plouffe W. (1978) An Asynchronous Programming Language and Computing Machine. Technical Report 114a, Department of Information and Computer Science, University of California at Irvine, CA

15. Arvind, Dahbura A.T., Caro A. (1997) Computer Architecture Research and the Real World. Technical Report MIT/CSG Memo 397, Laboratory for Computer Science, MIT, Cambridge, MA
16. Arvind, Nikhil R.S. (1987) Executing a Program on the MIT Tagged-Token Dataflow Architecture. Lecture Notes in Computer Science 259:1–29, Springer-Verlag, Berlin
17. Arvind, Thomas R.E. (1981) I-structures: An Efficient Data Type for Functional Languages. Technical Report MIT/LCS/TM-210, Laboratory for Computer Science, MIT, Cambridge, MA
18. Asada K., Terada H., Matsumoto S., Miyata S., Asano H., Miura H., Shimizu M., Komori S., Fukuhara T., Shima K. (1987) Hardware Structure of a One-Chip Data-Driven Processor: Q-p. In: Proceedings of the 16th International Conference on Parallel Processing (August), St.Charles, IL, 327–329
19. August D.I., Connors D.A., Mahlke S.A., Sias J.W., Crozier K.M., Cheng B.-C., Eaton P.R., Olaniran Q.B., Hwu W.W. (1998) Integrated Predicated and Speculative Execution in the IMPACT EPIC Architecture. In: Proceedings of the 25th Annual International Symposium on Computer Architecture (June-July), Barcelona, 227–237
20. August D.I., Hwu W.W., Mahlke S.A. (1997) A Framework for Balancing Control Flow and Predication. In: Proceedings of the 30th Annual International Symposium on Microarchitecture (December), Research Triangle Park, NC, 92–103
21. Bach P., Braun M., Formella A., Friedrich J., Grün T., Linchtenau C. (1997) Building the 4 Processor SB-PRAM Prototype. In: Proceedings of the 30th Hawaii International Symposium on System Science (January), 5:14–23
22. Barkhordarian S. (1987) RAMPS: A Realtime Structured Small-Scale Data Flow System for Parallel Processing. In: Proceedings of the 16th International Conference on Parallel Processing (August), St.Charles, IL, 610–613
23. Barroso L.A., Gharachorloo K., Bugnion E. (1998) Memory System Characterization of Commercial Workloads. In: Proceedings of the 25th Annual International Symposium on Computer Architecture (June-July), Barcelona, 3–14
24. Beck M., Ungerer T., Zehender E. (1993) Classification and Performance Evaluation of Hybrid Dataflow Techniques with Respect to Matrix Multiplication. In: Proceedings of the GI/ITG Workshop PARS (April), Dresden, 118–126
25. Becker M.C., Allen M.S., Moore C.R., Muhich J.S., Tuttle D.P. (1993) The PowerPC 601 Microprocessor. IEEE Micro 13:54–67 (October)
26. Bertin P., Roncin D., Vuillemin J. (1989) Introduction to Programmable Active Memories. In: McCanny J., McWhirther J., Swartslander E. (eds.) Systolic Array Processors. Prentice Hall, Englewood Cliffs, NJ, 300–309
27. Bhandarkar D., Ding J. (1997) Performance Characterization of the Pentium Pro Processor. In: Proceedings of the 3rd International Symposium on High-Performance Architecture (February), San Antonio, TX, 288–297
28. Bittner R., Athanas P. (1997) Wormhole Run-Time Reconfiguration. In: Proceedings of the ACM/SIGDA International Symposium on Field Programmable Gate Arrays (February), Monterey, CA, 79–85
29. Blanchard T.W., Tobin P.G. (1997) The PA 7300LC Microprocessor: A Highly Integrated System on a Chip. Hewlett-Packard Journal 43(3):43–47
30. Boekelheide K. (1979) A High-Level, Graphical, Data-Driven Language. In: Proceedings of the Workshop on Data Driven Languages and Machines (February), Toulouse
31. Böhm A.P.W., Gurd J.R., Teo Y.M. (1989) The Effect of Iterative Instructions in Dataflow Computers. In: Proceedings of the 18th International Conference on Parallel Processing (August), 201–208

32. Bolychevsky A., Jesshope C.R., Muchnik V.B. (1996) Dynamic Scheduling in RISC Architectures. IEE Proceedings Computers and Digital Techniques 143(5):309–317
33. Boothe R.F., Ranade A. (1992) Improved Multithreading Techniques for Hiding Communication Latency in Multiprocessors. In: Proceedings of the 19th Annual Symposium on Computer Architecture (May), Gold Coast, 214–223
34. Boothe R.F. (1993) Evaluation of Multithreading and Caching in Large Shared Memory Parallel Computers. Technical Report UCB/CSD-93-766, Computer Science Division, University of California, Berkeley, CA
35. Boughton G.A. (1994) Arctic Routing Chip. Lecture Notes in Computer Science 853:310–317, Springer-Verlag, Berlin
36. Brailsford D.F., Duckworth R.J. (1985) The MUSE Machine – an Architecture for Structured Data Flow Computation. New Generation Computing 3:181–195
37. Brehob M., Doom T., Enbody R., Moore W.H., Moore S.Q., Sass R., Severance C. (1996) Beyond RISC – The Post-RISC Architecture. Technical Report TR96-11, Michigan State University (March)
38. Brunvand E. (1993) The NSR Processor. In: Proceedings of the 26th Annual Hawaii International Conference on System Sciences, Maui, HI, 428–435
39. Burger D., Goodman J.R. (1997) Billion-Transistor Architectures (Guest Editor's Introduction). Computer 30:46–48 (September)
40. Burger D., Goodman J.R., Kägi A. (1996) Memory Bandwidth Limitations of Future Microprocessors. In: Proceedings of the 23rd Annual International Symposium on Computer Architecture (May), Philadelphia, PA, 78–89
41. Burger D., Goodman J.R., Kägi A. (1997) Limited Bandwidth to Affect Processor Design. IEEE Micro 17:55–62 (November/December)
42. Burger D., Kaxiras S., Goodman J.R. (1997) DataScalar Architectures. In: Proceedings of the 24th Annual International Symposium on Computer Architecture (June), Denver, CO, 338–349
43. Burgess B. (1998) Specialization: A Way of Life. Computer 31:43 (January)
44. Burgess B., Ullah N., Van Overen P., Ogden D. (1994) The PowerPC 603 Microprocessor. Communications of the ACM 37(6):34–42
45. Butler M., Yeh T.-Y., Patt Y.N., Alsup M., Scales H., Shebanow M. (1991) Single Instruction Stream Parallelism is Greater Than Two. In: Proceedings of the 18th Annual Symposium on Computer Architecture (May), Toronto, 276–286
46. Chang P.-Y., Hao E., Yeh T.-Y., Patt Y.N. (1994) Branch Classification: A New Mechanism for Improving Branch Predictor Performance. In: Proceedings of the 27th International Symposium on Microarchitecture (November-December), San Jose, CA, 22–31
47. Chen D.C., Rabaey J.M. (1992) A Reconfigurable Multiprocessor IC for Rapid Prototyping of Algorithmic-Specific High-Speed Datapaths. IEEE Journal of Solid-State Circuits 27(12)
48. Cho K.-R., Okura K., Asada K. (1992) Design of a 32-bit Fully Asynchronous Microprocessor (FAM). In: Proceedings of the 35th Midwest Symposium on Circuits and Systems (August), 1500–1503
49. Chrysos G.Z., Emer J.S. (1998) Memory Dependence Prediction Using Store Sets. In: Proceedings of the 25th Annual International Symposium on Computer Architecture (June-July), Barcelona, 142-153
50. Circello J., Edgington G., McCarthy D., Gay J., Schimke D., Sullivan S., Duerden R., Hinds C., Marquette D., Sood L., Crouch A., Chow D. (1995) The Superscalar Architecture of the MC68060. IEEE Micro 15:10–21 (April)
51. Colwell R.P. (1998) Maintaining a Leading Position. Computer 31:45–47 (January)

52. Colwell R.P., Nix R.P., O'Donnell J.J., Papworth D.B., Rodman P.K. (1987) A VLIW Architecture for a Trace Scheduling Compiler. In: Proceedings of the 2nd International Conference on Architectural Support for Programming Languages and Operating Systems (October), Palo Alto, CA, 180–192
53. Colwell R.P., Steck R.L. (1995) A 0.6 μm BiCMOS Processor with Dynamic Execution. In: Proceedings of the International Solid State Circuits Conference (February) 176–177
54. Cornish M., Hogan D.W., Jensen J.C. (1979) The Texas Instruments Distributed Data Processor. In: Proceedings of the Louisiana Computer Exposition, 189–193
55. Crisp R. (1997) Direct Rambus Technology: The New Main Memory Standard. IEEE Micro 17:18–28 (November/December)
56. Culler D.E. (1989) Managing Parallelism and Resources in Scientific Dataflow Programs. Ph.D. Thesis, MIT, Cambridge, MA
57. Culler D.E., Sah A., Schauser K.E., von Eicken T., Wawrzynek J. (1991) Fine-Grain Parallelism with Minimal Hardware Support: A Compiler-Controlled Threaded Abstract Machine. In: Proceedings of the 4th International Conference on Architectural Support for Programming Languages and Operating Systems (April), Santa Clara, CA, 164–175
58. Dally W.J., Fiske J., Keen J., Lethin R., Noakes M., Nuth P., Davison R., Fyler G. (1992) The Message-Driven Processor: A Multicomputer Processing Node with Efficient Mechanisms. IEEE Micro 12:23–39 (April)
59. David I., Ginosar R., Yoeli M. (1993) Self-Timed Architecture of a Reduced Instruction Set Computer. IFIP Transactions A-28:29–43
60. Davis A.L. (1978) The Architecture and System Method of DDM1: A Recursively Structured Data Driven Machine. In: Proceedings of the 5th Annual Symposium on Computer Architecture (April), Palo Alto, CA, 210–215
61. Dean M.E. (1992) Strip: A Self-Timed RISC Processor. Technical Report CSL-TR-92-543, Stanford University
62. DeHon A., Hartman D., Mirsky E. (1997) MATRIX: A Reconfigurable Computing Architecture with Configurable Instruction Distribution. In: Proceedings of the Symposium on High-Performance Chips – Hot Chips 9 (August), Stanford, CA
63. Dennis J.B., Gao G.R. (1994) Multithreaded Architectures: Principles, Projects, and Issues. In: Iannucci R.A., Gao G.R., Halstead R., Smith B. (eds.) Multithreaded Computer Architecture: A Summary of the State of the Art. Kluwer Academic Publishers, 1–74
64. Dennis J.B., Misunas D.P. (1975) A Preliminary Architecture for a Basic Data-Flow Processor. In: Proceedings of the 2nd Annual Symposium on Computer Architecture (January), Houston, TX, 126–132
65. Diefendorff, K., Allen M. (1992) Organization of the Motorola 88110 Superscalar RISC Microprocessor. IEEE Micro 12:40–63 (April)
66. Diefendorff K. (1994) History of the Power PC Architecture. Communications of the ACM 37(6):28–33
67. Diefendorff K., Oehler R., Hochsprung R. (1994) Evolution of the PowerPC Architecture. IEEE Micro 14:34–49 (April)
68. Diep T.A., Nelson C., Shen J.P. (1995) Performance Evaluation of the PowerPC 620 Microprocessor. In: Proceedings of the 22nd Annual International Symposium on Computer Architecture (June), Santa Margherita Ligure, 163–174
69. Dobberpuhl D.W., Witek R.T., Allmon R., Anglin R., Bertucci D., Britton S., Chao L., Conrad R.A., Dever D.E., Gieseke B., Hassoun S.M.N., Hoeppner G.W., Kuchler K., Ladd M., Leary B.M., Madden L., McLellan E.J.,

Meyer D.R., Montanaro J., Priore D.A., Rajagopalan V., Samudrala S., Santhanam S. (1992) A 200-MHz 64-bit Dual-Issue CMOS Microprocessor. Digital Technical Journal 4(4):35–50

70. Driesen K., Hoelzle U. (1998) Accurate Indirect Branch Prediction. In: Proceedings of the 25th Annual International Symposium on Computer Architecture (June-July), Barcelona, 167–177
71. Dulong C. (1998) The IA-64 Architecture at Work. Computer 31:24–31 (July)
72. Ebeling C., Cronquist D.C., Franklin P. (1996) RaPiD – Reconfigurable Pipelined Datapath. In: Proceedings of the Symposium on the Field Programmable Logic, Springer-Verlag, Berlin, 126–135
73. Ebeling C., Cronquist D.C., Franklin P. (1997) Configurable Computing: The Catalyst for High-Performance Architectures. In: Proceedings of the IEEE International Conference on Application-Specific Systems, Architectures and Processors (July), Zurich, 364–372
74. Edmondson J.H., Rubinfeld P., Preston R., Rajagopalan V. (1995) Superscalar Instruction Execution in the 21164 Alpha Microprocessor. IEEE Micro 15:33–43 (April)
75. Egan G.K. (1990) The CSIRAC II Dataflow Computer: Token and Node Definitions. Technical Report 31-009, School of Electrical Engineering, Swinburne Institute of Technology, Hawthorn, Australia
76. Eggers S.J., Emer J.S., Levy H.M., Lo J.L., Stamm R.M., Tullsen D.M. (1997) Simultaneous Multithreading: A Platform for Next-Generation Processors. IEEE Micro 17:12–19 (September/October)
77. Elston C.J., Christianson D.B., Findlay P.A., Steven G.B. (1995) Hades – Towards the Design of an Asynchronous Superscalar Processor. In: Proceedings of the Second Working Conference on Asynchronous Design Methodologies, London, 200–209
78. Endecott P.B. (1996) SCALP: A Superscalar Asynchronous Low-Power Processor. Ph.D. Thesis, University of Manchester
79. Espasa R., Valero M. (1997) Exploiting Instruction- and Data-Level Parallelism. IEEE Micro 17:20–27 (September/October)
80. Evers M., Chang P.-Y., Patt Y.N. (1996) Using Hybrid Branch Predictors to Improve Branch Prediction Accuracy in the Presence of Context Switches. In: Proceedings of the 23rd Annual International Symposium on Computer Architecture (May), Philadelphia, PA, 3–11
81. Evripidou P., Gaudiot J.-L. (1991) The USC Decoupled Multilevel Data-Flow Execution Model. In: Gaudiot J.L., Bic L. (eds.) Advanced Topics in Data-Flow Computing. Prentice Hall, Englewood Cliffs, NJ, 347–379
82. Exum M.R., Gaudiot J.-L. (1990) Network Design and Allocation Consideration in the Hughes Data-Flow Machine. Parallel Computing 13:17–34
83. Fillo M., Keckler S.W., Dally W.J., Carter N.P., Chang A., Gurevich Y., Lee W.S. (1995) The M-Machine Multicomputer. In: Proceedings of the 28th International Symposium on Microarchitecture (November/December), Ann Arbor, MI
84. Fisher J.A. (1979) The Optimization of Horizontal Microcode Within and Beyond Basic Blocks. Ph.D. Thesis, New York University
85. Fisher J.A. (1983) Very Long Instruction Word Architectures and the ELI-52. In: Proceedings of the 10th Annual Symposium on Computer Architecture (June), Stockholm, 140–150
86. Fisher J.A. (1997) Walk-Time Techniques: Catalyst for Architectural Change. Computer 30:40–42 (September)
87. Flynn M.J. (1995) Computer Architecture, Pipelined and Parallel Processor Design. Jones and Bartlett Publishers, Sudbury, MA

88. Flynn M.J. (1998) Computer Engineering 30 Years After the IBM Model 91. Computer 31:27–31 (April)
89. Foley P. (1996) The Mpact$^{(tm)}$ Media Processor Redefines the Multimedia PC. In: Proceedings of the 41st IEEE Computer Society International Conference COMPCON 96 (February), Santa Clara, CA, 311–318
90. Formella A., Keller J., Walle T. (1996) HPP: A High Performance PRAM. Lecture Notes in Computer Science 1123:425–434, Springer-Verlag, Berlin
91. Franklin M. (1993) The Multiscalar Architecture. Ph.D. Thesis, Computer Science Technical Report No. 1196, University of Wisconsin-Madison
92. Fromm R., Perissakis S., Cardwell N., Kozyrakis C.E., McGaughy B., Patterson D.A., Anderson T., Yelick K. (1997) The Energy Efficiency of IRAM Architectures. In: Proceedings of the 24th Annual International Symposium on Computer Architecture (June), Denver, CO, 327–337
93. Furber S.B., Day P., Garside J.D., Paver N.C., Woods J.V. (1993) A Micropipelined ARM. In: Proceedings of the IFIP TC10/WG 10.5 International Conference on Very Large Scale Integration (September), Grenoble, 5.4.1–10
94. Furber S.B., Day P., Garside J.D., Paver N.C., Temple S. (1996) AMULET2e. In: Proceedings of EMSYS'96 - OMI 6th Annual Conference (September), Berlin
95. Furber S.B., Garside J.D., Gilbert D.A. (1998) AMULET3: A High-Performance Self-Timed ARM Microprocessor. In: Proceedings of the IEEE International Conference on Computer Design (October), Austin, TX
96. Gabbay F., Mendelson A. (1997) Can Program Profiling Support Value Prediction? In: Proceedings of the 30th Annual International Symposium on Microarchitecture (December), Research Triangle Park, NC
97. Gabbay F., Mendelson A. (1998) The Effect of Instruction Fetch Bandwidth on Value Prediction. In: Proceedings of the 25th Annual International Symposium on Computer Architecture (June-July), Barcelona, 272-281
98. Gaudiot J.-L., Bic L. (1991) Advanced Topics in Dataflow Computing. Prentice Hall, Englewood Cliffs, NJ
99. Gaudiot J.-L., Vedder R.W., Tucker G.K., Campbell M.L., Finn D. (1985) A Distributed VLSI Architecture for Efficient Signal and Data Processing. IEEE Trans Comput 34(12):1072–1087
100. Gillingham P., Vogley B. (1997) SLDRAM: High-Performance, Open-Standard Memory. IEEE Micro 17:29–39 (November/December)
101. Glauert J.R.W., Gurd J.R., Kirkham C.C. (1985) Evolution of a Dataflow Architecture. In: Proceedings of the IFIP WG 10.3 Workshop on Hardware Supported Implementation on Concurrent Languages and Distributed Systems (March), Bristol, 1–18
102. Glück-Hiltrop E., Ramlow M., Schürfeld U. (1989) The Stollman Dataflow Machine. Lecture Notes in Computer Science 365:433–457, Springer-Verlag, Berlin
103. Gokhale M., Holmes W., Kosper A., Lucas S., Minnich R., Sweely D., Lopresti D. (1991) Building and Using a Highly Parallel Programmable Logic Array. Computer 24:81–89 (January)
104. Golston J. (1996) Single-Chip H.324 Videoconferencing. IEEE Micro 16:21–33 (August)
105. Gostelow K.P., Arvind (1982) The U-interpreter. Computer 15:42–49 (February)
106. Grafe V.G., Hoch J.E. (1990) The Epsilon-2 Multiprocessor System. Journal of Parallel and Distributed Computing 10:309–318
107. Grohoski G. (1998) Reining in Complexity. Computer 31:41–42 (January)
108. Grünewald W., Ungerer T. (1996) Towards Extremely Fast Context Switching in a Blockmultithreaded Processor. In: Proceedings of the 22nd Euromicro

Conference: Hardware and Software, Design Strategies (September), Prague, 592–599

109. Grünewald W., Ungerer T. (1997) A Multithreaded Processor Designed for Distributed Shared Memory Systems. In: Proceedings of the International Conference on Advances in Parallel and Distributed Computing (March), Shanghai, 206–213
110. Grunwald D., Klauser A., Manne S., Pleszkun A. (1998) Confidence Estimation for Speculation Control. In: Proceedings of the 25th Annual International Symposium on Computer Architecture (June-July), Barcelona, 122–131
111. Gulati M., Bagherzadeh N. (1996) Performance Study of a Multithreaded Superscalar Microprocessor. In: Proceedings of the 2nd International Symposium on High-Performance Computer Architectures (February), 291–301
112. Gwennap L. (1994) VLIW: The Wave of the Future? Microprocessor Report 8(2):18–21 (February 14)
113. Gwennap L. (1994) 620 Fills Out PowerPC Product Line. Microprocessor Report 8(14):12–16 (October 24)
114. Gwennap L. (1995) Intel's P6 Uses Decoupled Superscalar Design. Microprocessor Report 9(2):9–15 (February 16)
115. Gwennap L. (1996) Digital 21264 Sets New Standard. Microprocessor Report 10(14) (October 28)
116. Gwennap L. (1997) DanSoft Develops VLIW Design. Microprocessor Report 11(2):18–22 (February 17)
117. Gwennap L. (1997) Intel, HP Make EPIC Disclosure. IA-64 Instruction Set Goes Beyond Traditional RISC, VLIW. Microprocessor Report 11(14) (October 27)
118. Gwennap L. (1998) Merced Slips to Mid-2000. Microprocessor Report 12(8):1–2 (June 22)
119. Halstead R.H., Fujita T. (1988) MASA: A Multithreaded Processor Architecture for Parallel Symbolic Computing. In: Proceedings of the 15th Annual Symposium on Computer Architecture (May/June), Honolulu, HI, 443–451
120. Hammond L., Nayfeh B.A., Olukotun K. (1997) A Single-Chip Multiprocessor. Computer 30:79–85 (September)
121. Hammond L., Olukotun K. (1998) Considerations in the Design of Hydra: A Multiprocessor-on-Chip Microarchitecture. Technical Report CSL-TR-98-749, Computer Systems Laboratory, Stanford University
122. Hansen C. (1996) MicroUnity's MediaProcessor Architecture. IEEE Micro 16:34–41 (August)
123. Hartenstein R.W., Herz M., Hoffmann T., Nageldinger U. (1998) On Reconfigurable Co-Processing Units. Lecture Notes in Computer Science 1388:67, Springer-Verlag, Berlin
124. Hartenstein R.W., Hirschbiel A.G., Weber M. (1987) MoM - Map Oriented Machine. In: Proceedings of the International Workshop on Hardware Accelerators (September-October)
125. Hartenstein R.W., Hirschbiel A.G., Weber M. (1988) MoM (Map-oriented Machine), a Reconfigurable Procedurally Data-Driven Machine Architecture. In: Ambler T., Agraval P., Moore W. (eds.) Hardware Accelerators for Electrical CAD, Adam Hilger, Bristol, UK
126. Hartenstein R.W., Hirschbiel A.G., Weber M. (1989) The Map-oriented Machine (MoM), a Custom-designed Architecture Compared to Standard Designs. In: Proceedings of the CompEuro 89, Hamburg, 7–9
127. Hartenstein R., Kress R. (1995) A Datapath Synthesis System for the Reconfigurable Datapath Architecture. In: Proceedings of Asia and South Pacific Design Automation Conference, 479–484

128. Hartenstein R.W, Riedmüller M., Schmidt K., Weber M. (1991) A Novel ASIC Design Approach Based on a New Machine Paradigm. IEEE Journal of Solid-State Circuits 26(7)
129. Hauser J. R., Wawrzynek J. (1997) Garp: A MIPS Processor with a Reconfigurable Coprocessor. In: Proceedings of IEEE Workshop on FPGAs for Custom Computing Machines (April), Napa Valley, CA, 12–21
130. Heil T.H., Smith J.E. (1996) Selective Dual Path Execution. Technical Report, University of Wisconsin-Madison,
131. Hewlett-Packard (1990) PA-RISC 1.1 Architecture and Instruction Set Reference Manual (1st ed.). Part Number 09740-90039 (November)
132. Heller S., Traub K. (1985) Id Compiler User's Manual. Technical Report MIT/CSG Memo 248, Laboratory for Computer Science, MIT, Cambridge, MA
133. Hennessy J.L., Jouppi N., Baskett F., Gill J. (1981) MIPS: A VLSI Processor Architecture. In: Proceedings of the CMU Conference on VLSI Systems and Computation (October), Rockville, MD, 337–346
134. Hennessy J.L., Patterson D.A. (1996) Computer Architecture a Quantitative Approach (2nd ed.). Morgan Kaufmann, San Mateo, CA
135. Hill M.D. (1998) Multiprocessors Should Support Simple Memory-consistency Models. IEEE Micro 18:28–34 (August)
136. Hintz K.J., Tabak D. (1992) Microcontrollers: Architecture, Implementation, and Programming. McGraw-Hill, New York
137. Hiraki K., Shimada T., Nishida K. (1984) A Hardware Design of the SIGMA-1, a Data Flow Computer for Scientific Computations. In: Proceedings of the 13th International Conference on Parallel Processing (August), 524–531
138. Hirata H., Kimura K., Nagamine S., Mochizuki Y., Nishimura A., Nakase Y., Nishizawa T. (1992) An Elementary Processor Architecture with Simultaneous Instruction Issuing from Multiple Threads. In: Proceedings of the 19th Annual Symposium on Computer Architecture (May), Gold Coast, 136–145
139. Hofstee H.P., Dhong S.H., Meltzer D., Nowka K.J., Silberman J.A., Burns J.L., Posluszny S.D., Takahashi O. (1998) Designing for a Gigahertz. IEEE Micro 18:66–74 (May/June)
140. Hum H.H.J., Theobald K.B., Gao G.R. (1994) Building Multithreaded Architectures with Off-the-Shelf Microprocessor. In: Proceedings of the 8th International Parallel Processing Symposium (April), Cancún, 288–294
141. Hunt D. (1995) Advanced Performance Features for the 64-bit PA-8000. In: Proceedings of the 40th IEEE Computer Society International Conference COMPCON 95 (March), San Francisco, CA, 123–128
142. Hwu W.W. (1998) Introduction to Predicated Execution. Computer 31:49–50 (January)
143. Intel (1997) Pentium II Processor Developer's Manual. Intel Corporation 243502-001, October 1997
144. Intel (1997) Intel Architecture Software Developer's Manual. Vol. 1: Basic Architecture. Intel Corporation.
145. Intel (1998) P6 Family of Processors Hardware Developer's Manual. Intel Corporation 244001-001, September 1998
146. International Standard Organization (1996) ISO/IEC 13818-2 1996(E) Information Technology - Generic Coding of Moving Pictures and Associated Audio Information: Video.
147. Ito N., Sato M., Kuno E., Rokusawa K. (1986) The Architecture and Preliminary Evaluation Results of the Experimental Parallel Inference Machine PIM-D. In: Proceedings of the 13th Annual Symposium on Computer Architecture (June), Tokyo, 149–156

148. Jacobsen E., Rotenberg E., Smith J.E. (1996) Assigning Confidence to Conditional Branch Predictions. In: Proceedings of the 29th Annual International Symposium on Microarchitecture (December), Paris, 142–152
149. Jaggar D. (1997) ARM Architectures and Systems. IEEE Micro 17:9–11 (July/August)
150. Johnson M. (1991) Superscalar Microprocessor Design. Prentice Hall, Englewood Cliffs, NJ
151. Jouppi N. (1990) Improving Direct-Mapped Cache Performance by Addition of a Small Fully-Associative Cache and Prefetch Buffer. In: Proceedings of the 17th Annual Symposium on Computer Architecture (May), Seattle, WA, 364–373
152. Kalapathy P. (1997) Hardware-Software Interactions on Mpact. IEEE Micro 17:20–26 (March/April)
153. Kane G. (1995) PA-RISC 2.0 Architecture. Prentice Hall, Engelwood Cliffs, NJ
154. Kane G., Heinrich J. (1991) MIPS RISC Architecture. Prentice Hall, Engelwood Cliffs, NJ
155. Karp R.M., Miller R.E. (1966) Properties of a Model for Parallel Computation: Determinacy, Termination, Queueing. SIAM J Appl Math 14:1390–1411
156. Katayama Y. (1997) Trends in Semiconductor Memories. IEEE Micro 17:10–17 (November/December)
157. Kavi K.M., Hurson A.R., Patadia P., Abraham E., Shanmugam P. (1995) Design of Cache Memories for Multi-Threaded Dataflow Architecture. In: Proceedings of the 22nd Annual International Symposium on Computer Architecture (June), Santa Margherita Ligure, 253–264
158. Kavi K.M., Levine D.L., Hurson A.R. (1997) A Non-Blocking Multithreaded Architecture. In: Proceedings of the 5th International Conference on Advanced Computing (December), Madras, 171–177
159. Kawakami K., Gurd J.R. (1986) A Scalable Dataflow Structure Store. In: Proceedings of the 13th Annual Symposium on Computer Architecture (June), Tokyo, 243–250
160. Keeton K., Patterson D.A., He Y.Q., Raphael R.C., Baker W.E. (1998) Performance Characterization of a Quad Pentium Pro SMP Using OLTP Workloads. In: Proceedings of the 25th Annual International Symposium on Computer Architecture (June-July), Barcelona, 15–26
161. Killian E. (1998) Challenges, Not Roadblocks. Computer 31:44–45 (January)
162. Kishi M., Yasuhara H., Kawamura Y. (1983) DDDP: A Distributed Data Driven Processor. In: Proceedings of the 10th Annual Symposium on Computer Architecture (June), Stockholm, 236–242
163. Klauser A., Paithankar A., Grunwald D. (1998) Selective Eager Execution on the PolyPath Architecture. In: Proceedings of the 25th Annual International Symposium on Computer Architecture (June-July), Barcelona, 250–259
164. Kodama Y., Sakane H., Sato M., Yamana H., Sakai S., Yamaguchi Y. (1995) The EM-X Parallel Computer: Architecture and Basic Performance. In: Proceedings of the 22nd Annual International Symposium on Computer Architecture (June), Santa Margherita Ligure, 14–23
165. Kohn L., Maturana G., Tremblay M., Praghu A., Zyner G. (1995) The Visual Instruction Set (VIS) in UltraSPARC. In: Proceedings of the 40th IEEE Computer Society International Conference COMPCON 95 (March), San Francisco, CA, 462–469
166. Koren I., Mendelson B., Peled I., Silberman G.M. (1988) A Data-Driven VLSI Array for Arbitrary Algorithms. Computer 21:30–43 (October)

167. Kozyrakis C.E., Perissakis S., Patterson D.A., Anderson T., Asanović K., Cardwell N., Fromm R., Golbus J., Gribstad B., Keeton K., Thomas R., Treuhaft N., Yelick K. (1997) Scalable Processors in the Billion-Transistor Era: IRAM. Computer 30:75–78 (September)
168. Krishnan V., Torellas J. (1998) A Clustered Approach to Multithreaded Processors. In: Proceedings of the 1998 IPPS/SPDP Conference (March-April), Orlando, FL, 627–634
169. Kumar A. (1997) The HP PA-8000 RISC CPU. IEEE Micro 17:27–32 (March/April)
170. Lam M.S., Wilson R.P. (1992) Limits of Control Flow on Parallelism. In: Proceedings of the 18th Annual Symposium on Computer Architecture (May), Toronto, 46–57
171. Lamport L. (1979) How to Make a Multiprocessor Computer That Correctly Executes Multiprocess Programs. IEEE Trans Comput 28(9):690–691
172. Laudon J., Gupta A., Horowitz M. (1994) Interleaving: A Multithreading Technique Targeting Multiprocessors and Workstations. In: Proceedings of the 6th International Conference on Architectural Support for Programming Languages and Operating Systems (October), San Jose, CA, 308–318
173. Laudon J., Lenoski D. (1997) The SGI Origin: A ccNUMA Highly Scalable Server. In: Proceedings of the 24th Annual International Symposium on Computer Architecture (June), Denver, CO, 241–251
174. Lee D.C., Crowley P.J., Baer J.-L., Anderson T.E., Bershad B.N. (1998) Execution Characteristics of Desktop Applications on Windows NT. In: Proceedings of the 25th Annual International Symposium on Computer Architecture (June-July), Barcelona, 27–38
175. Lee J.K.F., Smith A.J. (1984) Branch Prediction Strategies and Branch Target Buffer Design. Computer 17:6–22 (January)
176. Lee R.B. (1995) Accelerating Multimedia with Enhanced Microprocessors. IEEE Micro 15:22–32 (April)
177. Lee R.B. (1996) Subword Parallelism with MAX-2. IEEE Micro 16:51–59 (August)
178. Lee W., Barua R., Frank M., Srikrishna D., Babb J., Sakar V., Amarasinghe S. (1998) Space-Time Scheduling of Instruction-Level Parallelism on a Raw Machine. In: Proceedings of the Eighth International Conference on Architectural Support for Programming Languages and Operating Systems (October), San Jose, CA, 46–57
179. Lesartre G., Hunt D. (1997) The Continuing Evolution of the PA-8000 Family. In: Proceedings of the 42nd IEEE Computer Society International Conference COMPCON 97 (February), San Jose, CA
180. Levine F., Thurber S. (1994) RISC System/6000 Power PC System Architecture. Morgan Kaufmann Publishers, San Francisco, CA
181. Li Z., Tsai J.-Y., Wang X., Yew P.-C., Zheng B. (1996) Compiler Techniques for Concurrent Multithreading with Hardware Speculation Support. Lecture Notes in Computer Science 1239:175–191, Springer-Verlag, Berlin
182. Lipasti M.H., Shen J.P. (1997) The Performance Potential of Value and Dependence Prediction. Lecture Notes in Computer Science 1300:1043–1052, Springer-Verlag, Berlin
183. Lipasti M.H., Shen J.P. (1997) Superspeculative Microarchitecture for Beyond AD 2000. Computer 30:59–66 (September)
184. Lipasti M.H., Wilkerson C.B., Shen J.P. (1996) Value Locality and Load Value Prediction. In: Proceedings of the 7th International Conference on Architectural Support for Programming Languages and Operating Systems (October), Cambridge, MA, 138–147

185. Liu Y.C., Gibson G.A. (1986) Microcomputer Systems: The 8086/8088 Family. Prentice Hall, Engelwood Cliffs, NJ

186. Lo J.L., Barroso L.A., Eggers S.J., Gharachorloo K., Levy H.M., Parekh S.S. (1998) An Analysis of Database Workload Performance on Simultaneous Multithreaded Processors. In: Proceedings of the 25th Annual International Symposium on Computer Architecture (June-July), Barcelona, 39–50

187. Loikkanen M., Bagherzadeh N. (1996) A Fine-Grain Multithreading Superscalar Architecture. In: Proceedings of the 1996 Conference on Parallel Architectures and Compilation Techniques (October), Boston, MA, 163–168

188. Mahlke S.A., Hank R.E., McCormick J.E., August D.I., Hwu W-M.W. (1995) A Comparison of Full and Partial Predicated Execution Support for ILP Processors. In: Proceedings of the 22nd Annual International Symposium on Computer Architecture (June), Santa Margherita Ligure, 138–149

189. Mahon M., Lee R., Miller T., Huck J., Bryg W. (1986) Hewlett-Packard Precision Architecture: The Processor. Hewlett-Packard Journal 37(8):4–21

190. Mangione-Smith W.H., Hutchings B., Andrews D., DeHon A., Ebeling C., Hartenstein R., Mencer O., Morris J., Palem K., Prasanna V.K., Spaanenburg H.A.E. (1997) Seeking Solutions in Configurable Computing. Computer 38:38–43 (December)

191. Mankovic T.E., Popescu V., Sullivan H. (1987) CHoPP Priciples of Operations. In: Proceedings of the 2nd International Supercomputer Conference (May), 2–10

192. Maquelin O.C., Hum H.H.J., Gao G.R. (1995) Costs and Benefits of Multithreading with Off-the-Shelf Processors. Lecture Notes in Computer Science 966:117–128, Springer-Verlag, Berlin

193. Martin A.J., Burns S.M., Lee T.K., Borkovic D., Hazewindus P.J. (1989) The Design of an Asynchronous Microprocessor. In: Proceedings of the Decennial Caltech Conference on VLSI, 351–373

194. Matsuoka H., Okamoto K., Hirono H., Sato M., Yokota T., Sakai S. (1998) Pipeline Design and Enhancement for Fast Network Message Handling in RWC-1 Multiprocessor. In: Proceedings of the Workshop on Multithreaded Execution, Architecture and Compilation (January-February), Las Vegas, NV

195. May C., Silha E., Simpson R., Wareen H. (1994) The PowerPC Architecture: A Specification for a New Family of RISC Processors. Morgan Kaufmann Publishers, San Francisco, CA

196. McFarling S. (1993) Combining Branch Predictors. WRL Technical Notes TN-36, Digital Western Research Laboratory

197. McGhan H., O'Connor M. (1998) PicoJava: A Direct Execution Engine for Java Bytecode. Computer 31:22–30 (October)

198. McHenry J. (1995) The WILDFIRE Custom Configurable Computer. In: Proceedings International Society of Optical Engineering: Field Programmable Gate Arrays (FPGAs) for Fast Board Development and Reconfigurable Computing, Philadelphia, PA, 189–199

199. Melear C. (1989) The Design of the 88000 RISC Family. IEEE Micro 9:26–38 (April)

200. Meyer J., Downing T. (1997) Java Virtual Machine. O'Reilly & Associates, Cambridge, MA

201. Mikschl A., Damm W. (1996) Msparc: A Multithreaded Sparc. Lecture Notes in Computer Science 1123:461–469, Springer-Verlag, Berlin

202. Milutinović V. (1997) Surviving the Design of a 200MHz RISC Microprocessor: Lessons Learned. IEEE Computer Society Press, Los Alamitos, CA

203. Milutinović V., Fura D., Helbig W. (1986) An Introduction to GaAs Microprocessor Architecture for VLSI. Computer 19:30–42 (March)

204. Milutinović V., Marković B., Tomašević M., Tremblay M. (1996) Split Temporal/Spatial Cache: Initial Performance Evaluation. In: Proceedings of the SCIzzL-5 Workshop (March), Santa Clara, CA, 72–78
205. Mirsky E., DeHon A. (1996) MATRIX: A Reconfigurable Computing Architecture with Configurable Instruction Distribution and Deployable Resources. In: Proceedings of 4th IEEE Workshop on FPGAs for Custom Computing Machines (April), 157–166
206. Morton S.V., Appleton S.S., Liebelt M.J. (1995) ECSTAC: A Fast Asynchronous Microprocessor. In: Proceedings of the Second Working Conference on Asynchronous Design Methodologies, London, 180–189
207. Müller S.M., Paul W.J. (1996) Making the Original Scoreboard Mechanism Deadlock Free. In: Proceedings of the 4th Israel Symposium on Theory of Computing and Systems, 92–99
208. Müller S.M., Paul W.J. (1998) On the Correctness of Hardware Scheduling Mechanisms for Out-of-Order Execution. Journal of Circuits, Systems and Computers 8(2):301–314
209. Nanya T., Ueno Y., Kagotani H., Kuwako M., Takamura A. (1994) TITAC: Design of a Quasy-Delay-Insensitive Microprocessor. IEEE Design and Test of Computers 11(2):50–63
210. Narasimhan V.L., Downs T. (1989) Operating System Features of a Dynamic Dataflow Array Processing System (PATTSY). In: Proceedings of the 3rd Annual Parallel Processing Symposium, 722–740
211. Nayfeh B.A., Hammond L., Olukotun K. (1996) Evaluation of Design Alternatives for a Multiprocessor Microprocessor. In: Proceedings of the 23rd Annual International Symposium on Computer Architecture (May), Philadelphia, PA, 67–77
212. Nunomura Y., Shimizu T., Tomisawa O. (1997) M32R/D – Integrating DRAM and Microprocessor. IEEE Micro 17:40–48 (November)
213. Nikhil R.S., Arvind (1989) Can Dataflow Subsume von Neumann Computing? In: Proceedings of the 16th Annual Symposium on Computer Architecture (May), Jerusalem, 262–272
214. Nikhil R.S., Papadopoulos G.M., Arvind (1992) *T: A Multithreaded Massively Parallel Architecture. In: Proceedings of the 19th Annual Symposium on Computer Architecture (May), Gold Coast, 156–167
215. Nishikawa H., Terada H., Komatsu K., Yoshida S., Okamoto T., Tsuji Y., Takamura S., Tokura T., Nishikawa Y., Hara S., Meichi M. (1987) Architecture of a One-Chip Data-Driven Processor: Q-p. In: Proceedings of the 16th International Conference on Parallel Processing (August), St.Charles, IL, 319–326
216. O'Connor J.M., Tremblay M. (1997) PicoJava-I: The Java Virtual Machine in Hardware. IEEE Micro 17:45–53 (March/April)
217. Oehler R.R., Groves R.D. (1990) IBM RISC System/6000 ProcessorArchitecture. IBM J Res Dev 34(1):23–36
218. Oehring H., Sigmund U., Ungerer Th. (1999) Simultaneous Multithreading and Multimedia. In: Proceedings of the Workshop on Multi-threaded Execution, Architecture and Compilation (MTEAC 99) in conjuction with HPCA-5 (January), Orlando, FL
219. Olukotun K., Nayfeh B.A., Hammond L., Wilson K., Chang K. (1996) The Case for a Single-Chip Multiprocessor. In: Proceedings of the 7th International Conference on Architectural Support for Programming Languages and Operating Systems (October), Cambridge, MA, 2–11
220. Oskin M., Chong F.T., Sherwood T. (1998) Active Pages: A Computation Model for Intelligent Memory. In: Proceedings of the 25th Annual International Symposium on Computer Architecture (June-July), Barcelona, 192–203

221. Page I. (1996) Reconfigurable Processor Architectures. Microprocessors and Microsystems 20:185–196
222. Palacharla S., Jouppi N.P., Smith J.E. (1997) Complexity-Effective Superscalar Processors. In: Proceedings of the 24th Annual International Symposium on Computer Architecture (June), Denver, CO, 206–218
223. Pan S.-T., So K., Rahmeh J.T. (1992) Improving the Accuracy of Dynamic Branch Prediction Using Branch Correlation. In: Proceedings of the 5th International Conference on Architectural Support for Programming Languages and Operating Systems (April), Boston, MA, 76–84
224. Papadopoulos G.M. (1988) Implementation of a General-Purpose Dataflow Multiprocessor. Technical Report TR-432, Laboratory for Computer Science, MIT, Cambridge, MA
225. Papadopoulos G.M., Culler D.E. (1990) Monsoon: An Explicit Token-Store Architecture. In: Proceedings of the 17th Annual Symposium on Computer Architecture (May), Seattle, WA, 82–91
226. Papadopoulos G.M., Traub K.R. (1991) Multithreading: A Revisionist View of Dataflow Architectures. In: Proceedings of the 18th Annual Symposium on Computer Architecture (May), Toronto, 342–351
227. Papworth D. (1996) Tuning the Pentium Pro Microarchitecture. IEEE Micro 16:8–15 (April)
228. Patnaik L.M., Govindarajan R., Ramadoss N.S. (1986) Design and Performance Evaluation of EXMAN: EXtended MANchester Data Flow Computer. IEEE Trans Comput 35:229–244
229. Patt Y.N., Patel S.J., Evers M., Friendly D.H., Stark J. (1997) One Billion Transistors, One Uniprocessor, One Chip. Computer 30:51–57 (September)
230. Patterson D.A., Anderson T., Cardwell N., Fromm R., Keeton K., Kozyrakis C., Thomas R., Yelick K. (1997) A Case for Intelligent RAM. IEEE Micro 17:34–43 (March/April)
231. Patterson D.A., Ditzel D.R. (1980) The Case for the Reduced Instruction Set Computer. ACM Computer Architecture News 8(6):25–33 (October)
232. Peleg A., Weiser U. (1996) MMX Technology Extension to the Intel Architecture. IEEE Micro 16:42–50 (August)
233. Peleg A., Wilkie S., Weiser U. (1997) Intel MMX for Multimedia PCs. Commun ACM 40(1):25–38
234. Plas A., Comte D., Gelly O., Syre J.C. (1976) LAU System Architecture: A Parallel Data Driven Processor Based on Single Assignment. In: Proceedings of the 6th International Conference on Parallel Processing (August), 293–302
235. Potter T., Vaden M., Young J., Ullah N. (1994) Resolution of Data and Control-Flow Dependencies in the PowerPC 601. IEEE Micro 14:18–29 (October)
236. Quénot G.M., Zavidovique B. (1991) A Data-Flow Processor for Real-Time Low-Level Image Processing. In: Proceedings of the IEEE Custom Integrated Circuits Conference (May), San Diego, CA, 1241–1244
237. Radin G. (1982) The 801 Minicomputer. In: Proceedings of the Symposium on Architectural Support for Programming Languages and Operating Systems (March), Palo Alto, CA, 39–47
238. Rathnam S., Slavenburg G. (1996) An Architectural Overview of the Programmable Multimedia Processors, TM-1. In: Proceedings of the 41st IEEE Computer Society International Conference COMPCON 96 (February), Santa Clara, CA, 319–326
239. Rau B.R., Fisher J.A. (1993) Instruction Level Parallel Processing: History, Overview, and Perspective. Journal on Supercomputing 7(1/2):9–50

240. Requa J.E., McGraw J.R. (1983) The Piecewise Data Flow Architecture: Architectural Concept. IEEE Trans Comput 32:425–438
241. Richardson W.F., Brunvand E. (1995) Fred: An Architecture for a Self-Timed Decoupled Computer. Technical Report UUCS-95-008, University of Utah
242. Robič B., Kolbezen P., Šilc J. (1992) Area Optimization of Dataflow-Graph Mappings. Parallel Computing 18:297–311
243. Robič B., Šilc J., Kolbezen P. (1987) Resource Optimization in Parallel Data Driven Architecture. In: Proceedings of the IASTED International Symposium Applied Informatics (February), Gründelwald, 86–89
244. Robič B., Vilfan B. (1996) Improved Schemes for Mapping Arbitrary Algorithms onto Processor Meshes. Parallel Computing 22:701–724
245. Roh L., Najjar W.A. (1995) Design of Storage Hierarchy in Multithreaded Architectures. In: Proceedings of the 28th International Symposium on Microarchitecture (November-December), Ann Arbor, MI, 271–278
246. Rotenberg E., Bennett S., Smith J.E. (1996) Trace Cache: A Low Latency Approach to High Bandwidth Instruction Fetching. In: Proceedings of the 29th Annual International Symposium on Microarchitecture (December), Paris, 24–34
247. Rotenberg E., Jacobson Q., Sazeides Y., Smith J. (1997) Trace Processors. In: Proceedings of the 30th Annual International Symposium on Microarchitecture (December), Research Triangle Park, NC, 138–148
248. Rowen C., Johnson M., Ries P. (1988) The MIPS R3010 Floating-Point Coprocessor. IEEE Micro 8:53–62 (June)
249. Rubinfeld P.I. (1998) Managing Problems at High Speed. Computer 31:47–48 (January)
250. Ryan B., Thompson T. (1994) PowerPC 604 Weighs In. Byte Magazine 19(6):265–266 (June)
251. Rychlik B., Faistl J., Krug B., Shen J.P. (1998) Efficiecy and Performance Impact of Value Prediction. In: Proceedings of the 1998 Conference on Parallel Architectures and Compilation Techniques (October), Paris, 148–154
252. Sakai S. (1995) Synchronization and Pipeline Design for a Multithreaded Massively Parallel Computer. In: Gao G.R., Bic L., Gaudiot J.-L. (eds.) Advanced Topics in Dataflow Computing and Multithreading. IEEE Computer Society Press, Los Alamitos, CA, 55–74
253. Sakai S., Okamoto K., Matsuoka K., Hirono H., Kodama Y., Sato M., Yokota T. (1993) Basic Features of a Massively Parallel Computer RWC-1. In: Proceedings of the 1993 Joint Symposium on Parallel Processing (May), Tokyo
254. Sakai S., Yamaguchi Y., Hiraki K., Kodama Y., Yuba T. (1989) An Architecture of a Dataflow Single Chip Processor. In: Proceedings of the 16th Annual Symposium on Computer Architecture (May), Jerusalem, 46–53
255. Santhanam S. (1996) StrongARM SA110, A 160MHz 32b 0.5W CMOS ARM Processor. In: Proceedings of the Symposium on High-Performance Chips – Hot Chips 8 (August), Stanford, CA
256. Saulsbury A., Pong P., Nowatzyk A. (1996) Missing the Memory Wall: The Case for Processor/Memory Integration. In: Proceedings of the 23rd Annual International Symposium on Computer Architecture (May), Philadelphia, PA, 90–101
257. Scott A.P., Burkhart K.P., Kumar A., Blumberg R.M., Ranson G.L. (1997) Four-Way Superscalar PA-RISC Processors. Hewlett-Packard Journal 48(4):8–15

258. Shriver B., Smith B. (1998) The Anatomy of a High-Performance Microprocessor, A Systems Perspective. IEEE Computer Society Press, Los Alamitos, CA
259. Sigmund U., Ungerer T. (1996) Evaluating a Multithreaded Superscalar Microprocessor Versus a Multiprocessor Chip. In: Proceedings of the 4th PASA Workshop on Parallel Systems and Algorithms (April), Jülich, 147–159
260. Sigmund U., Ungerer T. (1996) Identifying Bottlenecks in Multithreaded Superscalar Multiprocessors. Lecture Notes in Computer Science 1123:797–800, Springer-Verlag, Berlin
261. Šilc J., Robič B. (1989) Synchronous Dataflow-Based Architecture. Microprocessing and Microprogramming 27:315–322
262. Šilc J., Robič B. (1993) Program Partitioning for a Control/Data Driven Computer. Journal of Computing and Information Technology 1:47–55
263. Šilc J., Robič B., Ungerer T. (1998) Asynchrony in Parallel Computing: From Dataflow to Multithreading. Parallel and Distributed Computing Practices 1:57–83
264. Singh H., Lee M.-H., Lu G., Kurdahi F.J., Bagherzadeh N., Lang T., Heaton R., Filho E.M.C. (1998) MorphoSys: An Integrated Re-configurable Architecture. In: Proceedings of the NATO Symposium on System Concepts and Integration (April), Monterey, CA
265. Sites R.L. (1992) Alpha AXP Architecture. Digital Technical Journal 4(4)
266. Smith B.J. (1981) Architecture and Applications of the HEP Multiprocessor Computer System. SPIE Real-Time Signal Processing IV 298:241–248
267. Smith B.J. (1985) The Architecture of HEP. In: Kowalik J.S. (ed.) Parallel MIMD Computation: HEP Supercomputer and Its Applications. MIT Press, Cambridge, MA, 41–55
268. Smith J.E. (1981) A Study of Branch Prediction Strategies. In: Proceedings of the 8th Annual Symposium on Computer Architecture (May), Minneapolis, MI, 135–147
269. Smith J.E. (1998) Retrospective: A Study of Branch Prediction Techniques. In: 25 Years of the International Symposia on Computer Architecture. Selected Papers, ACM Press, 22–23
270. Smith J.E., Pleskun A.R. (1985) Implementation of Precise Interrupts in Pipelined Processors. In: Proceedings of the 12th Annual International Symposium on Computer Architecture (June), Boston, MA, 36–44
271. Smith J.E., Vajapeyam S. (1997) Trace Processors: Moving to Fourth-Generation Microarchitectures. Computer 30:68–74 (September)
272. Snelling D.F. (1993) The Design and Analysis of a Stateless Data-Flow Architectures. Technical Report UMCS-93-7-2, Department of Computer Science, University of Manchester, UK
273. Sohi G.S. (1997) Multiscalar: Another Fourth-Generation Processor. Computer 30:72 (September)
274. Sohi G.S., Breach S.E., Vijaykumar T.N. (1995) Multiscalar Processors. In: Proceedings of the 22nd Annual International Symposium on Computer Architecture (June), Santa Margherita Ligure, 414–425
275. Song P. (1997) HAL Packs SPARC64 onto Single Chip. Microprocessor Report 11(16) (December 8)
276. Song S.P., Denman M., Chang J. (1994) The PowerPC 604 RISC Microprocessor. IEEE Micro 14:8–17 (October)
277. Sproull R.F., Sutherland I.E., Molnar C.E. (1994) Counterflow Pipeline Processor Architecture. IEEE Design and Test of Computers 11(3):48–59
278. Srini V.P. (1986) An Architectural Comparison of Dataflow Systems. Computer 19:68–88 (March)

279. Stiller A. (1999) Cindy und Bert – Architekturen von Katmai (Pentium III) und Sharptooth (K6-III). ct 5:124–129 (March)
280. Strohschneider J., Klauer B., Zickenheimer S., Waldschmidt K. (1994) Adarc: A Fine Grain Dataflow Architecture with Associative Communication Network. In: Proceedings of the 20th Euromicro Conference: System Architecture and Integration (September), Liverpool, 445–450
281. Suessmith B.W., Paap III G. (1994) PowerPC 603 Power Management. Commun ACM 37(6):43–46
282. Sun Microsystems (1997) picoJava-I Microprocessor Core Architecture. Hardware and Networking Microelectronics Whitepapers, WPR-0014-01
283. Sun Microsystems (1997) Sun Unveils Its First Java Processor, microJava-701 Looks to Post Industry's Highest Caffeinemarks. Sun Microsystems, News and Events, Press Releases (October 15)
284. Sun Microsystems (1998) microJava 701 Extending the Java Paradigm. Microelectronics Whitepapers (March), http://www.sun.com/ microelectronics/whitepapers/
285. Sutherland I.E. (1989) Micropipelines. Commun ACM 32(6):720–738
286. Tabak D. (1995) Advanced Microprocessors. McGraw-Hill, New York
287. Takahashi N., Yoshida M., Amamiya M. (1983) A Data Flow Processor Array System: Design and Analysis. In: Proceedings of the 10th Annual Symposium on Computer Architecture (June), Stockholm, 243–250
288. Takamura A., Kuwako M., Imai M., Fujii T., Ozawa M., Fukasaku I., Ueno Y., Nanya T. (1997) TITAC-2: An Asyncronous 32-bit Microprocessor Based on Scalable-Delay-Insensitive Model. In: Proceedings of the IEEE International Conference on Computer Design (October), 288–294
289. Tau E., Chen D., Eslick I., Brown J., DeHon A. (1995) A First Generation DPGA Implementation. In: Proceedings of the 3rd Canadian Workshop of Field-Programmable Devices (May-June), Montreal, 138–143
290. Temma T., Iwashita M., Matsumoto K., Kurokawa H., Nukiyama T. (1985) Data Flow Processor Chip for Image Processing. IEEE Trans Electronic Devices 32:1784–1791
291. Texas Instruments (1994) TMS320C80 Technical Brief. Texas Instruments, Houston, TX
292. Texas Instruments (1998) TMX320C6201 Digital Signal Processor Advance Information. Texas Instruments, Houston, TX, March.
293. Texas Instruments (1998) Fixed- and Floating-Point DPS - One Architecture. DSP Products 'C6x information. Texas Instruments, Houston, TX
294. Thistle M., Smith B.J. (1988) A Processor Architecture for Horizon. In: Proceedings of the Supercomputing '88 (November), Orlando, FL, 35–41
295. Thornton J.E. (1961) Parallel Operation in the Control Data 6600. In: Proceedings of the Fall Joint Computers Conference, 26:33–40
296. Thornton J.E. (1970) Design of a Computer: The Control Data 6600. Glenview, IL
297. Tierno J.A., Martin A.J., Borkovic D., Lee T.K. (1994) A 100-MIPS GaAs Asynchronous Microprocessor. IEEE Design and Test of Computers 11(2):43–49
298. Tomasulo R.M. (1967) An Efficient Algorithm for Exploiting Multiple Arithmetic Units. IBM J Res Dev 11(1):25–33
299. Treleaven P.C., Brownbridge D.R., Hopkins R.P. (1982) Data-Driven and Demand-Driven Computer Architectures. ACM Computing Surveys 14:93–143
300. Tremblay M. (1998) Increasing Work, Pushing the Clock. Computer 31:40–41 (January)
301. Tremblay M., O'Connor J.M. (1996) UltraSPARC I: A Four-Issue Processor Supporting Multimedia. IEEE Micro 16:42–50 (April)

302. Tsai J.-Y., Yew P.-C. (1996) The Superthreaded Architecture: Thread Pipelining with Run-Time Data Dependence Checking and Control Speculation. In: Proceedings of the 1996 Conference on Parallel Architectures and Compilation Techniques (October), Boston, MA, 35–46
303. Tullsen D.M., Eggers S.J., Levy H.M. (1995) Simultaneous Multithreading: Maximizing On-Chip Parallelism. In: Proceedings of the 22nd Annual International Symposium on Computer Architecture (June), Santa Margherita Ligure, 392–403
304. Tullsen D.M., Eggers S.J., Emer J.S., Levy H.M., Lo J.L., Stamm R.L. (1996) Exploiting Choice: Instruction Fetch and Issue on an Implementable Simultaneous Multithreading Processor. In: Proceedings of the 23rd Annual International Symposium on Computer Architecture (May), Philadelphia, PA, 191–202
305. Tyson G., Lick K., Farrens M. (1997) Limited Dual Path Execution. Technical Report CSE-TR 346-97, University of Michigan
306. Uht A.K., Sindagi V. (1995) Disjoint Eager Execution: An Optimal Form of Speculative Execution. In: Proceedings of the 28th International Symposium on Microarchitecture, (November-December), Ann Arbor, MI, 313–325
307. Uht A.K., Sindagi V., Somanathan S. (1997) Branch Effect Reduction Techniques. Computer 30:71–81 (May)
308. Unger A., Ungerer T., Zehendner E. (1998) A Compiler Technique for Speculative Execution of Alternative Program Paths Targeting Multithreaded Architectures. In: Proceedings of the Yale Multithreaded Programming Workshop (June), New Haven, CT
309. Unger A., Ungerer T., Zehendner E. (1998) Static Speculation, Dynamic Resolution. In: Proceedings of the 7th Workshop on Compilers for Parallel Computers (June /July), Linkoping, 243–253
310. Vajapeyam S., Mitra T. (1997) Improving Superscalar Instruction Dispatch and Issue by Exploiting Dynamic Code Sequences. In: Proceedings of the 24th Annual International Symposium on Computer Architecture (June), Denver, CO, 1–12
311. Vedder R.W., Campbell M.L., Tucker G.K. (1985) The Hughes Data Flow Multiprocessor. In: Proceedings of the 5th International Conference on Distributed Computing Systems (May), 2–9
312. Veen A.H., van den Born R. (1990) The RC Compiler for the DTN Dataflow Computer. Journal of Parallel and Distributed Computing 10:319–322
313. Villasenor J., Mangione-Smith H. (1997) Configurable Computing. Sci Am 276(6):54–59
314. Villasenor J., Schoner B., Chia K.-N., Zapata C., Kim H.J., Jones C., Lansing S., Mangione-Smith B. (1996) Configurable Computing Solutions for Automatic Target Recognition. In: Proceedings of 4th IEEE Workshop on FPGAs for Custom Computing Machines (April), 70–79
315. Wadge W.W., Ashcroft E.A. (1985) LUCID, The Dataflow Programming Language. Academic Press, San Diego, CA
316. Waingold E., Taylor M., Srikrishna D., Sarkar V., Lee W., Lee V., Kim J., Frank M., Finch P., Barua R., Babb J., Amarasinghe S., Agarwal A. (1997) Baring it All to Software: Raw Machines. Computer 30:86–93 (September)
317. Wall D.W. (1991) Limits of Instruction-level Parallelism. In: Proceedings of the 4th International Conference on Architectural Support for Programming Languages and Operating Systems (April), Santa Clara, CA, 176–188
318. Wallace S., Calder B., Tullsen D.M. (1998) Threaded Multiple Path Execution. In: Proceedings of the 25th Annual International Symposium on Computer Architecture (June-July), Barcelona, 238–249

319. Wallace S., Bagherzadeh N. (1998) Modeled and Measured Instruction Fetching Performance for Superscalar Microprocessors. IEEE Transactions on Parallel and Distributed Systems 9(6):570–578
320. Watson I., Gurd J.R. (1979) A Prototype Data Flow Computer with Token Labelling. In: Proceedings of the National Computer Conference, AFIPS Proceedings 48:623–628
321. Wayner P. (1996) Sun Gambles on Java Chips. Byte Magazine 21(11):79–88 (November).
322. Weiss S., Spilinger I.Y., Silberman G.M. (1993) Architectural Improvements for a Data-Driven VLSI Processing Array. Journal of Parallel and Distributed Computing 19:308–322
323. Weiss S., Smith J.E. (1994) Power and PowerPC: Principles, Architecture, Implementation. Morgan Kaufmann, San Francisco, CA
324. Wilkes M.V. (1995) Computing Perspectives. Morgan Kaufmann, San Francisco, CA
325. Wirthlin M.J., Hutchings B.L. (1995) A Dynamic Instruction Set Computer. In: Proceedings of 3rd IEEE Workshop on FPGAs for Custom Computing Machines (April), 99–107
326. Wulf W.A., McKee S.A. (1995) Hitting the Memory Wall: Implications of the Obvious. ACM Computer Architecture News 23(1):20–24 (March)
327. Xilinx (1997) XC6200 Field Programmable Gate Arrays. Xilinx data book (April 24), http://www.xilinx.com/
328. Yamauchio T., Hammond L., Olukotun K. (1997) A Single Chip Multiprocessor Integrated with DRAM. Technical Report CSL-TR-97-731, Computer Systems Laboratory, Stanford University
329. Yao Y. (1996) Chromatic's Mpact 2 Boosts 3D. Microprocessor Report 10(15):53–58 (November 18)
330. Yeager K.C. (1996) The MIPS R10000 Superscalar Microprocessor. IEEE Micro 16:28–40 (April)
331. Yeh T.-Y., Patt Y.N. (1992) Alternative Implementation of Two-Level Adaptive Branch Prediction. In: Proceedings of the 19th Annual Symposium on Computer Architecture (May), Gold Coast, 124–134
332. Yeh T.-Y., Patt Y.N. (1993) A Comparison of Dynamic Branch Predictors that use Two Levels of Branch History. In: Proceedings of the 20th Annual International Symposium on Computer Architecture (May), San Diego, CA, 257–266
333. Young C., Smith M. (1994) Improving the Accuracy of Static Branch Prediction Using Branch Correlation. In: Proceedings of the 6th Annual International Symposium on Architectural Support for Programming Languages and Operating Systems (October), San Jose, CA, 232–241
334. Zehendner E., Ungerer T. (1987) The ASTOR Architecture. In: Proceedings of the 7th International Conference on Distributed Computing Systems (September), Berlin, 424–430
335. – (1998) SIA Releases Technology Roadmap. IEEE Micro 18:2 (January/February)
336. – (1997) The National Technology Roadmap for Semiconductors. SEMATECH, Austin, TX
337. – (1997) 1600-MIPS-DSP unter der Lupe. Design & Elektronik 8:6–8 (April 22)

Index

Printing (computer to plate): Mercedes-Druck, Berlin
Binding: Stürtz AG, Würzburg

Printed in the United States
130677LV00002B/31/P